HANDBOOK ON MODELLING FOR DISCRETE OPTIMIZATION

T0180180

Recent titles in the **INTERNATIONAL SERIES IN OPERATIONS RESEARCH & MANAGEMENT SCIENCE**
Frederick S. Hillier, Series Editor, *Stanford University*

* A list of the early publications in the series is at the end of the book *

HANDBOOK ON MODELLING
FOR DISCRETE OPTIMIZATION

Edited by

GAUTAM APPA
Operational Research Department
London School of Economics

LEONIDAS PITSOULIS
Department of Mathematical and Physical Sciences
Aristotle Universlty of Thessaloniki

H.PAUL WILLIAMS
Operational Research Department
London School of Economics

 Springer

Gautam Appa
London School of Economics
United Kingdom

Leonidas Pitsoulis
Aristotle University of Thessaloniki
Greece

H. Paul Williams
London School of Economics
United Kingdom

ISBN-13: 978-1-4419-4107-7

ISBN-10: 0-387-32942-0 (e-book)
ISBN-13: 978-0387-32942-0 (e-book)

Printed on acid-free paper.

9 8 7 6 5 4 3 2 1

springer.com

Contents

List of Figures

List of Tables

Contributing Authors

Gautam Appa

Department of Operational Research
London School of Economics, London
United Kingdom

g.appa@lse.ac.uk

Stanislav Busygin

Department of Industrial and Systems
Engineering, University of Florida
303 Weil Hall, Gainesville FL 32611
USA

busygin@ufl.edu

Jean-François Cordeau

Canada Research Chair in Distribution
Management and GERAD, HEC Montreal
3000 chemin de la Côte-Sainte-Catherine
Montreal, Canada H3T 2A7

cordeau@crt.umontreal.ca

Warren D'Souza

University of Maryland School of Medicine
22 South Green Street
Baltimore, MD 21201
USA

wdsou001@umaryland.edu

Michael C. Ferris

Computer Sciences Department
University of Wisconsin
1210 West Dayton Street, Madison
WI 53706,
USA

ferris@cs.wisc.edu

Vasilis Friderikos

Centre for Telecommunications Research
King's College London
United Kingdom

vasilis.friderikos@kcl.ac.uk

Sabino M. Gadaleta

Numerica
PO Box 271246
Fort Collins, CO 80527-1246
USA

smgadaleta@numerica.us

Ignacio E. Grossmann

Department of Chemical Engineering
Carnegie Mellon University, Pittsburgh
PA 15213,
USA

ig0c@andrew.cmu.edu

Susara A. van den Heever

Department of Chemical Engineering
Carnegie Mellon University, Pittsburgh
PA 15213,
USA

susara@andrew.cmu.edu

John N. Hooker

Graduate School of Industrial Administration
Carnegie Mellon University, PA 15213
USA

jh38@andrew.cmu.edu

Giusseppe Lancia
Dipartimento di Matematica e Informatica
Universita di Udine
Via delle Scienze 206, 33100 Udine
Italy
lancia@dei.unipd.it

Gilbert Laporte
Canada Research Chair in Distribution
Management and GERAD, HEC Montreal
3000 chemin de la Côte-Sainte-Catherine
Montreal, Canada H3T 2A7
gilbert@crt.umontreal.ca

Dimitris Magos
Department of Informatics
Technological Educational Institute of Athens
12210 Athens, Greece
dmagos@teiath.gr

Christos T. Maravelias
Department of Chemical and Biological
Engineering, University of Wisconsin
1415 Engineering Drive,
Madison, WI 53706-1691,
USA
christos@engr.wisc.edu

Robert R. Meyer
Computer Sciences Department
University of Wisconsin
1210 West Dayton Street, Madison
WI 53706,
USA
rrm@cs.wisc.edu

Gautam Mitra
CARISMA
School of Information Systems,
Computing and Mathematics,
Brunel University, London
United Kingdom
gautam.mitra@brunel.ac.uk

Ioannis Mourtos
Department of Economics
University of Patras, 26500 Rion, Patras
Greece
imourtos@upatras.gr

Katerina Papadaki
Department of Operational Research
London School of Economics, London
United Kingdom
k.p.papadaki@lse.ac.uk

Panos Pardalos
Department of Industrial and Systems
Engineering, University of Florida
303 Weil Hall, Gainesville FL 32611
USA
pardalos@ufl.edu

Leonidas Pitsoulis
Department of Mathematical and
Physical Sciences, Aristotle University
of Thessaloniki, 54124 Thessaloniki,
Greece
pitsouli@gen.auth.gr

Chandra Poojari
CARISMA
School of Information Systems,
Computing and Mathematics
Brunel University, London
United Kingdom
chandra.poojari@brunel.ac.uk

Aubrey Poore
Department of Mathematics
Colorado State University
Fort Collins, 80523,
USA
aubrey.poore@colostate.edu

Oleg A Prokopyev

*Department of Industrial and Systems
Engineering, University of Florida
303 Weil Hall, Gainesville FL 32611
USA*

oap4ripe@ufl.edu

Suvrajeet Sen

*Department of Systems and Industrial
Engineering , University of Arizona,
Tuscon, AZ 85721
USA*

sen@sie.arizona.edu

Douglas R. Shier

*Department of Mathematical Sciences
Clemson University, Clemson,*

*SC 29634-0975
USA*

shierd@clemson.edu

Benjamin J. Slocumb

*Numerica
PO Box 271246
Fort Collins, CO 80527-1246
USA*

bjslocumb@numerica.us

H. Paul Williams

*Department of Operational Research
London School of Economics, London
United Kingdom*

h.p.williams@lse.ac.uk

Preface

The primary reason for producing this book is to demonstrate and communicate the pervasive nature of Discrete Optimisation. It has applications across a very wide range of activities. Many of the applications are only known to specialists. Our aim is to rectify this.

It has long been recognized that "modelling" is as important, if not more important, a mathematical activity as designing algorithms for solving these discrete optimisation problems. Nevertheless solving the resultant models is also often far from straightforward. Although in recent years it has become viable to solve many large scale discrete optimisation problems some problems remain a challenge, even as advances in mathematical methods, hardware and software technology are constantly pushing the frontiers forward.

The subject brings together diverse areas of academic activity as well as diverse areas of applications. To date the driving force has been Operational Research and Integer Programming as the major extention of the well-developed subject of Linear Programming. However, the subject also brings results in Computer Science, Graph Theory, Logic and Combinatorics, all of which are reflected in this book.

We have divided the chapters in this book into two parts, one dealing with general methods in the modelling of discrete optimisation problems and one with specific applications. The first chapter of this volume, written by Paul Williams, can be regarded as a basic introduction of how to model discrete optimisation problems as Mixed Integer Programmes, and outlines the main methods of solving them.

Chapter 2, written by Pardalos et al., deals with the intriguing relationship between the continuous versus the discrete approach to optimisation problems. The authors in chapter 2 illustrate how many well known hard discrete optimisation problems can be modelled and solved by continuous methods, thereby giving rise to the question of whether or not the discrete nature of the problem is the true cause of its computational complexity or the presence of nonconvexity.

Another subject of great relevance to modelling is Logic. This is covered in chapter 3. The author, John Hooker, describes the relationship with an alternative solution (and modelling) approach known as Constraint Satisfaction or, as it is sometimes called, Constraint Logic Programming. This approach has emerged more from Computer Science than Operational Research. However, the possibility of "hybrid methods" based on combining the approaches is on the horizon, and has been realized with some problem specific implementations.

In chapter 4 Appa et al. illustrate how discrete optimisation modelling and solution methods can be applied to answer questions regarding a problem arising from combinatorial mathematics. Specifically the authors present various optimisation formulations of the mutually orthogonal latin squares problem, from constraint programming (which is covered in detail in chapter 3) to mixed integer programming formulations and matroid intersection, all of which can be used to answer existence questions for the problem.

It has long been established that Networks can model most of today's complex systems such as transportation systems, telecommunication systems, and computer networks to name a few, and network optimisation has proven to be a valuable tool in analyzing the behavior of these systems for design purposes. Chapter 5 by Shier enhances further the applicability of network modelling by presenting how it can also be applied to less apparent systems ranging from genomics, sports and artificial intelligence.

Chapter 6 is the last chapter in the methods part of the book, where Cordeau and Laporte discuss a class of problems known as vehicle routing problems. Vehicle routing problems enjoy a plethora of applications in the transportation and logistics sector, and the authors in chapter 6 present the state of the art with respect to exact and heuristic methods for solving them.

In the second part of the book various real life applications are presented, most of them formulated as mixed integer linear or nonlinear programming problems. Chapter 7 by Papadaki and Friderikos, is concerned with the solution of optimization problems arising in resource management problems in wireless cellular systems by employing a novel approach, the so called approximate dynamic programming.

Most of the discrete optimisation models presented in this book are of deterministic nature, that is the values of the input data are assumed to be known with certainty. There are however real life applications where such an assumption is inapplicable, and stochastic models need to be considered. This is the subject of chapter 8, by Mitra et al. where stochastic mixed integer programming models are discussed for supply chain management problems.

In chapters 9 and 10 Grossmann et al. present how discrete optimisation modeling can be efficiently applied to two specific application areas. In chapter 9 mixed integer linear programming models are presented for the problem of scheduling regulatory tests of new pharmaceutical and agrochemical products, while in chapter 10 a mixed integer nonlinear model is presented for the optimal planning of offshore oilfield infrastructures. In both chapters the authors also present solution techniques.

Optimization models to radiation therapy for cancer patients is the subject discussed in chapter 11 by Ferris and Meyer. They show how the problem of irradiating patients for treatment of cancerous tumors can be formulated as a discrete optimisation problem and can be solved as such.

In chapter 12 the data association problem that arises in target tracking is considered by Poore et al. The objective in this chapter is to partition the data that is created by multiple sensors observing multiple targets into tracks and false alarms, which can be formulated as a multidimensional assignment problem, a notoriously difficult integer programming problem which generalizes the well known assignment problem.

Finally chapter 13 is concerned with the life sciences, and Lancia shows how some challenging problems of Computational Biology can now be solved as discrete optimisation models.

Assembling and planning this book has been much more of a challenge than we at first envisaged. The field is so active and diverse that it has been difficult covering the whole subject. Moreover the contributors have themselves been so deeply involved in practical applications that it has taken longer than expected to complete the volume.

We are aware that, within the limits of space and the time of contributors we have not been able to cover all topics that we would have liked. For example we have been unable to obtain a contributor on Computer Design, an area of great importance, or similarly on Computational Finance and Air Crew Scheduling. By way of mitigation we are pleased to have been able to bring together some relatively new application areas.

We hope this volume proves a valuable work of reference as well as to stimulate further successful applications of discrete optimisation.

GAUTAM APPA, LEONIDAS PITSOULIS, H. PAUL WILLIAMS

Acknowledgments

We would like to thank the contributing authors of this volume, and the many anonymous referees who have helped us review the chapters all of which have been thoroughly refereed. We are also thankful to the staff of Springer, in particular Gary Folven and Carolyn Ford, as well as the series editor Fred Hillier.

Paul Williams acknowledges the help which resulted from Leverhulme Research Fellowship RF&G/9/RFG/2000/0174 and EPSRC Grant EP/C530578/1 in preparing this book.

I

METHODS

Chapter 1

THE FORMULATION AND SOLUTION OF DISCRETE OPTIMISATION MODELS

H. Paul Williams

Department of Operational Research
London School of Economics, London
United Kingdom

h.p.williams@lse.ac.uk

Abstract   This introductory chapter first discusses the applicability of Discrete Optimisa-
                tion and how Integer Programming is the most satisfactory method of solving
                such problems. It then describes a number of modelling techniques, such as
                linearisng products of variables, special ordered sets of variables, logical condi-
                tions, disaggregating constraints and variables, column generation etc. The main
                solution methods are described, i.e. Branch-and-Bound and Cutting Planes. Fi-
                nally alternative methods such as Lagrangian Relaxation and non-optimising
                methods such as Heuristics and Constraint Satisfaction are outlined.

Keywords: Integer Programming, Global Optima, Fixed Costs,Convex Hull, Reformulation,
                Presolve, Logic, Constraint Satisfaction.

1.   The Applicability of Discrete Optimisation

The purpose of this introductory chapter is to give an overview of the scope
for discrete optimisation models and how they are solved. Details of the modelling necessary is usually problem specific. Many applications are covered in
other chapters of this book. A fuller coverage of this subject is given in [25]
together with many references.

For solving discrete optimization models, when formulated as (linear) Integer Programmes (IPs), much fuller accounts, together with extensive references, can be found in Nemhauser and Wolsey [19] and Williams [24]. Our
purpose, here, is to make this volume as self contained as possible by describing the main methods.

Limiting the coverage to *linear* IPs should not be seen as too restrictive in view of the reformulation possibilities described in this chapter.

It is not possible, totally, to separate modelling from the solution methods. Different types of model will be appropriate for different methods. With some methods it is desirable to modify the model in the course of optimisation. These two considerations are addressed in this chapter.

The modelling of many *physical* systems is dominated by *continuous* (as opposed to *discrete*) mathematics. Such models are often a simplification of reality, but the discrete nature of the real systems is often at a microscopic level and continuous modelling provides a satisfactory simplification. What's more, continuous mathematics is more developed and unified than discrete mathematics. The calculus is a powerful tool for the optimisation of many continuous problems. There are, however, many systems where such models are inappropriate. These arise with physical systems (e.g. construction problems, finite element analysis etc) but are much more common in decision making (operational research) and information systems (computer science). In many ways, we now live in a 'discrete world'. Digital systems are tending to replace analogue systems.

2.   Integer Programming

The most common type of model used for discrete optimisation is an **Integer Programme** (IP) although Constraint Logic Programmes (discussed in the chapter by Hooker) are also applicable (but give more emphasis to obtaining feasible rather than optimal solutions). An IP model can be written:

$$\text{Maximise/Minimise} \quad c' x + d' y \tag{1.1}$$

$$\text{Subject to:} \quad A x + B y \leq b \tag{1.2}$$

$$x \in \Re, y \in Z \tag{1.3}$$

x are (generally non-negative) continuous variables and y are (generally non-negative and bounded) integer variables.

If there are no integer variables we have a **Linear Programme** (LP). If there are no continuous variables we have a **Pure IP** (PIP). In contrast the above model is known as a **Mixed IP** (MIP). Most practical models take this form. It should be remarked that although the constraints and objective of the above model are linear this is not as restrictive, as it might seem. In IP, non-linear expressions can generally be linearised (with possibly some degree of approximation) by the addition of extra integer variables and constraints. This is explained in this chapter. Indeed IP models provide the most satisfactory way of solving, general non-linear optimisation problems, in order to obtain a **global** (as opposed to local) **optimum**. This again is explained later in this chapter.

LPs are 'easy' to solve in a well defined sense (they belong to the complexity class P). In contrast, IPs are difficult, unless they have a special structure, (they are in the NP – complete complexity class). There is often considerable scope for formulating, or reformulating, IP models in a manner which makes them easier to solve.

3. The Uses of Integer Variables

The most obvious use of integer variables is to model quantities which can only take integer values e.g. numbers of cars, persons, machines etc. While this may be useful, if the quantities will always be small integers (less than, say, 10), it is not particularly common. Generally such quantities would be represented by continuous variables and the solutions rounded to the nearest integer.

3.1 0-1 Variables

More commonly integer variables are restricted to two values, 0 or 1 and known as **0-1 variables** or **binary variables**. They then represent Yes/No decisions (e.g. investments). Examples of their use are given below. Before doing this it is worth pointing out that any bounded integer variable can be represented as a sum of 0-1 variables. The most compact way of doing this is to use a binary expansion of the coefficients. Suppose e.g. we have an integer variable y such that:

$$0 \leq y \leq U \tag{1.4}$$

Where U is an upper bound on the value of y. Create 0-1 variables:

$$y_0 , y_1 , y_2 , \quad . \quad . \quad . \quad y_{\lfloor \log_2 U \rfloor}$$

then y can be represented by:

$$y_0 + 2y_1 + 4y_2 + \ldots + 2^{\lfloor \log_2 U \rfloor} y_{\lfloor \log_2 U \rfloor} \tag{1.5}$$

3.1.1 Fixed Charge Problems. A very common use of 0-1 variables is to represent investments or decisions which have an associated **fixed cost** (f). If carried out, other continuous variables $x_1, x_2, \ldots, x_n$ bringing in e.g. **variable profits** $p_1, p_2, \ldots, p_n$ come into play. This situation would be modelled as:

$$\text{Maximise} \quad \sum_j p_j x_j - f y \tag{1.6}$$

$$\text{Subject to:} \quad \sum_j x_j - M y \leq 0 \tag{1.7}$$

(and other constraints)

where M represents an upper bound on the combined level of the continuous activities.

If $y = 0$ (investment not made) all the x_j are forced to zero. On the other hand if $y = 1$ (investment is made) the fixed cost f is incurred but the x_j can take positive values and profit is made.

There are, of course, many extensions to this type of model where alternative or additional investments can be made, etc. The basic idea is, however, the same.

3.1.2   Indicator Variables.

A very powerful modelling technique is to use 0-1 variables to distinguish situations which are ('discretely') different. If there is a simple dichotomy then the two possible values of a 0-1 variable can be used to make the distinction. For example, suppose we want to model the condition

$$\sum_j a_{1j}x_j \leq b_1 \quad \textbf{or} \quad \sum_j a_{2j}x_j \leq b_2 \tag{1.8}$$

Such a condition is known as a *disjunction* of constraints (in contrast to the usual LP *conjunction* of constraints). Let $y = 0$ impose the first constraint in the disjunction and $y= 1$ impose the second constraint. This is achieved by the following *conjunction of* constraints

$$\sum_j a_{1j}x_j - M_1 y \leq b_1 \tag{1.9}$$

$$\sum_j a_{2j}x_j + M_2 y \leq M_2 + b_2 \tag{1.10}$$

M_1 is an upper bound on the value of the expression $\sum_j a_{1j}x_j - b_1$ and M_2 an upper bound on $\sum_j a_{2j}x_j - b_2$. (If either of the expressions has no upper bound then the disjunction can only be modelled in special conditions.) There is an alternative and preferable way of modelling this condition which is considered later in this chapter.

Should a disjunction of more than two constraints be needed then more than one $0 - 1$ variable can be used. For example the situation

$$\sum_j a_{ij}x_j \leq b_1 \quad \textbf{or} \quad \sum_j a_{2j}x_j \leq b_2 \quad \textbf{or} \ldots \textbf{or} \quad \sum_j a_{nj}x_j \leq b_n \tag{1.11}$$

can be modelled as

$$\sum_j a_{ij}x_j - M_1 y_1 \leq b_1 \tag{1.12}$$

$$\sum_j a_{2j}x_j - M_2 y_2 \leq b_2 \tag{1.13}$$

$$\sum_j a_{nj}x_j - M_n y_n \leq b_n \tag{1.14}$$

$$y_1 + y_2 + \cdots + y_n \leq n - 1 \tag{1.15}$$

The situation where at least k of a given set of constraints $(1 < k < n)$ hold can easily be modelled by amending (1.15).

There are many situations where one wishes to model such disjunctions. For example:

- This operation must finish before the other starts **or** vice versa;

- We must either not include this ingredient **or** include it above a given threshold level.

- At least (most) k of a given set of possible warehouses should be built.

In fact all of IP could be reduced to disjunctions of linear constraints. This is only a useful paradigm for some problems. However the modelling of disjunctive conditions has been thoroughly analysed in Balas [2].

3.2   Special Ordered Sets of Variables (Discrete Entities)

If a quantity can take a number of discrete values then a convenient method of modelling is a *Special Ordered Set of Type 1* (S1 set). Suppose we wish to model a quantity which can take values $a_1, a_2, \ldots, a_n$. We would represent this quantity by

$$x = a_1 y_1 + a_2 y_2 + \cdots + a_n y_n \tag{1.16}$$

and with the constraint

$$y_1 + y_2 + \cdots + y_n \;=\; 1$$

with the stipulation that

Exactly one member of the set $\{y_1, y_2, \cdots, y_n\}$ *can be non-zero*

Note that it is not necessary to stipulate that the y_j variables be integer and 0-1.

An S1 set (as opposed to the individual variables in it) should be regarded as a *discrete entity*. It can be considered as a generalisation of a general integer variable where values are not evenly spaced. S1 sets are regarded as entities analogous to integer variables when the Branch-and Bound algorithm, described later in this chapter, is used.

A variant of S1 sets are *Special Ordered sets of Type 2* (S2 sets). These are widely used to model *non-linear functions*. In order to model a non-linear function it must be separated into the sum of non-linear functions of a single variable. This can generally be done by introducing new variables and constraints. (The possibility of doing this goes back to one of Hilbert's problems.)

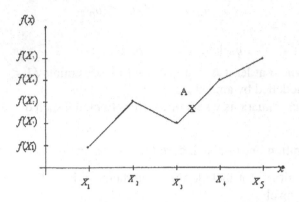

Figure 1.1.   A piecewise linear approximation to a non-linear function

Each non-linear function of a single variable is approximated by a *piecewise linear function* i.e. a series of linear segments as illustrated in Figure 1.1.   A grid value is defined for the relevant values of the argument x (not necessarily evenly spaced) and the function $f(x)$ defined at those values. We define an S2 set $\{y_1, y_2, \dots y_5\}$ with the constraints

$$x = X_1 y_1 + X_2 y_2 + X_3 y_3 + X_4 y_4 + X_5 y_5 \qquad (1.17)$$

$$f(x) = f(X_1)y_1 + f(X_2)y_2 + f(X_3)y_3 + f(X_4)y_4 + f(X_5)y_5 \qquad (1.18)$$

$$y_1 + y_2 + y_3 + y_4 + y_5 = 1 \qquad (1.19)$$

together with the stipulation that

At most two adjacent y_5 can be non-zero

Note that the y_j are continuous. If, for example, $y_3 = \frac{1}{4}, y_4 = \frac{3}{4}$ (implying the other $y_j=0$) this corresponds to the point A in Figure 1.1. By this means, only the points on one of the straight-line segments in Figure 1.1 are represented by $(x, f(x))$. Clearly a more refined grid can give a more accurate approximation to a function (at the expense of more variables).

The concept of Special Ordered Sets are described by [3] with an extension to deal with functions of more than one variable in [4]. There are, of course, alternative methods of modelling condition (1.11) using 0-1 variables since we again have a disjunction of possibilities. However, using Special Ordered Sets has computational advantages when using the Branch-and-Bound algorithm.

4.   The Modelling of Common Conditions

4.1   Products of Variables

It has already been remarked that non-linear expressions can be reformulated using linear constraints involving integer variables. A common non-
linearity is products of integer variables. As explained above (bounded) general integer variables can always be expressed using 0-1 variables. Therefore
we confine our attention to products of 0-1 variables.

Consider the product $y_1 y_2$ which we will represent by a new variable z
where y_1 and y_2 are 0-1 variables. This product itself can only take values 0 or
1 depending on the values of y_1 and y_2. If either is zero the product itself is 0.
Otherwise it is 1.

Logically we have

$$y_1 = 0 \ \textbf{or} \ y_2 = 0 \ \text{(or both)} \ \textbf{implies} \ z = 0$$

This can be modelled by

$$y_1 + y_2 - 2z \geq 0 \tag{1.20}$$

(There is a 'better' way of modelling this which will be given later in this
chapter.) We also have to model

$$y_1 = 1 \ \textbf{and} \ y_2 = 1 \ \textbf{implies} \ z = 1.$$

This can be modelled by

$$y_1 + y_2 - z \leq 1 \tag{1.21}$$

If we have a product of 3 or more $0-1$ variables then we can repeat the above
formulation procedure.

A product of a continuous variable x and a 0-1 variable y can be represented
by a continuous variable z. The variable z will then be equal to x or to zero
depending on whether y is 1 or 0. Either way we have

$$z - x \leq 0 \tag{1.22}$$

If $y = 1$ then we want the condition $z = x$. This may be done by modelling
the condition

$$y = 1 \ \textbf{implies} \ z \geq x$$

since, together with (1.22), this would imply $z = x$. We can model the condition as

$$x - z + My \leq M \tag{1.23}$$

M must be chosen as a suitably large number (an upper bound) which x (and
therefore z) will not exceed.

If we wish to model the product of an 0-1 variable and an *expression* this
formulation can be extended in an economical manner, as done by Glover [10].

4.2   Logical Conditions

The common use of 0-1 variables in IP models suggests the analogy with modelling of True/False propositions which forms the subject matter of the Propositional Calculus (Boolean Algebra). This close relationship is explored further in this book in Chapter 3 by Hooker. Here we give a taste for the power of IP to model the relationships of the Propositional Calculus.

In IP (and LP) our propositions are constraints which may be true or false. (An LP model consists of a conjunction of constraints i.e. they must all be true). We can represent the truth or falsity of a constraint by the setting of a 0-1 variable. If, for example, we have the constraint

$$\sum_j a_j\, x_j \leq b \qquad (1.24)$$

then we can incorporate the 0-1 variable y into the constraint to give

$$\sum_j a_j x_j + My \leq M + b \qquad (1.25)$$

where M is an upper bound on the expression

$$\sum_j a_j x_j - b \qquad (1.26)$$

If $y = 1$ the constraint (1.24) is forced to hold. If $y = 0$ it becomes vacuous. Sometimes we may wish to model the condition

$$\sum_j a_j x_j \leq b \ \textbf{implies} \ \ y = 1$$

This can be represented by its equivalent *contrapositive* statement

$$y = 0 \ \textbf{implies} \ \sum_j a_j x_j > b \qquad (1.27)$$

Since it is conventional only to use non-strict inequalities we can replace

$$\sum_j a_j x_j > b$$

by

$$\sum_j a_j x_j \geq b + \epsilon \qquad (1.28)$$

where ϵ is a suitable small positive number. Now (1.27) can be modelled (by analogy with (1.25)) as

$$\sum_j a_j x_j + my \geq b + \epsilon \qquad (1.29)$$

where m is a lower bound on the expression

$$\sum_j a_j x_j - b$$

It is now possible to model the standard connectives of the Propositional Calculus applied to constraints by means of (in)equalities between 0-1 variables.

We have already modelled the **'or'** condition by constraints (1.9) and (1.10) and (1.12) to (1.15). The **'and'** condition is simply modelled by repeating the constraints (as in LP). The **'implies'** condition is modelled thus:

$$y_1 = 1 \text{ \textbf{implies} } y_2 = 1$$

is represented by $y_1 - y_2 \leq 0$

We now have all the machinery for modelling logical conditions which will be explained in greater depth in Hooker's chapter (see chapter 3).

5.   Reformulation Techniques

Despite advances in methods of solving Discrete Optimisation problems, as well as the dramatic increase in the speed and storage capacity of computers, many such problems still remain very difficult to solve. There is, however, often great flexibility in the way such problems are modelled as Integer Programmes. It is frequently possible to remodel a problem in a way which makes it easier to solve. This is the subject of this section.

5.1   The Convex Hull of Integer Solutions

The constraints of a Linear Programme (LP) restrict the feasible solutions to a set which can be represented by a Polytope in multidimensional space. For Integer Programmes the set is further restricted to the lattice of integer points within a Polytope. This is illustrated, in Figure 1.2, by the constraints of a 2-Variable IP, which can therefore be represented in 2-dimensional space.

$$
\begin{align}
-x_1 + 2x_2 &\leq 7 &\text{(1.30)}\\
x_1 + 3x_2 &\leq 15 &\text{(1.31)}\\
7x_1 - 3x_2 &\leq 23 &\text{(1.32)}\\
x_1, x_2 &\geq 0 \text{ and integer} &\text{(1.33)}
\end{align}
$$

The boundaries of constraints (1.30), (1.31) and (1.32) are represented by the lines AB, BC and CD respectively and the non-negativity conditions by OA and OD. However, the integrality restrictions on x_1 and x_2 restrict us further to the lattice of points marked. If we were to ignore the integrality conditions then, given an objective function, we would have an LP model. An optimal

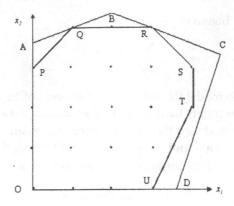

Figure 1.2.   The convex hull of a pure IP

solution would lie at one of the vertices O, A, B, C, or D (if there were alternative solutions some of them would be between these vertices). The LP got by ignoring the integrality conditions is known as the "LP Relaxation" of the model.

However, if we were to represent the constraints whose boundaries are the bold lines we would have a more constrained LP problem whose solutions would be O, P, Q, R, S, T or U which would satisfy the integrality requirements. The region enclosed in the bold lines is known as the *Convex Hull of Feasible Integer Solutions.* It is the smallest convex set containing the feasible integer solutions. For this example the constraints defining the convex hull are:

$$-x_1 + x_2 \leq 3 \tag{1.34}$$

$$x_2 \leq 4 \tag{1.35}$$

$$x_1 + x_2 \leq 7 \tag{1.36}$$

$$x_1 \leq 4 \tag{1.37}$$

$$2x_1 - x_2 \leq 6 \tag{1.38}$$

$$x_1, x_2 \geq 0 \tag{1.39}$$

Computationally LP models are much easier to solve than IP models. Therefore reformulating an IP by appending or substituting these *facet defining constraints* would appear to make a model easier to solve. With some types of model it is fairly straightforward to do this. In general, however, the derivation of facet defining constraints is extremely difficult and there is no known systematic way of doing it. What is more, there are often an astronomic number of facets making it impossible to represent them all with limited computer storage.

Before describing ways of partially using the concept above, to our advantage, we point out that the example above is a Pure Integer Programme. The concept still applies to Mixed Integer Programmes. Suppose, for example, that constraints (1.33) have been modified to only stipulate that x_2 should be integer. The feasible solutions would be as represented by the horizontal lines in Figure 1.3. The convex hull of feasible integer solutions is described by the

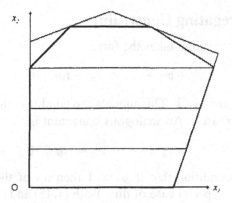

Figure 1.3. The convex hull of a mixed IP

bold lines and the facet defining constraints are:

$$-x_1 + x_2 \leq 3 \tag{1.40}$$
$$x_2 \leq 4 \tag{1.41}$$
$$7x_1 + 11x_2 \leq 65 \tag{1.42}$$
$$7x_1 - 3x_2 \leq 23 \tag{1.43}$$
$$x_1, x_2 \geq 0 \tag{1.44}$$

We now describe ways in which some commonly arising constraints can be reformulated to make the associated LP Relaxation 'tighter', even if we do not produce facet defining constraints for the complete model.

5.2   'Big M' coefficients

These arise when a particular set of activities (continuous or integer) can only take place if another, $0 - 1$ variable, takes the value 1. An example of this is the constraint (1.7) in the Fixed Charge problem. The condition is correctly modelled so long as the value of M is sufficiently large not to place a spurious restriction on the value of $\sum_j x_j$ (or some other set of activities). But if M is larger than it need be, the associated LP relaxation will be less constrained than it need be i.e. the associated polytype of the LP relaxation will be larger than necessary. Therefore when such a situation arises it is desirable to make M a strict upper bound on the appropriate expression, if possible. It may be worth maximising the appropriate expression subject to the other constraints (as an LP) in order to find as small a value of M to be acceptable. A number of packages try to reduce the values of 'Big M coefficients' automatically.

5.3   Disaggregating Constraints

A common PIP constraint takes the form

$$x_1 + x_2 + \ldots + x_n - ny \leq 0 \qquad (1.45)$$

where all variables are $0 - 1$. This models the condition that if any of x_j take the value 1 then so must y. An analogous constraint is

$$x_1 + x_2 + \ldots + x_n - ny \geq 0 \qquad (1.46)$$

which models the condition that if $y = 1$ then all of the x_j must be one. Constraint (1.20) is a special case of this. Both (1.45) and (1.46) can be disaggregated into

$$x_j - y \leq 0 \quad \text{for all} \quad j \qquad (1.47)$$

and

$$x_j - y \geq 0 \quad \text{for all} \quad j \qquad (1.48)$$

respectively. The groups of constraints (1.47) and (1.48) both have the same set of feasible integer solutions as (1.45) and (1.46) respectively but have more constrained LP relaxations. e.g. the solution $x_1 = \frac{1}{2}, x_2 = \frac{3}{4}, x_3 = \ldots = x_n = 0, y = \frac{1}{n}$ satisfies (1.45) and (1.46) but violates (1.47) and (1.48). Hence this fractional solution cannot arise from an LP relaxation using (1.47) and (1.48). However, any fractional solution which satisfies (1.47) and (1.48) also satisfies (1.45) and (1.46) respectively. There are many other, more complicated, constraints and groups of constraints which have been analysed to give tighter LP relaxations. See for example Wolsey [26] and Nemhauser and Wolsey [19].

5.4   Splitting Variables

In order to disaggregate constraints it is sometime possible to split variables up into component variables which can then appear in different constraints. A general way of doing this is described by Balas [2] and Jeroslow [15] under the title 'Disjunctive Constraints'.

We have already shown how it is possible to model a disjunction of constraints (1.11) by (1.12) to (1.15). There is another formulation for which the LP relaxation is tighter.

Suppose we have a disjunction of groups of constraints which can be written as

$$A_1 x \geq b_1 \text{ or } A_2 x \geq b_2 \text{ or } \ldots \text{ or } A_n x \geq b_n \qquad (1.49)$$

where x and b_i are vectors. We also assume that the constraints in each of the 'clauses' above have the same *recession directions*. This means that if they are open polytopes, as represented in Figure 4, the directions of the extreme rays are the same. The polytopes in Figure 4 have different recession directions since direction DS does not form part of the left hand polytope. If, as is the

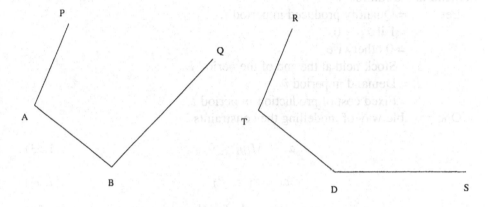

Figure 1.4. Polytopes with different recession directions

case with the constraints in (1.11), the polytopes are all closed then the problem does not arise since there are no recession directions. In practice this will often be the case as we will place 'big M' coefficients as upper bounds on the variables. For disjunction (1.49), however, we make no such assumption. In order to model (1.49) we split the vector x into n component vectors x_1, x_2, ..., x_n i.e.

$$x = x_1 + x_2 + \ldots + x_n \qquad (1.50)$$

We then specify the following constraints

$$A_j x_j \quad \geq b_j y_j \text{ for all } j \qquad (1.51)$$

$$y_1 + y_2 + \ldots + y_n = 1 \qquad (1.52)$$

where y_j are $0-1$ integer variables. Balas [2] shows that if it is possible to represent an IP as a disjunction of polytopes in this way then we can obtain a convex hull representation (in the higher dimensional space of $x = x_1 + x_2 + \ldots + x_n$)

5.5   Multicommodity Flows

If a MIP model contains variables representing flows of material it may be advantageous to disaggregate these flows into different commodities. This is illustrated by the **lot sizing problem** (see e.g. Wolsey [26]). In this problem it has to be decided, in different periods, whether to produce a certain quantity (a 'lot') of material to satisfy future demand. There is a fixed charge associated with producing any at all. Hence an optimal policy will be only to produce in certain periods but retain stock from this production to satisfy some of the demand in the future.

Let x_i = Quantity produced in period i

y_i = 1 if $x_i \geq 0$

= 0 otherwise

z_i = Stock held at the end of the period i

d_i = Demand in period i

f_i = Fixed cost of production in period i

One possible way of modelling the constraints is

$$x_i - M_i y_i \leq 0 \tag{1.53}$$

$$x_1 - z_1 = d_1 \tag{1.54}$$

$$z_{i-1} + x_i - z_i = d_i \text{ for all } i \neq 1 \tag{1.55}$$

Notice the use of the 'big-M' coefficient in constraints (1.53) representing upper bounds on the amount produced in each period.

An alternative, and more constrained, way of modelling this problem is to split the variables into components representing different 'commodities'. We distinguish between the different quantities produced in a period to meet demand in *different* future periods.

Let x_{ij} = Quantity produced in period i to meet demand in period j

z_{ij} = Stock held at end of period i to meet demand in period j

The constraints can now be reformulated as

$$x_{ij} - d_j y_i \leq 0 \text{ for all } i, j \tag{1.56}$$

$$x_{1j} - z_{1j} = d_j \text{ for all } j \tag{1.57}$$

$$\sum_{i<j} (z_{i-1,j} + x_{ij} - z_{ij}) = d_j \text{ for all } j \tag{1.58}$$

Notice that it is no longer necessary to use the 'big-M' coefficients since the demand in period j places an upper bound on the value of each x_{ij} variable.

If all the constraints were of the form (1.56), (1.57) and (1.58) it can be shown that the LP relaxation would produce the integer optimal solution. In practice there will probably be other constraints in the model which prevent this happening. However, this second formulation is more constrained (although larger).

5.6 Reducing the Number of Constraints

One of the effects of most of the reformulations described in the previous section has been to increase the number of constraints. While there is great advantage to be gained in tightening the LP relaxation in this way the increased size of the models will, to some extent, counter the reduction in overall solution time. Therefore, in this section, we look at ways of reducing the number of constraints in an IP model.

5.6.1 Avoiding Constraints by Virtue of Optimality.
Sometimes certain constraints will automatically be satisfied at the optimal solution. For example, with the formulation of the fixed charge problem constraint (1.7) does not rule out the 'stupid' solution $x_j = 0$ for all j and $y = 1$. However, it is not necessary to put in extra constraints to avoid this since such a solution would be non-optimal.

However, there are situations in which it is desirable to put in constraints which are not necessary, by virtue of optimality, but which aid the solution process using the Branch and Bound algorithm described below. Hooker et al [14] discusses such an example in relation to modelling a chemical process and terms the extra constraints as 'logic cuts'.

5.6.2 Avoiding Constraints by Introducing Extra Variables.
A number of formulations of important Combinatorial Optimisation problems involve an **exponential** number of constraints. The best known example of this is the famous **Travelling Salesman Problem** discussed in eg Lawler et al [17]. The problem is to route a Salesman around a number of cities, returning to the beginning and covering the minimum total distance. It can be formulated as a PIP model as follows:

$$x_{ij} = 1 \quad \text{if tour goes from } i \text{ to } j \text{ directly}$$
$$= 0 \quad \text{otherwise}$$

Given that the distance from i to j is c_{ij} the model is

$$\text{Minimise} \quad \sum_{ij} c_{ij} x_{ij} \quad (1.59)$$

$$\text{subject to} \quad \sum_i x_{ij} \;=\; 1 \text{ for all } j \qquad\qquad (1.60)$$

$$\sum_j x_{ij} \;=\; 1 \text{ for all } i \qquad\qquad (1.61)$$

$$\sum_{i,j \in S} x_{ij} \;\leq\; |S| - 1 \quad \text{for all } S \subset \{2, 3, \ldots, n\} \quad (1.62)$$

where $2 \leq |S| \leq |V| - 2$. Constraints (1.60) and (1.61) guarantee that each city is entered exactly once and left exactly once respectively. However, constraints (1.62) are needed to prevent 'subtours' i.e. going around disconnected subsets of the cities. There are $2^{n-1} - n - 1$, such constraints: i.e. an **exponential** number. It would be impossible to fit all these constraints in a computer for even modest values of n. In practice such constraints are usually added (with others) on an 'as-needed' basis in the course of optimisation.

However, it is possible to avoid an exponential number of constraints by introducing extra variables (a polynomial number). There are a number of ways of doing this which are surveyed by Orman and Williams [20]. We present one such formulation here. It relies on a Network Flow formulation associated with the network of arcs between cities.

$$y_{ij} = \text{Flow (of some commodity) between } i \text{ and } j$$

$$y_{ij} - (n-1)\, x_{ij} \leq 0 \quad \text{for all } i, j \qquad\qquad (1.63)$$

$$\sum_j y_{1j} = n - 1 \qquad\qquad (1.64)$$

$$\sum_j y_{ij} - \sum_k y_{ki} = -1 \quad \text{for all } i \neq 1 \qquad\qquad (1.65)$$

together with constraints (1.60) and (1.61).

Constraints (1.63) only allow flow in an arc if it exists (similar to the 'big M' constraints discussed earlier). Constraint (1.64) introduces a flow of $n-1$ units of the commodity into city 1 and constraint (1.65) takes 1 unit of the commodity out of all the other $(n-1)$ arcs. In order to dispose of all the commodity the underlying network must be connected, so ruling out subtours.

Hence the exponential number of constraints (1.62) has been replaced by constraints (1.63), (1.64) and (1.65): a total of n^2. In addition we have doubled the number of variables from $n(n-1)$ to $2n(n-1)$.

Unfortunately this formulation does not perform as well as the conventional formulation, for reasons described below, if a linear programming based IP algorithm is used. Rather surprisingly this formulation can be drastically improved by a *multicommodity* flow formulation analogous to that described for the lot sizing problem. This results in a model whose LP relaxation has equal strength to the conventional formulation, but has only a polynomial number of constraints.

5.7   Column Generation

Besides the advantages of introducing extra variables to avoid an excess number of constraints it may be very difficult (or impossible) to capture some complicated conditions by means of (linear) constraints. In such circumstances it may only be practical to enumerate all, or some of, the possible solutions to the complicating conditions. These can then be generated in the course of optimization if deemed worthwhile.

The classic example of this is the Cutting Stock Problem (see e.g. Gilmore and Gomory [9]) where it is wished to cut given orders for widths of material (e.g. wallpaper) out of standard widths. The objective is to minimise total waste (which is equivalent to minimising the total number of rolls used). An equivalent problem is sometimes referred to as the 'Bin Packing Problem'. Here we wish to pack items into bins using the minimum number of bins.

A possible formulation would be to include variables with the following interpretation:

$$x_{ij} = \quad \text{Quantity of order } i \text{ (e.g. a given width) taken from roll } j$$

Constraints could then be stated to:

(i) limit the widths taken from each standard roll to be within the standard
       width

(ii) meet total demand

This would be a somewhat cumbersome model with a large amount of unnecessary symmetry (i.e. many equivalent solutions would be investigated in the course of optimisation).

The usual way of solving such problems is to enumerate some of the solutions to (i). This is done by giving possible patterns of widths which may fit into the standard widths.

We illustrate this important technique by a small example. A plumber has a stock of 13 metre copper pipes and needs to cut off the following orders:

10 lengths of 4 metres
10 lengths of 5 metres
23 lengths of 6 metres

How should he cut up the 13 metre pipes in order to minimise the number he needs to use ? There are a number of possible cutting patterns he could adopt,

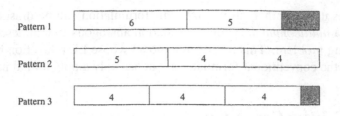

Figure 1.5.   A cutting pattern

for example three are shown in Figure 1.5. In fact, in this example, there are
only six possible patterns (leaving out obviously redundant ones, which are
'dominated' by others). However we will begin by restricting ourselves to the
above three.

Variables are introduced with the following interpretation:

$$x_j = \text{ Number of standard pipes cut using pattern } j \text{ in our model}$$

The model becomes:
$$\text{Minimise } x_1 + x_2 + x_3 \qquad (1.66)$$

$$\text{Subject to: } \quad 2x_2 + 3x_3 \; \geq \; 10 \qquad (1.67)$$
$$x_1 + x_2 \; \geq \; 10 \qquad (1.68)$$
$$x_1 \; \geq \; 23 \qquad (1.69)$$
$$x_1, x_2, x_3 \; \geq \; 0 \text{ and integer}$$

Notice that we have modelled the demands for 4, 5 and 6 metre pipes respectively by constraints (1.67) to (1.69) i.e. we have modelled constraints (ii).
However we have predefined some solutions to constraints (i). If we solve the
above model we obtain the solution

$$x_1 = 23, x_2 = 0, x_3 = 4$$

i.e. we use 23 of pattern 1 and 4 of pattern 3 making a total of 27 standard
lengths. However, we have not incorporated some patterns in the model. Notice that each possible pattern gives a *column* of the model e.g. the column for
x_2 has coefficients 2,1,0 since pattern 2 has 2 of length 4, 1 of length 5 and 0
of length 5.

If the above model is solved as a *Linear Programme* (LP) we obtain *dual
values* for the constraints (this concept is explained in any standard text on LP).
In this case they are $\frac{1}{3}, 0, 1$. Suppose we were to consider another pattern with
a_1 of length 4, a_2 of length 5 and a_3 of length 6 we must have

$$4a_1 + 5a_2 + 6a_3 \leq 13 \qquad (1.70)$$

The variable corresponding to this pattern would have a *reduced cost* of

$$1 - \frac{1}{3}a_1 - a_3$$

To make it a desirable variable to enter the solution we wish to make this quantity as small as possible.

$$\text{i.e. Maximise} \quad \frac{1}{3}a_1 + a_3 \tag{1.71}$$

subject to constraint (1.70). This type of IP with one constraint is known as a *Knapsack Problem* and efficiently solved by *Dynamic Programming* (see e.g. Bellman [5]). Details of this procedure are beyond the scope of this chapter. For this example the optimal solution is $a_1, a_2 = 0, a_3 = 2$ i.e. we should consider using the pattern in Figure 1.6.

Pattern 4   6   6

Figure 1.6. Optimal pattern

Note that the analysis has been based on the *LP Relaxation* and not the IP solution to (1.66)-(1.69). Therefore it is not necessarily the case that the new pattern will be beneficial. However, in this case we append a new variable x_4 to the model, representing pattern 4. This gives

$$\text{Minimise} \quad x_1 + x_2 + x_3 + x_4 \tag{1.72}$$

$$\text{Subject to:} \quad 2x_2 + 3x_3 \geq 10 \tag{1.73}$$

$$x_1 + x_2 \geq 10 \tag{1.74}$$

$$x_1 + 2x_4 \geq 23 \tag{1.75}$$

$$x_1, x_2, x_3, x_4 \geq 0 \text{ and integer}$$

This revised model has optimal solution

$$x_1 = 5, \quad x_2 = 5, \quad x_3 = 0, \quad x_4 = 9$$

i.e. we use 5 of pattern 1, 5 of pattern 2 and 9 of pattern 4 making a total of 19 standard lengths. Clearly this is an improvement of the previous solution. The procedure can be repeated generating new patterns only when worthwhile. In fact, for this example, this is the optimal solution demonstrating that only 3 of the six possible patterns are ever needed.

Column Generation techniques, such as this, are common in LP and IP to avoid excessive numbers of variables. The exact formulations of the applications are problem specific and beyond the scope of this chapter. One of the most economically important applications of the technique is to the Air Crew Scheduling problem.

6.   Solution Methods

There is an important class of problems for which Linear Programming (LP) automatically gives integer solutions making specialist IP methods unnecessary. These are models which are presented as, or can be converted to, *Network Flow* models.

It is important to recognise that IP models are generally much more difficult to solve than LP models. Theoretically they lie in the NP-complete complexity class (unless they have a special structure) whereas LP models lie in the P complexity class (see for example Garey and Johnson [8]). This theoretical distinction is borne out by practical experience. Some IP models can prove very difficult to solve optimally in a realistic period of time. It is often much more difficult (than for LP models) to predict how long it will take to solve them. However, as discussed earlier, reformulation can sometimes have a dramatic effect on solution success.

Historically the first method used to solve IPs was known as *'Cutting Planes'* and is due to Gomory [11] and [13]. In practice, however, another, simpler, method known as Branch-and-Bound due to Land and Doig [16] and Dakin [7] was used and variants formed the basis of commercial systems. These two methods are described in the next two sections.

Commercial systems now, however, use a combination of both methods. While Branch-and-Bound forms the basic solution framework, Cutting Planes are used within this framework (Branch-and-Cut). This combined approach is also described here. In addition heuristic choices are made within Branch-and-Bound to guide the, somewhat arbitrary, tree search.

Reformulation methods are also incorporated in some systems (usually known as Presolve).

Finally variables (columns) are sometimes generated in the course of optimization. This is usually referred to as 'Branch-and-Price' for the reason described in the previous section.

6.1   Cutting Planes

In order to illustrate this approach we will refer to the example illustrated in figure 1.2. Each of the new facet defining constraints, given there (inequalities (1.34), (1.35), (1.36), (1.37) and (1.38)) can be regarded as *cutting planes*. If appended to the model they may aid the solution process. Rather than define them all, prior to solving the model, the usual approach is to append them, in the course of optimisation, if needed.

We illustrate this process by considering the model with an objective function i.e.

$$\text{Maximise } x_1 + x_2 \tag{1.76}$$

$$\text{Subject to: } -x_1 + 2x_2 \leq 7 \tag{1.77}$$

$$x_1 + 3x_2 \leq 15 \qquad (1.78)$$

$$7x_1 - 3x_2 \leq 23 \qquad (1.79)$$

$$x_1, x_2 \geq 0 \text{ and integer} \qquad (1.80)$$

Figure 1.7 illustrates this. This first step is to solve the *Linear Programming*

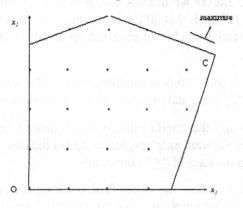

Figure 1.7. An integer programme

Relaxation of the model. That is to ignore the integrality requirements and to solve the model as an (easier) LP.

The concept of a '*Relaxation*' is broader than this and used widely in the solution of IP models. The essential idea

$$\text{Subject to:} \qquad 2x_2 + 3x_3 \geq 10 \qquad (1.81)$$

$$x_1 + x_2 \geq 10 \qquad (1.82)$$

$$x_1 + 2x_4 \geq 23 \qquad (1.83)$$

$$x_1, x_2, x_3, x_4 \geq 0 \text{ and integer}$$

is to leave out some of the conditions or constraints on a (difficult) IP model and solve the resultant (easier) model. By leaving out constraints we may well obtain a '*better*' solution (better objective value), but one which does not satisfy all the original constraints. This solution is then used to advantage in the subsequent process.

For the above example the optimal solution of the LP Relaxation is

$$x_1 = 4\frac{3}{4}, x_2 = 3\frac{5}{12}, \quad \text{Objective} = 8\frac{1}{6}$$

We have obtained the optimal fractional solution at C. The next step is to define a *cutting plane* which will *cut-off* the fractional solution at C, without removing any of the feasible integer solutions (represented by bold dots). This is known as the *separation* problem. Obviously a number of cuts are possible. The facet defining constraints, are possible cuts as are "shallower" cuts which are not facets. A major issue is to create a *systematic* way of generating those cuts (ideally facet defining) which *separate* C from the feasible integer solutions. Before we discuss this, however, we present a major result due to Chvátal [6]. This is that all the facet defining constraints (for a PIP) can be obtained by a finite number of repeated applications of the following two procedures.

(i) *Add together constraints* in suitable multiples (when all in the same form eg "$\leq$" or "$\geq$") and add or subtract "$=$" constraints in suitable multiples.

(ii) Divide through the coefficients by their greatest common divisor and *round the right-hand-side coefficient* up (in the case of "$\geq$" constraints) or down (in the case of "$\leq$" constraints).

We illustrate this by deriving all the facet defining constraints for the example above. However, we emphasise that our choice of which constraints to add, and in what multiples, is ad-hoc. It is a valid procedure but not systematic. This aspect is discussed later.

1. Add
$$-x_1 + 2x_2 \leq 7$$

$$-x_1 \leq 0$$

to give $-2x_1 + 2x_2 \leq 7$

Divide by 2 and round to give

$$-x_1 + x_2 \leq \left\lfloor \frac{7}{2} \right\rfloor = 3 \tag{1.84}$$

2. Add
$$-x_1 + 2x_2 \leq 7$$

$$-x_1 + 3x_2 \leq 15$$

to give $5x_2 \leq 22$

Divide by 5 and round to give

$$x_2 \leq \left\lfloor \frac{22}{5} \right\rfloor = 4 \tag{1.85}$$

3. Add
$$x_1 + 3x_2 \leq 15$$

$$-7x - 3x_2 \leq 23$$

to give $8x_1 \leq 38$

Divide by 2 and round to give

$$x_1 \leq \left\lfloor \frac{38}{8} \right\rfloor = 4 \qquad (1.86)$$

4. Add $\qquad x_1 + 3x_2 \leq 15$

$$x_1 \leq 4$$

to give $3x_1 + 3x_2 \leq 23$

Divide by 3 and round to give

$$-x_1 + x_2 \leq \left\lfloor \frac{23}{3} \right\rfloor = 7 \qquad (1.87)$$

5. Add $\qquad 2 \times (7x_1 - 3x_2 \leq 23)$

$$-x_1 \leq 0$$

to give $14x_1 - 7x_2 \leq 46$

Divide by 7 and round to give

$$-2x_1 - x_2 \leq \left\lfloor \frac{46}{7} \right\rfloor = 6 \qquad (1.88)$$

Constraints (1.84) to (1.88), together with the original constraints (1.80) make up all the facet defining constraints for this model. Note that all the new facet defining constraints, apart from constraint (1.87), have been derived using *one* rounding operation, after adding some of the *original* constraints in suitable multiples. Such constraints are known as **rank-1 cuts**. In contrast constraint (1.87) was derived using constraint (1.86), which was itself derived as a rank-1 cut. Therefore constraint (1.87) is known as a **rank-2 cut**. The rank of a cut is a measure of the complexity of its derivation. For some problems (eg. The Travelling Salesman Problem) there is no limit to the rank of the facet defining constraints, giving a further indication of their computational difficulty.

As emphasised above, however, in general the derivation of Chvátal cuts is arbitrary. It is a valid 'proof' procedure for showing cuts to be valid but we have not given a systematic way of doing this. In general there is no such systematic way. If, however, we use LP to obtain the solution of the relaxed problem we can use this solution to derive some Chvátal cuts.

In the example we began by obtaining the fractional solution at C. At C constraints (1.78) and (1.79) are 'binding' i.e. they cannot be removed without altering the LP optimal solution. Also x_1 and x_2 take positive values (in LP parlance they are 'basic' variables). If we add 'slack' variables to constraints (1.78) and (1.79) to make them equations we obtain

$$x_1 + 3x_2 + u_2 \quad = \quad 15 \tag{1.89}$$

$$7x_1 - 3x_2 \quad + \quad u_3 = 23 \tag{1.90}$$

We can eliminate x_1 from equation (1.90) (using equation (1.89)) and then x_2 from equation (1.90) (using equation (1.90)) by Gaussian Elimination. Making x_1 and x_2 the subjects of their respective equations gives:

$$x_1 = \frac{19}{4} - \frac{u_2}{8} - \frac{u_3}{8} \tag{1.91}$$

$$x_2 = \frac{41}{12} - \frac{7u_2}{24} + \frac{u_3}{24} \tag{1.92}$$

Setting u_1 and u_2 ('non-basic' variables) to zero gives the fractional solution, we have already obtained, to the LP Relaxation. However we want an *integer* solution to (1.78), (1.79) and other constraints (1.79) and (1.80). Clearly since x_1 and x_2 can only take integer values this is true for u_1 and u_2 as well. (1.91) and (1.92) can be rewritten separating out their integer and fractional parts and also ensuring that the expression on the right is non-negative. This gives:

$$x_1 + u_2 + u_3 - 4 = \frac{3}{4} + \frac{7}{8}u_2 + \frac{7}{8}u_3 \tag{1.93}$$

$$x_2 + u_2 - 3 = \frac{5}{12} + \frac{17}{24}u_2 + \frac{u_3}{24} \tag{1.94}$$

Since the expressions on the right must be both *integer* and *strictly positive* we have

$$\frac{3}{4} + \frac{7}{8}u_2 + \frac{7}{8}u_3 \geq 1 \tag{1.95}$$

$$\frac{5}{12} + \frac{17}{24}u_2 + \frac{u_3}{24} \geq 1 \tag{1.96}$$

Using expressions (1.89) and (1.90) to substitute for u_2 and u_3 and eliminating fractions gives

$$7(15 - x_1 - 3x_2) + 7(23 - 7x_1 + 3x_2) \geq 2$$

i.e.

$$7x_1 \leq 33 \tag{1.97}$$

and

$$17\,(15 - x_1 - 3x_2) + (23 - 7x_1 + 3x_2) \geq 14$$

i.e.

$$x_1 + 2x_2 \leq 11 \tag{1.98}$$

The cuts which we have derived are known as **Gomory Cuts** and are illustrated in figure 1.8. Notice that these cuts are shallower than Chvátal cuts but they

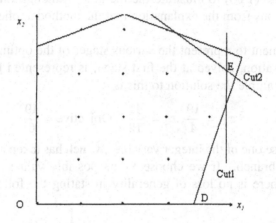

Figure 1.8.   An integer programme with Gomory cuts

can be generated systematically if the Simplex Algorithm is used to solve the LP Relaxation.

We could append either or both of cuts (1.97) and (1.98) to the model and resolve. If both cuts are appended we obtain another fractional solution at E.

$$x_1 = 4\frac{11}{17}, x_2 = 3\frac{3}{7}, \quad \text{Objective} = 7\frac{14}{17}$$

This solution can be used to generate further cuts. Ultimately, in a finite number of steps, we will obtain the optimal integer solution at S (figure 1.2).

$$x_1 = 4, x_2 = 3, \quad \text{Objective} = 7$$

If, however, we had appended the Chvátal cuts (1.87) and (1.88) at the first stage we could have obtained the optimal integer solution in one step. As we have emphasised before, however, there is no known systematic way of generating Chvátal cuts.

Further discussion of the many methods of generating cutting planes is beyond the scope of this introductory chapter but can be found in eg. [19]. We should, however, mention that the generation of Gomory Cuts can be generalised to deal with the (much more common) Mixed Integer (MIP) case. This is done by Gomory [12]. Although such cuts are usually "very weak", in practice they often prove very effective.

6.2   Branch-and-Bound

This approach again starts by solving the LP Relaxation. We again use the model (1.76) to (1.80) to illustrate the method. Although this model is a PIP it will be obvious from the explanation that the method applies equally well to MIPs.

It is convenient to represent the various stages of the optimisation by a *tree*. The LP Relaxation, solved at the first stage, is represented by node 0 of the tree. In our example the solution to this is

$$x_1 = \frac{19}{4}, x_2 = \frac{41}{12}, \text{ Objective} = \frac{49}{6}$$

We then choose one of the integer variables, which has taken a fractional value, and create a branch. If we choose x_1 its possible values are illustrated in figure 1.9. There is no loss of generality in stating the following *dichotomy*

Figure 1.9.   Possible values of an integer variable

$x_1 \leq 4$ **or** $x_1 \geq 5$. Whats more this rules out the current fractional solution. The choice of which integer variable to *branch-on* is heuristic. In general if a variable x takes a fractional value $N + f$ where $0 < f < 1$ then we create the dichotomy $x \leq N$ **or** $x \geq N + 1$. We append each of the alternatives extra constraints separately to the original model and represent the two new problems as two new nodes of the solution tree, illustrated in figure 1.10.

The situation is illustrated in Figure 1.11.

The solution at node 1 corresponds to the point S_1. For convenience the new nodes are numbered in order of their generation. At each node we write the solution of its LP relaxation. Notice that

(a) the objective value gets worse (*strictly* no better) as we go down a branch (smaller for a maximisation or larger for a minimisation).

(b) the LP relaxation of a new node may become infeasible.

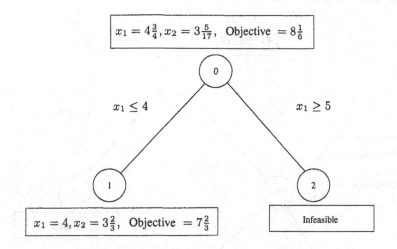

$$x_1 = 4\tfrac{3}{4}, x_2 = 3\tfrac{5}{17}, \quad \text{Objective } = 8\tfrac{1}{6}$$

0

$x_1 \le 4$ $x_1 \ge 5$

1   2

$$x_1 = 4, x_2 = 3\tfrac{2}{3}, \quad \text{Objective } = 7\tfrac{2}{3}$$ Infeasible

Figure 1.10.   The first branch of a solution tree

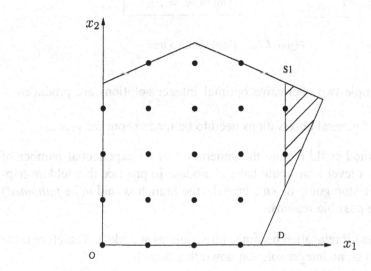

Figure 1.11.   Solution space of the first branch

The process can now be continued by branching on variable x_2 to produce the solution tree in Figure 1.12.

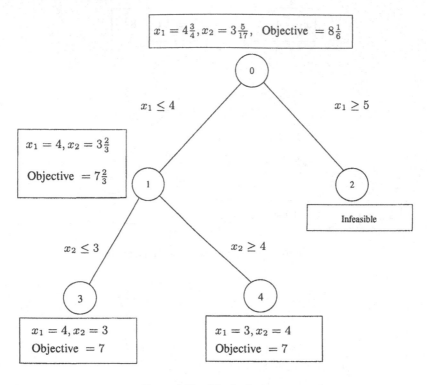

$$x_1 = 4\tfrac{3}{4}, x_2 = 3\tfrac{5}{17}, \quad \text{Objective} = 8\tfrac{1}{6}$$

0

$x_1 \leq 4$　　　　　　$x_1 \geq 5$

$$x_1 = 4, x_2 = 3\tfrac{2}{3}$$

$$\text{Objective} = 7\tfrac{2}{3}$$

1

2

Infeasible

$x_2 \leq 3$　　　　　　$x_2 \geq 4$

3

4

$$x_1 = 4, x_2 = 3$$
$$\text{Objective} = 7$$

$$x_1 = 3, x_2 = 4$$
$$\text{Objective} = 7$$

Figure 1.12.　Final solution tree

In this example two alternative optimal integer solutions are produced at nodes 3 and 4.

A number of general observations need to be made about the process.

1 The method could require the generation of an exponential number of nodes. At level r we could have 2^r nodes. In practice this seldom happens. We stop going down a branch (the branch is said to be *fathomed*) for three possible reasons.

   a. The LP relaxation becomes infeasible as at node 2. Therefore there can be no integer solution down this branch.

   b. We obtain an integer solution as at node 3.

   c. The objective value of the LP relaxation is worse than that of a known integer solution. Such a branch is said to be *bounded* because the objective value of the known integer solution provides a bound (lower bound for a maximisation or upper bound for a min-imisation) on the values of the optimal integer solution. This has not happened in this case.

2 In principle this process might not terminate if the feasible region of the LP relaxation was 'open'. However, in practice, this seldom happens and can be avoided by giving the integer variables upper and lower (usually zero) bounds.

3 When a node does not lie at the bottom of a branch which has been fathomed it is known as a *Waiting Node*. Once there are no more waiting nodes (as in Figure 1.12) the best integer solution must be an optimal solution. Should no integer solution be found the model is shown to be (integer) infeasible.

4 The choices of which variables to branch on and which waiting nodes to develop are *heuristic*. How these choices are made in practice are beyond the scope of this chapter (and sometimes specific to the com-mercial packages used). There are, however, general strategic choices to be made regarding which waiting nodes to develop. A *depth-first* strat-egy involves going down individual branches of the tree in the hope of finding an integer solution quickly. Such a solution may be satisfactory (although not proven to be optimal) or used as a bound in order to fathom other future branches. A *breadth-first* strategy involves solving the LP relaxation of all waiting nodes at a particular level before proceeding to the next level. For example we solved the LP relaxation at node 2 before developing node 1 further. Which strategy is best adopted depends on the nature of the problem and whether one wants a (good) integer solu-tion or the proven optimum. Such considerations are discussed in [24]. There are also computational considerations as to how one reoptimises the LP relaxation between nodes (eg. The *Primal* or *Dual Simplex* algo-rithms). These are beyond the scope of this chapter but discussed in eg. [19].

6.2.1   Other Branches.   In order to illustrate the Branch-and-Bound method we branched on *integer variables* using the *dichotomy* that if an integer variable x takes the fractional value $N + f$ where $0 < f < 1$ we have

$$x \leq N \text{ or } x \geq N + 1$$

Other dichotomies are possible

$$\text{eg. } x = 0 \text{ or } x > 0$$

6.2.2   Branching on Entities.   In section 1.3.2 we explained the concept of *Special Ordered Sets* and emphasised they should be regarded as *entities* rather than just as individual variables. This is to allow the concept of branching on the entity. If for example we have a S2 set $\{x_0, x_1, x_2, \ldots, x_n\}$ we require that *at most two adjacent x_j can be non-zero* (In practice we usually also stipulate the *convexity* constraint that the variables sum to 1, but this is not necessary to the concept). The branching dichotomy that we use is

$$x_0 + x_1 + \ldots + x_{j-1} = 0 \text{ or } x_{j+1} + x_{j+2} + \ldots + x_n = 0$$

The rational behind this is that the order of the variables is significant. Therefore, if a number of variables currently take non-zero values, we choose a variable 'in the middle' of these, x_j, and, without loss of generality, force all the variables to the left to zero or all the variables to the right to be zero. Such a branching can radically reduce the size of the tree search from branching on individual variables.

6.3   Branch-and-Cut

The use of cutting planes to solve IP models has already been discussed. There is often great merit in using them within the Branch-and-Bound method. If, for example we were to apply the Gomory Cuts (1.97) and (1.98) to the LP relaxation (at node 0) we would start with the fractional solution at E in Figure 1.8 and the Branch-and-Bound tree would take the same course as before. In general, however, one might expect it to reduce this search tree. Such cuts, incorporated into the solution process, would be known as *global cuts* since, being applied at node 0, they would be valid for the whole model.

Alternatively, or additionally, we could apply cuts at particular nodes of the tree. Such cuts would be *local cuts* since they might only be valid at that particular node and all nodes descending from it. For example, at node 1 in Figure 1.10 we have the solution represented by point S_1 in Figure 1.11. We could obtain Gomory Cuts to the LP relaxation at node 1 and apply them. One such Gomory Cut obtainable here is (1.98) (In this example it is also a global cut). Appending this cut gives the solution

$$x_1 = 4, x_2 = 3\frac{1}{2}, \text{ Objective } = 7\frac{1}{2}$$

In this, trivially small, example there is no obvious advantage in applying cutting planes in the tree search. However for larger examples it is often worthwhile.

6.4   Lagrangian Relaxation

The methods described so far have relied on the *LP Relaxation* of an IP to help in the solution process. It has already been pointed out that this is not the only possible relaxation. Indeed many relaxations are possible. One such, that has proved particularly valuable for many problems, is *Lagrangian Relaxation*. The idea here is to remove some of the constraints of the model and incorporate them in the objective function. There will then be a penalty (degradation of the objective function) in violating them but violation will not render the relaxed model infeasible. If the constraints so removed are complicated ones to satisfy then the resultant relaxed model may be easier to solve.

Because the success of the method depends on which constraints are removed and incorporated in the objective function its use is problem dependent. We illustrate the method by means of our previous small example.

$$\text{Maximise} \quad x_1 + x_2 \tag{1.99}$$

$$\text{Subject to:} \quad -x_1 + 2x_2 \le 7 \tag{1.100}$$

$$x_1 + 3x_2 \le 15 \tag{1.101}$$

$$7x_1 - 3x_2 \le 23 \tag{1.102}$$

$$x_1, x_2 \ge 0, \text{ and integer} \tag{1.103}$$

Suppose we were to remove constraint (1.102) but penalise breaking it. The amount by which it is broken could be written as

$$V = 7x_1 - 3x_2 - 23 \tag{1.104}$$

As we wish to minimise this quantity we could subtract $7x_1 - 3x_2$ in a suitable multiple, from the objective function. The multiple we choose can be based on the dual value of the constraint (if we start by solving the LP relaxation). The usual method is to use a technique called *subgradient optimisation* which is beyond the scope of this chapter but described in [19]. This multiplier can be regarded as a *Lagrange Multiplier.* Hence the use of the name "*Lagrangian Relaxation*" For the purposes of this illustration we will take the multiplier as 0.1. This transforms the problem to :

$$\text{Maximise} \quad \frac{3}{10}x_1 + \frac{13}{10}x_2 \tag{1.105}$$

$$\text{Subject to:} \quad -x_1 + 2x_2 \le 7 \tag{1.106}$$

$$x_1 + 3x_2 \le 15 \tag{1.107}$$

$$x_1, x_2 \ge 0, \tag{1.108}$$

Now it might be the case, for problems with a structure, that the model above was easier to solve *as an IP* than the original model.

An example of this is if we were left with one constraint. We would then have a *Knapsack Problem* which we could solve quickly by Dynamic Programming (see Bellman [5]). Another example is with the Travelling Salesman Problem where we are left with a Spanning Tree Problem for which there exists a fast simple algorithm. It should be noted, however, that if the relaxed problem is easier to solve as an IP because it is an LP which gives an integer solution e.g. a Network Flow problem, then there is no virtue to be gained in *Lagrangian Relaxation.*

Returning to our example, solving the model above (by whatever means) yields the optimal integer solution.

$$x_1 = 3, \ x_2 = 4 \qquad\qquad (1.109)$$

Note that the constraint removed (1.102) has been satisfied automatically whereas if it had been removed and given a smaller penalty it might not have been satisfied. This is clearly one of the optimal solutions to the original problem.

6.5   Heuristics

Heuristics are quick, easy-to-understand methods which are widely used to solve combinatorial problems quickly but non-optimally. They are usually problem specific. Therefore for those two reasons they are not discussed in this general chapter on *optimisation* (There is considerable discussion in Chapter 6 by Cordeau and Laporte in relation to Distribution problems). However it is very valuable to incorporate heuristics inside the Branch-and-Bound method. The overall structure of Branch-and-Bound guarantees an *optimal* solution, if carried out to completion, but the incorporation of heuristics within this structure can greatly speed the process.

We have already mentioned that heuristics can be valuable in the choices of branching variables and waiting node. They can, however, also be applied at a waiting node to try to obtain an *integer* solution before developing the node. Such an integer solution, if found, will give a *bound* on the value of the final optimal integer solution. This bound may be used, to good effect, in bounding other waiting nodes. It may also, of course, be a valuable solution in itself.

6.6   Constraint Satisfaction

This, alternative, approach to solving some Discrete Optimization problems can be very powerful. It is discussed at length in Chapter 3 by Hooker. An alternative name for this approach is Constraint Logic Programming. We outline the method here.

One of the merits of Constraint Satisfaction is that it allows a richer set of modelling expressions than simply the linear expressions of IP. The essential idea of Constraint Satisfaction is that every variable has a finite domain of possible values it may take. The predicates within this modelling language relate the variables. Therefore when a variable is assigned a particular value in its domain this may further restrict the possible values which related variables can take in their domains. For example in the *Assignment Problem* we wish to assign n objects $\{a_1, a_2, \ldots, a_n\}$ to n other objects $\{b_1, b_2, \ldots, b_n\}$ so that each object is uniquely assigned. The usual IP model would contain 0-1 variables.

$$x_{ij} = 1 \text{ iff } a_i \text{ is assigned to } b_j$$

$$= 0 \text{ otherwise}$$

For Constraint Satisfaction we would have different (less) variables

$$x_i = \text{index of object to which } a_i \text{ is assigned}$$

together with a predicate

$$\textbf{all\_different}\{x_1, x_2, \ldots, x_n\}$$

meaning that each of $a_1, a_2, \ldots, a_n$ must be assigned to a different object b_1, $b_2, \ldots, b_n$. Each x_i would initially have the domain $\{1, 2, \ldots, n\}$. When however a particular x_i is assigned to a particular value b_j (either permanently or temporarily) its index would be taken out of the domains of the other x_i (either permanently or temporarily). A tree search is performed to systematically explore the possible feasible solutions. Applying the search intelligently can result in the obtaining of a feasible solution more quickly than for IP.

As the above example demonstrates Constraint Satisfaction models are usually much more compact than IP models. The successive *domain reduction* of the variables mirrors the *Presolve* facility of many commercial IP systems as was mentioned previously.

However Constraint Satisfaction does not rely on the concept of an LP relaxation which can prove very valuable in reducing a tree search. In fact for some models (eg. the Assignment problem mentioned above) the LP relaxation yields a feasible integer solution immediately. For some, other problems with a strong *combinatorial* flavour, however, the LP relaxation is of little value.

Constraint Satisfaction sometimes is a very good way of finding feasible solutions quickly to 'needle-in-a-haystack' type problems where the finding of any feasible solution is difficult. It is less successful at finding *optimal* solutions to problems with many feasible solutions (eg. the Travelling Salesman problem). In such cases successively improved objective values have to be used to try to find solutions which are better.

There is considerable promise in designing *hybrid* systems which combine both the methods of IP and Constraint Satisfaction (see chapter 4. To date these have been most successful for specialist problems where the structure of the model can be exploited. So far there has been less progress in general hybrid systems.

6.7   Other Methods

In view of its immense practical importance, combined with the major difficulties of computation, there has been much research in alternative methods. To date these have not resulted in systems which can compete with Branch-and-Bound (augmented by Cutting Planes, Presolve etc). Logic methods have been applied to some pure 0-1 models. Here the linear constraints of IP are replaced by expressions in the *Propositional Calculus (Boolean Algebra)* and Logic methods (eg. *Resolution*) used to solve the resultant model. Chapter 3 by Hooker discusses some of these methods.

Another class of methods (not covered in this volume) use analogous but *discrete* concepts to those of linear algebra (Hilbert and Gröbner bases and lattice reduction, etc). A reference to such methods is Aardal, Weismantel and Wolsey [1].

The method of Projection (Fourier-Motzkin elimination) has been extended to IP models by Williams [22] and [23] and is well covered in Martin [18] with a discussion of the solution of Diophantine equations. Finally Schrijver [21] gives a very full account of the theoretical aspects of solving IP models with very many references.

References

[1] K. AARDAL, R. WEISMANTEL, AND L. A. WOLSEY. Non-Standard Approaches to Integer Programming, **Discrete Applied Mathematics**, 123 (2002), 5-74

[2] E. BALAS. Disjunctive Programming: Properties of the Convex Hull of feasible points, **Discrete Applied Mathematics**, 89 (1998), 1-46

[3] E.M.L. BEALE and J.A. TOMLIN. Special facilities in a general mathematical programming system for non-convex problems using ordered sets of variables, in J. Lawrence [Ed], **Proceedings of the 5TH INTERNATIONAL CONFERENCE ON OPERATIONS RESEARCH** 1969 Tavistock, London

[4] E.M.L. BEALE. Branch and Bound methods for numerical optimisation of non-convex functions, in M.M. Barritt and D. Wishart (Eds) COMPSTAT 80: **PROCEEDINGS IN COMPUTATIONAL STATISTICS 1975** pp 11-20, Physica Verlag, Wien

[5] R. BELLMAN. *Dynamic Programming.* 1957 Princeton University Press

[6] V. CHVÁTAL. Edmonds Polytopes and a Hierarchy of Combinatorial Problems, **Discrete Mathematics**, 4 (1973) 305-337

[7] R. J. DAKIN. A Tree Search Algorithm for Mixed Integer Programming Problems, **Computer Journal**, 8 (1965), 250-255

[8] M. R. GAREY AND D. S. JOHNSON. Computers and Interactibility: A Guide to the Theory of NP-Completeness, 1979, Freeman

[9] P.C. GILMORE AND R.E. GOMORY. A Linear Programming Approach to the Cutting Stock Problem Part I, **Operations Research**, 9 (1961) 849 – 859

[10] F. GLOVER. Improved Linear Integer Programming Formulations of Nonlinear Integer Problems, **Management Science** 224 (1975), 455-459

[11] R. E. GOMORY. Outline of an Algorithm for Integer Solutions to Linear Programs, **Bulletin of the American Mathematical Society**, 64 (1958), 275-278

[12] R. E. GOMORY. An Algorithm for the Mixed Integer Problem, Research Report, RM-2597 (1960), The Rand Corporation

[13] R. E. GOMORY. An Algorithm for Integer Solutions to Linear Programs, **Recent Advances in Mathematical Programming**, R. Graves and P. Wolf (Eds), 1983, McGraw-Hill pp. 269-302

[14] J. N. HOOKER, H. YAN, I. GROSSMANN AND R. RAMAN. Logic cuts processing networks with fixed charges, **Computers and Operations Research**, 21 (1994) 265-279

[15] R. JEROSLOW. Logic-based decision support: Mixed integer model formulation, Annals of Discrete Mathematics 40, 1989, North Holland, Amsterdam

[16] A. H. LAND AND A. G. DOIG. An Automatic Method for Solving Discrete Programming Problems, **Econometrics**, 28 (1969) 497-520

[17] E.L. LAWLER, J.K. LENSTRA, A.H.G. RINNOOY KAN AND D.B. SHMOYS (Eds). **The Travelling Salesman Problem**. 1995, Wiley, Chichester

[18] R. K. MARTIN. **Large Scale Linear and Integer Optimization**, 1999, Kluwer

[19] G. L. NEMHAUSER AND L.A. WOLSEY. *Integer and Combinatorial Optimisation*. 1988, Wiley, New York

[20] A.J. ORMAN AND H.P. WILLIAMS. **A Survey of Different Integer Programming Formulations of the Travelling Salesman Problem**, Working Paper LSEOR 04.67, 2004, London School of Economics.

[21] A. SCHRIJVER. **Theory of Linear and Integer Programming**, 1986, Wiley

[22] H. P. WILLIAMS. Fourier-Motzkin Elimination Extension to Integer Programming Problems, **Journal of Combinatorial Theory (A)**, 21 (1976), 118-123

[23] H. P. WILLIAMS. A Characterisation of all Feasible Solutions to an Integer Program, **Discrete Applied Mathematics**, 5 (1983) 147-155

[24] H. P. WILLIAMS. *Model Solving in Mathematical Programming*. Wiley, 1993

[25] H. P. WILLIAMS. *Model Building in Mathematical Programming*. 4th Edition, Wiley, 1999

[26] L. A. WOLSEY. Strong formulations for mixed integer programming: a survey, **Mathematical Programming,** 45 (1989) 173-191

Chapter 2

CONTINUOUS APPROACHES FOR SOLVING DISCRETE OPTIMIZATION PROBLEMS*

Panos M Pardalos, Oleg A Prokopyev and Stanislav Busygin
Department of Industrial and Systems Engineering
University of Florida
303 Weil Hall, Gainesville FL 32611
USA
{ pardalos,oap4ripe,busygin } @ufl.edu

Abstract   This chapter contains short expository notes on applying continuous approaches for solving discrete optimization problems. We discuss some general aspects of the connection between integer programming and continuous optimization problems, along with several specific examples. The considered problems include maximum clique, satisfiability, the Steiner tree problem, minimax and semidefinite programming.

Keywords: Discrete optimization, global optimization, non-convex optimization, semidefinite programming, linear complementarity problem, max clique, satisfiability, minimax theory, Steiner problem

1.   Introduction

Discrete (or combinatorial) optimization problems are usually defined as problems with a discrete feasible domain and/or a discrete domain objective function. These types of problems model a great variety of applications in different areas of human activities.

Solution methods for discrete optimization problems are generally classified into *combinatorial and continuous approaches*. Continuous approaches are based on different characterizations, or reformulations of the considered problem in a continuous space. These characterizations include equivalent continuous nonconvex formulations, or continuous relaxations, which embed the

---

*This work was partially supported by grants from NSF, NIH, CRDF and AirForce.

initial discrete domain into a larger continuous space. While combinatorial approaches represent a more natural and typical way of addressing discreteness of the considered problems, continuous methods may provide new insight into the problem structure and properties, and allow one to develop more efficient algorithms for computing optimal and approximate solutions.

In this chapter we give a brief overview of some continuous approaches for solving discrete problems. This chapter is a completely revised and rewritten version of [54].

2.   Equivalence of Mixed Integer and Complementarity Problems

Given matrices $A_{n \times n}$, $B_{n \times l}$ and a vector $b \in \mathbb{R}^n$ with rational entries, *the mixed integer feasibility problem* is to find (x, z) such that $x \in \mathbb{R}^n$, $x \geq 0$, $z \in \{0, 1\}^l$ that satisfy

$$Ax + Bz = b.$$

Given a matrix $M_{n \times n}$ and a vector $q \in \mathbb{R}^n$ with rational entries, *the linear complementarity problem (LCP)* is to find $x \in \mathbb{R}^n$ such that

$$x \geq 0, \ Mx + q \geq 0, \ x^T(Mx + q) = 0,$$

or prove that such x does not exist.

THEOREM 1 *[53] The mixed integer feasibility problem can be reduced to the solution of a linear complementarity problem.*

Proof.   The condition $z_i \in \{0, 1\}$ can be equivalently expressed as:

$$z_i + w_i = 1, \ z_i \geq 0, \ w_i \geq 0, \ z_i w_i = 0,$$

where z_i are continuous variables and for each z_i a new continuous variable w_i is introduced. Let $s, t \in \mathbb{R}^n$ be such that

$$s = Ax + Bz - b \geq 0, \ t = -Ax - Bz + b \geq 0.$$

These two inequalities are satisfied if and only if $s = t = 0$, which implies that $Ax + Bz = b$. Therefore, the mixed integer feasibility problem is reduced to solution of the following LCP(M, q): Find v, y such that

$$v \geq 0, y \geq 0, v^T y = 0, v = My + q,$$

where

$$y = \begin{pmatrix} z \\ x \\ \theta \end{pmatrix}, v = \begin{pmatrix} w \\ s \\ t \end{pmatrix}, M = \begin{pmatrix} -I & 0 & 0 \\ B & A & 0 \\ -B & -A & 0 \end{pmatrix}, q = \begin{pmatrix} e \\ b \\ -b \end{pmatrix},$$

where $\theta \in R^n$ and $e \in R^l$ is the vector of all 1's.

□

LCP generalizes or unifies a great variety of important problems in optimization. Furthermore, complementarity is a fundamental tool in continuous optimization since it expresses optimality conditions [37, 38]. For more detailed information on LCP and related problems we also refer to [13, 45, 51, 53]. Some recent results on successive convex relaxation methods for LCP and related problems are discussed in [46].

Let $S = \{x : Mx + q \geq 0, \ x \geq 0\}$ be the domain defined by the inequality constraints in LCP, then the following theorem can be formulated [38]:

THEOREM 2 *Let* $f_1(x) = x^T(Mx+q)$ *and* $f_2(x) = \sum_{i=1}^{n} \min\{(Mx+q)_i, x_i\}$, *where* $(Mx+q)_i$ *is the i–th component of* $Mx + q$, *and* x_i *the i–th component of* x. *Then each solution* x^* *of the LCP satisfies*

(i) $f_1(x^*) = \min\{f_1(x) : x \in S\} = 0$,

(ii) $f_2(x^*) = \min\{f_2(x) : x \in S\} = 0$,

and conversely.

This result implies that LCP is equivalent to the quadratic minimization problem in (i), or to the problem stated in (ii), where the objective function is piecewise linear concave. Therefore, the mixed integer feasibility problem can be reformulated as equivalent continuous optimization problems (i) and (ii) with known optimal objective function values.

Moreover, it can be also shown that the LCP can be always solved by solving a specific mixed integer feasibility problem [53, 55]. Given the LCP(M, q) consider the following mixed zero-one integer problem:

$$0 \leq My + \alpha q \leq e - z,$$
$$\alpha \geq 0, 0 \leq y \leq z, \qquad (2.1)$$
$$z \in \{0, 1\}^n.$$

THEOREM 3 *Let* (α^*, y^*, z^*) *with* $\alpha^* > 0$ *be a feasible point of the above mixed integer problem (which is always feasible). Then* $x^* = y^*/\alpha^*$ *solves the LCP. If* $\alpha^* = 0$ *is the only feasible value of* α, *then the LCP has no solution.*

We have shown above that LCP can be reduced to a specific mixed integer feasibility problem and vice versa, any mixed integer feasibility problem can be reformulated as an instance of LCP. It means that these two problems are equivalent. The equivalence of mixed integer programming and linear complementarity problems strongly supports the point of view that discrete problems can be viewed as continuous global optimization problems.

3.   Continuous Formulations for 0–1 Programming Problems

Consider a classical general linear 0–1 programming problem:

$$\min \quad c^T x$$

$$\text{s.t.} \quad Ax \le b, \; x \in \{0,1\}^n$$

where A is a real $m \times n$ matrix, $c \in \mathbb{R}^n$ and $b \in \mathbb{R}^m$. We can reduce this problem to an equivalent concave minimization problem:

$$\min \quad f(x) = c^T x + \mu x^T (e - x)$$

$$\text{s.t.} \quad Ax \le b, \; 0 \le x \le e$$

where μ is a sufficiently large positive number. The function $f(x)$ is concave since $-x^T x$ is concave. The equivalence of the two problems is based on the facts that a concave function attains its minimum at a vertex, and that $x^T (x - e) = 0, 0 \le x \le e$, implies $x_i = 0$ or 1 for $i = 1, \ldots, n$. Note that a vertex of the feasible domain is not necessarily a vertex of the unit hypercube $0 \le x \le e$, but the global minimum is attained only when $x^T (e - x) = 0$, provided that μ is large enough.

Applying similar techniques, general nonlinear 0–1 problems can be reduced to equivalent concave minimization problems. For example, consider the quadratic 0–1 problem of the following form:

$$\min \quad f(x) = c^T x + x^T Q x$$

$$\text{s.t.} \quad x \in \{0,1\}^n$$

where Q is a real symmetric $n \times n$ matrix. Given any real number μ, let $\bar{Q} = Q + \mu I$ where I is the $n \times n$ identity matrix, and $\bar{c} = c - \mu e$. Then the above quadratic 0–1 problem is equivalent to the problem:

$$\min \quad \bar{f}(x) = \bar{c}^T x + x^T \bar{Q} x$$

$$\text{s.t.} \quad x \in \{0,1\}^n$$

because of the equality $\bar{f}(x) = f(x)$. If we choose μ such that $\bar{Q} = Q + \mu I$ becomes a negative semidefinite matrix (e.g., $\mu = -\lambda$, where λ is the largest eigenvalue of Q), then the objective function $\bar{f}(x)$ becomes concave and the constraints can be replaced by $0 \le x \le e$. Therefore, the problem is equivalent to the minimization of a quadratic concave function over the unit hypercube.

For a general 0–1 problem, in order to be formulated as an equivalent continuous problem, function $f(x)$ must possess some specific properties. The following result was proved in [23]:

THEOREM 4 *Consider the problem*

$$\min f(x), \tag{2.2}$$

$$g(x) \geq 0, \tag{2.3}$$

$$x \in \{0, 1\}^n, \tag{2.4}$$

and the problem

$$\min[f(x) + \mu x^T(e - x)], \tag{2.5}$$

$$g(x) \geq 0, \tag{2.6}$$

$$0 \leq x \leq e \tag{2.7}$$

and suppose that f is bounded on $\bar{X} = \{x \in \mathbb{R}^n; 0 \leq x \leq e\}$ and Lipschitz continuous on an open set $A \supseteq \{0, 1\}^n$. Then $\exists \mu_0 \in \mathbb{R}$ such that $\forall \mu > \mu_0$ problems (2.2)-(2.4) and (2.5)-(2.7) are equivalent.

Moreover, it can be also shown that under some additional conditions the objective function of the continuous equivalent problem is concave:

THEOREM 5 *If $f(x)$ is $C_2(\bar{X})$, with $\bar{X} = \{x \in \mathbb{R}^n; 0 \leq x \leq e\}$ there is a $\mu_1 \in \mathbb{R}$ such that for any $\mu \geq \mu_1$ problem (2.5)-(2.7), equivalent to (2.2)-(2.4), has a concave objective function.*

For more detailed description of different relations between general integer programming and continuous optimization problems we refer the reader to [23].

4. The Maximum Clique and Related Problems

In the previous two sections we discussed some theoretical aspects of relations between discrete and continuous optimization problems. In this section we describe a specific example of applying continuous global optimization methods to solving *the maximum clique problem*. The maximum clique problem is a well-known hard combinatorial optimization problem which finds many important applications in different domains.

Let $G(V, E)$ be a simple undirected graph, $V = \{1, 2, \ldots, n\}$. The *adjacency matrix* of G is a matrix $A_G = (a_{ij})_{n \times n}$, such that $a_{ij} = 1$ if $(i, j) \in E$, and $a_{ij} = 0$ if $(i, j) \notin E$. The *complement graph* of G is the graph $\bar{G} = (V, \bar{E})$, where $\bar{E}$ is the complement of E.

A *clique* Q is a subset of V any two vertices of which are adjacent, i.e., for any distinct $i \in Q$ and $j \in Q$, $(i, j) \in E$. A clique is called *maximal* if there is no other vertex in the graph adjacent to all vertices of Q. A clique of the largest size in the graph is called a *maximum clique*. The *maximum clique problem* asks to find a maximum clique. Its cardinality is called the *clique number* of the graph and is denoted by $\omega(G)$.

Along with the maximum clique problem, we can consider the *maximum independent set problem*. A subset $I \subseteq V$ is called an *independent set* if the edge set of the subgraph induced by I is empty. The maximum independent set problem is to find an independent set of maximum cardinality. The *independent number* $\alpha(G)$ (also called the *stability number*) is the cardinality of a maximum independent set of G.

It is easy to observe that I is an independent set of G if and only if I is a clique of $\bar{G}$. Therefore, maximum independent set problem for some graph G is equivalent to solving maximum clique problem in the complementary graph $\bar{G}$ and vice versa.

Next, we may associate with each vertex $i \in V$ of the graph a positive number w_i called the vertex *weight*. This way, along with the adjacency matrix A_G, we consider the vector of vertex weights $w \in \mathbb{R}^n$. The total weight of a vertex subset $S \subseteq V$ will be denoted by

$$W(S) = \sum_{i \in S} w_i.$$

The *maximum weight clique problem* asks for a clique Q of the largest $W(Q)$ value. This value is denoted by $\omega(G, w)$. Similarly, we can define *maximum weight independent set problem*.

The maximum cardinality and the maximum weight clique problems along with maximum cardinality and the maximum weight independent set problems are NP-hard [22], so it is considered unlikely that an exact polynomial time algorithm for them exists. Approximation of large cliques is also hard. It was shown in [36] that unless $NP = ZPP$ no polynomial time algorithm can approximate the clique number within a factor of $n^{1-\epsilon}$ for any $\epsilon > 0$. In [42] this margin was tightened to $n/2^{(\log n)^{1-\epsilon}}$.

The approaches to the problem offered include such common combinatorial optimization techniques as sequential greedy heuristics, local search heuristics, methods based on simulated annealing, neural networks, genetic algorithms, tabu search, etc. However, there are also methods utilizing various formulations of the clique problem as a continuous optimization problem. An extensive survey on the maximum clique problem can be found in [5].

The simplest integer formulation of the maximum weight independent set problem is the following so called *edge formulation*:

$$\max f(x) = \sum_{i=1}^{n} w_i x_i, \tag{2.8}$$

subject to

$$x_i + x_j \leq 1, \ \forall \ (i, j) \in E, \ x \in \{0, 1\}^n. \tag{2.9}$$

In [60] Shor considered an interesting continuous formulation for the maximum weight independent set by noticing that edge formulation (2.8)-(2.9) is equivalent to the following multiquadratic problem:

$$\max f(x) = \sum_{i=1}^{n} w_i x_i, \tag{2.10}$$

subject to

$$x_i x_j = 0, \ \forall \ (i,j) \in E, \ x_i^2 - x_i = 0, \ i = 1, 2, ..., n. \tag{2.11}$$

Applying dual quadratic estimates, Shor reported good computational results [60]. Lagrangian bounds based approaches for solving some related discrete optimization problem on graphs is discussed in [61].

Next two continuous polynomial formulations were proposed in [34, 35].

THEOREM 6 *If x^* is the solution of the following (continuous) quadratic program*

$$\max f(x) = \sum_{i=1}^{n} x_i - \sum_{(i,j) \in E} x_i x_j = e^T x - 1/2 x^T A_G x$$

subject to

$$0 \leq x_i \leq 1 \text{ for all } 1 \leq i \leq n$$

then, $f(x^*)$ equals the size of the maximum independent set.

THEOREM 7 *If x^* is the solution of the following (continuous) polynomial program*

$$\max f(x) = \sum_{i=1}^{n} (1 - x_i) \prod_{(i,j) \in E} x_j$$

subject to

$$0 \leq x_i \leq 1 \text{ for all } 1 \leq i \leq n$$

then, $f(x^*)$ equals the size of the maximum independent set.

The reader is referred to [1] for description of several heuristic algorithms based on the results of the above two theorems.

In 1965, Motzkin and Straus formulated the maximum clique problem as a quadratic program over a simplex [50].

THEOREM 8 (MOTZKIN–STRAUS) *The global optimum value of the quadratic program*

$$\max f(x) = \frac{1}{2} x^T A_G x \tag{2.12}$$

subject to

$$\sum_{i \in V} x_i = 1, \ x \geq 0 \tag{2.13}$$

is

$$\frac{1}{2}\left(1 - \frac{1}{\omega(G)}\right). \tag{2.14}$$

A recent direct proof of this theorem can be found in [1]. Similar formulations with some heuristic algorithms were proposed in [8, 24].

In [7], the formulation was generalized for the maximum weight clique problem in the following natural way. Let $w_{\min}$ be the smallest vertex weight existing in the graph and a vector $d \in \mathbb{R}^n$ be such that

$$d_i = 1 - \frac{w_{\min}}{w_i}.$$

Consider the following quadratic program:

$$\max f(x) = x^T(A_G + \text{diag}(d_1, \ldots, d_n))x \tag{2.15}$$

subject to

$$\sum_{i \in V} x_i = 1, \ x \geq 0. \tag{2.16}$$

THEOREM 9 *The global optimum value of the program (2.15)-(2.16) is*

$$1 - \frac{w_{\min}}{\omega(G, w)}. \tag{2.17}$$

Furthermore, for each maximum weight clique Q^* of the graph $G(V, E)$ there is a global maximizer x^* of the program (2.15, 2.16) such that

$$x_i^* = \begin{cases} w_i/\omega(G, w), & \text{if } i \in Q^* \\ 0, & \text{if } i \in V \setminus Q^*. \end{cases} \tag{2.18}$$

Obviously, when all $w_i = 1$, we have the original Motzkin–Straus formulation.

Properties of maximizers of the Motzkin–Straus formulation were investigated in [25]. Generally, since the program (2.12) is not concave, its arbitrary (local) maximum does not give us the clique number. Furthermore, if x^* is some maximizer of (2.12), its nonzero components do not necessarily correspond to a clique of the graph. However, S. Busygin showed in [7] that nonzero components of any global maximizer of (2.15)-(2.16) correspond to a complete multipartite subgraph of the original graph and any maximal clique of this subgraph is the maximum weight clique of the graph. Therefore, it is not hard to infer a maximum weight clique once the maximizer is found.

Performing the variable scaling $x_i \rightarrow \sqrt{w_i} x_i$, one may transform the formulation (2.15)-(2.16) to

$$\max f(x) = x^T A_G^{(w)} x \qquad (2.19)$$

subject to

$$z^T x = 1, \; x \geq 0, \qquad (2.20)$$

where $A_G^{(w)} = (a_{ij}^{(w)})_{n \times n}$ is such that

$$a_{ij}^{(w)} = \begin{cases} w_i - w_{\min}, & \text{if } i = j \\ \sqrt{w_i w_j}, & \text{if } (i,j) \in E \\ 0, & \text{if } i \neq j \text{ and } (i,j) \notin E, \end{cases} \qquad (2.21)$$

and

$$z \in \mathbb{R}^n : \; z_i = \sqrt{w_i} \qquad (2.22)$$

is the vector of square roots of the vertex weights. The attractive property is this rescaled formulation is that maximizers corresponding to cliques of the same weight are located at the same distance from the zero point. Making use of this, S. Busygin developed a heuristic algortihm for the maximum weight clique problem, called QUALEX-MS, which is based on the formulation (2.19)-(2.20) and shows a great practical efficiency [7]. Its main idea consists in approximating the nonnegativity constraint by a spherical constraint, which leads to a trust region problem known to be polynomially solvable. Then, stationary points of the obtained trust region problem show correlation with maximum weight clique indicators with a high probability.

A good upper bound on $\omega(G, w)$ can be obtained using semidefinite programming. The value

$$\vartheta(G, w) = \max_{X \in \mathcal{S}_n^+} z^T X z, \qquad (2.23)$$

$$\text{s.t. } x_{ij} = 0 \; \forall (i,j) \in E, \quad \text{tr}(X) = 1,$$

where z is defined by (2.22), $\mathcal{S}_n^+$ is the cone of positive semidefinite $n \times n$ matrices, and $\text{tr}(X) = \sum_{i=1}^n x_{ii}$ denotes the *trace* of the matrix X, is known as the *(weighted) Lovász number* (ϑ-function) of a graph. It bounds from above the (weighted) independence number of the graph. Hence, considering it for the complementary graph, we obtain an upper bound on $\omega(G, w)$. It was shown in [9] that unless $P = NP$, there exists no polynomial-time computable upper bound on the independence number provably better than the Lovász number (i.e., such that whenever there was a gap between the independence number and the Lovász number, that bound would be closer to the independence number than the Lovász number). It implies that the Lovász number of the complementary graph is most probably the best possible approximation from above for the clique number that can be achieved in polynomial time in the worst case. For a survey on the Lovász number and its remarkable properties we refer the reader to [44].

It is worth mentioning that the semidefinite program (2.23) can be also used for deriving large cliques of the graph. Burer, Monteiro, and Zhang employed a low-rank restriction upon this program in their *Max-AO* algorithm and obtained good computational results on a wide range of the maximum clique problem instances [6].

5.   The Satisfiability Problem

The satisfiability problem (SAT problem) is one of the classical hard combinatorial optimization problems. This problem is central in mathematical logic, computing theory, artificial intelligence, and many industrial application problems. It was the first problem proved to be NP-complete [12, 22]. More formally, this problem is defined as follows [32]. Let $x_1, \ldots, x_m$ be a set of Boolean variables whose values are either 0 (false), or 1 (true), and let $\bar{x}_i$ denote the negation of x_i. A *literal* is either a variable or its negation. A *clause* is an expression that can be constructed using literals and the logical operation *or* ($\vee$). Given a set of n clauses $C_1, \ldots, C_n$, the problem is to check if there exists an assignment of values to variables that makes a Boolean formula of the following *conjunctive normal form (CNF)* satisfiable:

$$C_1 \wedge C_2 \wedge \ldots \wedge C_n,$$

where $\wedge$ is a logical *and* operation.

SAT problem can be treated as a constrained decision problem. Another possible heuristic approach based on optimization of a non-convex quadratic programming problem is described in [40, 41]. SAT problem was formulated as an integer programming feasibility problem of the form:

$$B^T w \leq b, \tag{2.24}$$

$$w \in \{-1, 1\}^n, \tag{2.25}$$

where $B \in \mathbb{R}^{n \times m}$, $b \in \mathbb{R}^m$ and $w \in \mathbb{R}^n$. It is easy to observe that this integer programming feasibility problem is equivalent to the following non-convex quadratic programming problem:

$$\max \ w^T w \tag{2.26}$$

$$B^T w \leq b, \tag{2.27}$$

$$-e \leq w \leq e. \tag{2.28}$$

Let $A = [B, I, -I]$, where I is an $n \times n$ identity matrix, and let $c^T = (b^T, 1, \ldots, 1)$. Then (2.26)-(2.28) can be rewritten as

$$\max \ w^T w \tag{2.29}$$

$$A^T w \leq c. \tag{2.30}$$

Kamath et al. applied an interior point method for solving problem (2.29)-(2.30) (a nice review on interior point methods for combinatorial optimization can be found in [49]). The proposed algorithm was based on minimization of the following potential function:

$$\min \ \psi(w) = \log \sqrt{m - w^T w} - \frac{1}{n} \sum_{k=1}^{n} \log(c_k - a_k^T w)$$

subject to

$$A^T w \leq c.$$

Furthermore, Kamath et al. applied their approach to solving SAT instances obtained from the inductive inference problem, which plays an important role in artificial intelligence and machine learning applications [41].

In another more recent approach, universal satisfiability models (*UniSAT*) can be formulated transforming a discrete SAT problem on Boolean space $\{0,1\}^m$ into a continuous SAT problem on real space $\mathbb{R}^m$. Thus, this decision problem is transformed into a global optimization problem which can be addressed applying global optimization techniques [29, 30].

The main idea of UniSAT model is to transform Boolean connectors $\vee$ and $\wedge$ in CNF formulas into $\times$ and $+$ of ordinary addition and multiplication operations, respectively. The *true* value of the CNF formula is converted to the global minimum value, i.e. 0, of the objective function, which can be chosen to be a multivariable polynomial in exponential or logarithmic form.

Given a CNF formula $F(\mathbf{x})$ from $\{0,1\}^m$ to $\{0,1\}$ with n clauses $C_1, \ldots, C_n$, we define a real function $f(\mathbf{y})$ from $\mathbb{R}^m$ to $\mathbb{R}$ that transforms the SAT problem into an unconstrained global optimization problem.

Nondifferentiable Unconstrained Global Optimization:

$$\min_{\mathbf{y} \in \mathbb{R}^m} f(\mathbf{y}) \tag{2.31}$$

where

$$f(\mathbf{y}) = \sum_{i=1}^{n} c_i(\mathbf{y}). \tag{2.32}$$

A clause function $c_i(\mathbf{y})$ is a product of m *literal functions* $q_{ij}(y_j)$ $(1 \leq j \leq m)$:

$$c_i = \prod_{j=1}^{m} q_{ij}(y_j), \tag{2.33}$$

where

$$q_{ij}(y_j) = \begin{cases} |y_j - 1| & \text{if literal } x_j \text{ is in clause } C_i \\ |y_j + 1| & \text{if literal } \bar{x}_j \text{ is in clause } C_i \\ 1 & \text{if neither } x_j \text{ nor } \bar{x}_j \text{ is in } C_i \end{cases} \tag{2.34}$$

The correspondence between **x** and **y** is defined as follows (for $1 \leq i \leq m$):

$$x_i = \begin{cases} 1 & \text{if } y_i = 1 \\ 0 & \text{if } y_i = -1 \\ undefined & \text{otherwise} \end{cases}$$

Clearly, $F(\mathbf{x})$ is true iff $f(\mathbf{y})=0$ on the corresponding $\mathbf{y} \in \{-1, 1\}^m$.

Polynomial Unconstrained Global Optimization:

$$\min_{\mathbf{y} \in \mathbb{R}^m} f(\mathbf{y}), \qquad (2.35)$$

where

$$f_1(\mathbf{y}) = \sum_{i=1}^{n} c_i(\mathbf{y}). \qquad (2.36)$$

A clause function $c_i(\mathbf{y})$ is a product of m *literal functions* $q_{ij}(y_j)$ $(1 \leq j \leq m)$:

$$c_i = \prod_{j=1}^{m} q_{ij}(y_j), \qquad (2.37)$$

where

$$q_{ij}(y_j) = \begin{cases} (y_j - 1)^{2p} & \text{if } x_j \text{ is in clause } C_i \\ (y_j + 1)^{2p} & \text{if } \bar{x}_j \text{ is in clause } C_i \\ 1 & \text{if neither } x_j \text{ nor } \bar{x}_j \text{ is in } C_i \end{cases} \qquad (2.38)$$

where p is a positive integer.

The correspondence between **x** and **y** is defined as follows (for $1 \leq i \leq m$):

$$x_i = \begin{cases} 1 & \text{if } y_i = 1 \\ 0 & \text{if } y_i = -1 \\ undefined & \text{otherwise} \end{cases}$$

Clearly, $F(\mathbf{x})$ is true iff $f(\mathbf{y})=0$ on the corresponding $\mathbf{y} \in \{-1, 1\}^m$.

These models transform SAT problem from a discrete, constrained decision problem into an unconstrained global optimization problem. A good property of the transformation is that these models establish a *correspondence between the global minimum points of the objective function and the solutions of the original SAT problem*. A CNF $F(\mathbf{x})$ is true *if and only if* $f(\mathbf{y})$ takes the global minimum value 0 on the corresponding **y**. Extensive computational testing and comparisons with classical techniques such as the Davis-Putnam algorithm [14] and other methods indicate the significance of the global optimization approaches [29, 30, 31].

Finally, an extensive survey on different solution methods for SAT problem including continuous optimization based methods can be found in [32]. For more information on SAT and related problems we refer the reader to [15].

6. The Steiner Problem in Graphs

Let $G = (\mathcal{N}, \mathcal{A}, \mathcal{C})$ be an undirected graph, where $\mathcal{N} = \{1, \ldots, n\}$ is a set of nodes, $\mathcal{A}$ is a set of undirected arcs (i, j) with each arc incident to two nodes, and $\mathcal{C}$ is a set of nonnegative costs c_{ij} associated with undirected arcs (i, j). The *Steiner Problem in Graphs (SPG)* is defined as follows: Given a graph $G = (\mathcal{N}, \mathcal{A}, \mathcal{C})$, and a node subset $\mathcal{R} \in \mathcal{N}$, find the minimum cost tree on G such that it would connect all the vertices in $\mathcal{R}$.

The *Steiner Problem in Directed Graphs (SPDG)* is the directed graph version of the Steiner problem. Consider $G^d = (\mathcal{N}, \mathcal{A}^d, \mathcal{C}^d)$, where $\mathcal{N} = \{1, \ldots, n\}$ is a set of nodes, $\mathcal{A}^d$ is a set of directed arcs (i, j) incident to nodes in $\mathcal{N}$, and $\mathcal{C}^d = \{c_{ij}^d\}$ is a set of costs associated with $\mathcal{A}^d$. Let $\mathcal{R}$ (*a set of regular nodes*) be a subset of $\mathcal{N}$, and r (*a root node*) be an arbitrary node in $\mathcal{R}$. We define a *directed Steiner tree* T rooted at node r, with respect to $\mathcal{R}$ on G, as a directed tree, where all nodes in $\mathcal{R} \setminus \{r\}$ can be reached from r via directed paths in T, and all leaf nodes are in $\mathcal{R}$.

The SPDG is defined as follows: Given a directed graph $G^d = (\mathcal{N}, \mathcal{A}^d, \mathcal{C}^d)$, a node subset $\mathcal{R} \in \mathcal{N}$, and a root node $r \in \mathcal{R}$, find the minimum cost directed Steiner tree on G with respect to $\mathcal{R}$ that has node r as the root node.

It is easy to observe that the SPDG is a generalization of the SPG. The SPG can be transformed to a special case of SPDG, where the arc cost structure is symmetric, in the following way. Given an undirected graph $G = (\mathcal{N}, \mathcal{A}, \mathcal{C})$, consider the corresponding directed graph $G^d = (\mathcal{N}, \mathcal{A}^d, \mathcal{C}^d)$, where for each undirected arc $(i, j) \in \mathcal{A}$ we create two oppositely directed arcs (i, j) and $(j, i) \in \mathcal{A}^d$ with the same cost, i.e. $c_{ij}^d = c_{ji}^d = c_{ij}$.

SPG is NP-hard, so it is considered unlikely that an exact polynomial time algorithm for it exists [21, 22]. Moreover, this problem remains NP-hard for several restricted cases such as unweighted graphs, bipartite graphs and planar graphs [22]. For more information on the Steiner tree and related problems, models, applications and algorithms we refer the reader to [11, 33, 39, 65]. A variant of the Steiner tree problem on a set of points in a metric space will be discussed in §2.8.

Next, we present a continuous formulation for SPDG proposed in [43]. In this formulation, SPDG is ϵ-approximated by a series of concave optimization problems. It can be also proved that the limit of the series with ϵ going to zero is the fixed charge formulation of SPDG.

Let ϵ be a positive real number and y_{ij} be a flow variable corresponding to arc (i, j). We can define the following minimum cost flow problem with a separable concave objective function $\mathcal{F}_\epsilon$:

$$(\mathcal{F}_\epsilon): \quad \min f(\epsilon, Y) = \sum_{(i,j) \in \mathcal{A}^d} c_{ij} \frac{y_{ij}}{y_{ij} + \epsilon} \qquad (2.39)$$

subject to

$$AY = B, \ Y \geq 0, \tag{2.40}$$

where A is an $|\mathcal{N}| \times |\mathcal{A}^d|$ node-arc incidence matrix, Y is an $|\mathcal{N}| \times 1$ vector of y_{ij}'s, B is an $\mathcal{N} \times 1$ vector whose entries b_i are associated with the set of nodes $\mathcal{N}$ as follows: $b_i = 0$ if $i \in \mathcal{N} \backslash \mathcal{R}$, $b_i = -1$ if $i \in \mathcal{R} \backslash \{r\}$, and $b_r = |\mathcal{R}| - 1$. It is not difficult to observe that due to the strict concavity of $f_\epsilon(Y)$, the unimodularity of A and integrality of B, the optimal solution of $\mathcal{F}_\epsilon$ is always integer.

Let $x_{ij} = \frac{y_{ij}}{y_{ij}+\epsilon}$. Next we can note that if ϵ goes to zero then the limit of $\mathcal{F}_\epsilon$ becomes the network formulation of SPDG of the form:

$$(\mathcal{F}): \qquad \min f(Y) = \sum_{(i,j) \in \mathcal{A}^d} c_{ij} x_{ij} \tag{2.41}$$

subject to

$$AY = B, \ Y \geq 0, \tag{2.42}$$

$$0 \leq y_{ij} \leq (|\mathcal{R}| - 1)x_{ij} \ \ \forall (i,j) \in \mathcal{A}^d, \tag{2.43}$$

$$x_{ij} \in \{0,1\} \ \ \forall (i,j) \in \mathcal{A}^d. \tag{2.44}$$

The following theorem gives the worst-case bound on the error of the SPDG approximation by the solution of $\mathcal{F}_\epsilon$ [43].

THEOREM 10 *The error on approximating f^* by $f^*(\epsilon)$ is less than*

$$\frac{\epsilon}{\epsilon+1} \sum_{(i,j) \in \mathcal{A}^d} c_{ij},$$

i.e.,

$$0 < f^* - f^*(\epsilon) < \frac{\epsilon}{\epsilon+1} \sum_{(i,j) \in \mathcal{A}^d} c_{ij}.$$

7.   Semidefinite Programming Approaches

Semidefinite programming (SDP) is a special case of convex programming and an extension of linear programming. More precisely, a semidefinite program is the problem of optimizing a linear function of a symmetric matrix subject to linear equality constraints and the constraint that the matrix is positive semidefinite. Since the set of positive semidefinite matrices constitutes a convex cone, semidefinite programming is also called linear programming over cones or cone-LP [2].

Let $\mathcal{S}^n$ be the family of symmetric matrices of order n. For any two matrices $A, \ B \in \mathcal{S}^n$, denote by $A \bullet B$ the inner product of A and B, i.e., $A \bullet B =$

$Tr(A^T B)$. If $A - B$ is positive semidefinite then we write $A \succeq B$. Formally, SDP is defined as the problem of the form [57]:

$$\min U \bullet Q_0 \qquad (2.45)$$

$$\text{s.t.} \quad U \bullet Q_i = c_i, \quad i = 1, \ldots, m \qquad (2.46)$$

$$U \succeq 0 \qquad (2.47)$$

Here, the given matrices $Q_i \in \mathcal{S}^n$ are required to be linearly independent, vector $c \in \mathbb{R}^n$, and unknown $U \in \mathcal{S}^n$.

The semidefinite programming is a rapidly growing area in optimization. For the last decade there has been a lot of research activity in semidefinite programming. The significance of semidefinite programming is that it allows one to formulate tighter continuous relaxations than linear programming relaxations. Since SDP can be approximately solved in polynomial time that leads to polynomial-time approximation algorithms for many discrete optimization problems.

Next we consider an example:

Max-Bisection (MB): Given an undirected graph $G = (V, E)$ with non-negative weights w_{ij} for each edge in E (and $w_{ij} = 0$ if $(i,j) \notin E$), find a partition (V_1, V_2) of vertex set V to maximize the weighted sum of edges between V_1 and V_2 under condition that $|V_1| = |V_2| = n/2$.

Assigning for each node i a corresponding binary variable x_i, MB can be formulated as the following quadratic binary programming problem:

$$\max_{x \in \{-1,1\}^n} \frac{1}{4} \sum_{1 \le i,j \le n} w_{ij}(1 - x_i x_j) \qquad (2.48)$$

$$\text{s.t.} \quad \sum_{i=1}^{n} x_i = 0, \qquad (2.49)$$

where constraint (2.49) ensures that $|V_1| = |V_2| = n/2$. The semidefinite relaxation of (2.48)-(2.49) can be expressed as:

$$\max \frac{1}{4} \sum_{i,j} w_{ij}(1 - X_{ij}) \qquad (2.50)$$

$$\text{s.t.} \quad ee^T \bullet X = 0, \ X_{jj} = 1, \ j = 1, \ldots, n, \ X \succeq 0, \qquad (2.51)$$

where $X \in \mathbb{R}^{n \times n}$ is a symmetric matrix. Obviously, (2.50)-(2.51) is a relaxation of (2.48)-(2.49), since for any feasible solution x of (2.48)-(2.49), $X = xx^T$ is feasible for (2.50)-(2.51).

Using this semidefinite programming relaxation Frieze and Jerrum [20] applied Goemans and Williamson's approach for Max-Cut [28] and developed

0.651-approximation algorithm for MB. This bound was improved to 0.699 by Ye in [66].

Furthermore, the result for the Max-Cut problem was generalized for a general boolean quadratic programming problem,

$$\max\{q(x) = x^T Q x \mid x \in \{\pm 1\}^n\},$$

to obtain an approximation result as follows

$$\bar{q} - q(x) \leq \frac{4}{7}(\bar{q} - \underline{q}),$$

where $\underline{q}$ and $\bar{q}$ denote their minimal and maximal objective values, respectively [52, 67].

An excellent short survey on SDP can be found in [63]. For a more complete list of references, surveys, and additional detailed information the reader may refer to [3, 4, 47, 56, 62] and an annotated bibliography on SDP by H. Wolkowicz [64]. Surveys on applying SDP in combinatorial optimization are to be found in [27, 58].

8.   Minimax Approaches

A minimax problem is usually formulated as

$$\min_{x \in X} \max_{y \in Y} f(x, y),$$

where $f(x, y)$ is a function defined on the product of sets X and Y [16]. Minimax theory initiated by Von Neumann plays a very important role in many areas including but not limited to game theory, optimization, scheduling, location, allocation, packing and computational complexity. A nice survey on minimax problems in combinatorial optimization is given in [10]. In [17] Du and Pardalos presented a short review on continuous minimax approaches for solving discrete problems. More detailed information on minimax theory and applications can be found in [16, 59].

We have discussed the Steiner tree problem in graphs in an earlier section. Similarly we can define the Steiner tree problem in a metric space: Given a set of points in a metric space, find a shortest network interconnecting the points in the set. Such shortest network is called the *Steiner minimum tree (SMT)* on the point set.

A *minimum spanning tree* on a set of points is defined as the shortest network interconnecting the given points with all edges between the points. The *Steiner ratio* in a metric space is the largest lower bound for the ratio between lengths of a minimum Steiner tree and a minimum spanning tree for the same set of points in the given metric space. Gilbert and Pollak conjectured that in the Euclidean plane this ratio is equal to $\sqrt{3}/2$ [26]. In 1990 Du and Hwang finally proved this result transforming the Steiner ratio problem to a continuous minimax problem [18]. The central part of the proof is a new theorem proved for the following minimax problem:

$$\min_{x \in X} \max_{i \in I} f_i(x),$$

where X is a polytope in the n-dimensional Euclidean space $\mathbb{R}^n$, I is a finite index set, and $f_i(x)$'s are continuous functions over X.

THEOREM 11 (MINIMAX THEOREM) *Let $g(x) = \max_{i \in I} f_i(x)$. If every f_i is a concave function, then the global minimum value of $g(x)$ over the polytope X is achieved at some critical point, namely, a point satisfying the following condition: there exists an extreme subset Y of X such that $x \in Y$ and the index set $I(x)(= \{i|g(x) = f_i(x)\})$ is maximal over Y.*

A continuous version of this theorem was proved by Du and Pardalos in [19].

In the next example we consider the problem of *packing circles in a square*. What is the maximum radius of n equal circles that can be packed into a unit square? This problem can be equivalently stated in the following way: How should n points be arranged into a unit square such that the minimum distance between them is greatest? Denote by r_n the maximum radius in the first formulation and by d_n the max-min distance in the second problem. It is easy to show that

$$r_n = \frac{d_n}{2(1 + d_n)}.$$

The second problem can be formulated as the following minimax problem:

$$\min_{x_i \in [0,1] \times [0,1]} \max_{1 \leq i < j \leq n} -\|x_i - x_j\|.$$

For fixed i and j the function $\|x_i - x_j\|$ is convex and, therefore, the above stated minimax theorem can be applied. In [48] Maranas et al. obtained some new results for $n = 15, 28, 29$ using the described minimax approach.

References

[1] J. Abello, S. Butenko, P.M. Pardalos, and M.G.C. Resende, Finding independent sets in a graph using continuous multivariable polynomial formulations, *Journal of Global Optimization* 21/4 (2001), pp. 111–137.

[2] F. Alizadeh, Optimization Over Positive Semi-Definite Cone: Interior-Point Methods and Combinatorial Optimization, in: P.M. Pardalos, ed., *Advances in Optimization and Parallel Computing*, North-Holland, 1992, pp. 1–25.

[3] F. Alizadeh, Interrior-Point Methods in Semidefinite Programming with Applications to Combinatorial Optimization, *SIAM J. Opt.* Vol. 5 (1995), pp. 13–51.

[4] S.J. Benson, Y. Ye, X. Zhang, Solving large-scale semidefinite programs for combinatorial optimization, *SIAM J. Optim.* Vol. 10/2 (2000) pp. 443–461.

[5] I.M. Bomze, M. Budinich, P.M. Pardalos, and M. Pelillo, The maximum clique problem, in: D.-Z. Du and P.M. Pardalos, eds., *Handbook of Combinatorial Optimization* (Supplement Volume A), Kluwer Academic (1999), 1–74.

[6] S. Burer, R.D.C. Monteiro, and Y. Zhang, Maximum stable set formulations and heuristics based on continuous optimization, *Mathematical Programming* **94**:1 (2002) 137–166.

[7] S. Busygin, A new trust region technique for the maximum weight clique problem, *Combinatorial Optimization 2002*, Paris, France, 2002. Submitted to *CO02 Special Issue of Discrete Applied Mathematics*. Available at http://www.busygin.dp.ua/npc.html.

[8] S. Busygin, S. Butenko, P.M. Pardalos, A Heuristic for the Maximum Indepepndent Set Problem Based on Optimization of a Quadratic Over a Sphere, *Journal of Combinatorial Optimization* Vol. 6, pp. 287–297, 2002.

[9] S. Busygin and D. Pasechnik, On $\bar{\chi}(G) - \alpha(G) > 0$ gap recognition and $\alpha(G)$-upper bounds, *ECCC Report TR03-052* (2003).

[10] F. Cao, D.-Z. Du, B. Gao, P.-J. Wan, P.M. Pardalos, Minimax Problems in Combinatorial Optimization, in: D.-Z. Du, P.M. Pardalos, eds., *Minimax and Applications*, Kluwer Academic Publishers, 1995, pp. 269–292.

[11] X. Cheng, Y. Li, D.-Z. Du, H.Q. Ngo, Steiner Trees in Industry, in: D.-Z. Du, P.M. Pardalos,eds., *Handbook of Combinatorial Optimization*, Suppl. Vol. B, pp. 193–216, 2005.

[12] S. Cook. The complexity of theorem-proving procedures, in: *Proc. 3rd Ann. ACM Symp. on Theory of Computing*, Association for Computing Machinery, pp. 151158, 1971.

[13] R.W. Cottle, J.S. Pang, R.E. Stone, *The Linear complementaruty problem.* Academic Press (1992).

[14] M. Davis, H. Putnam, A computing procedure for quantification theory, *Journal of ACM*, 7, pp. 201–215, 1960.

[15] D.-Z. Du, J. Gu, P.M. Pardalos, eds., *Satisfiability Problem: Theory and Applications*, DIMACS Series Vol. 35, American Mathematical Society (1998).

[16] D.-Z. Du, P.M. Pardalos, eds., *Minimax and Applications*, Kluwer Academic Publishers, 1995.

[17] D.-Z. Du, P.M. Pardalos, Global Minimax Approaches for Solving Discrete Problems, Lecture Notes in Economics and Mathematical Systems, vol. 452, Springer-Verlag (1997), pp. 34–48.

[18] D.Z. Du and F.K. Hwang, An approach for proving lower bounds: solution of Gilbert-Pollak's conjecon Steiner ratio, *Proceedings 31th FOCS* (1990), pp. 76–85.

[19] D.Z. Du and P.M. Pardalos, A continuous version of a result of Du and Hwang, *Journal of Global Optimization* 4 (1994), 127–129.

[20] Improved approximation algorithms for max k-cut and max bisection, *Proc. 4th IPCO Conference*, pp. 1–13, 1995.

[21] M. Garey and D. Johnson, The rectilinear Steiner tree problem is NP-complete, *SIAM J. Appl. Math.* 32, pp. 826–834, 1977.

[22] M. Garey and D. Johnson, *Computers and Intractability: A Guide to the Theory of NP-Completeness* (Freeman & Co., 1979).

[23] F. Giannesi and F. Niccolucci, Connections between nonlinear and integer programming problem, in: *Symposia Mathematica* Vol XIX, Instituto Nazionale di Alta Mathematica, Acad. Press. N.Y., pp. 161–176, 1976.

[24] L.E. Gibbons, D.W. Hearn, P.M. Pardalos, and M.V. Ramana, A Continuous based heuristyic for the maximum clique problem, in: D.S. Johnson and M.A. Trick, eds., *Cliques, Coloring and Satisfiability: Second DIMACS Implementation Challenge*, DIMACS Series, American Math Society, Vol. 26, pp. 103–124, 1996.

[25] L.E. Gibbons, D.W. Hearn, P.M. Pardalos, and M.V. Ramana, Continuous characterizations of the maximum clique problem, *Math. Oper. Res.* 22 (1997) 754–768.

[26] E.N. Gilbert, H.O. Pollak, Steiner minimal trees, *SIAM J. Appl. Math,* 16 (1968), pp. 1–29.

[27] M.X. Goemans, Semidefinite programming in combinatorial optimization, *Math. Programming* 79 (1997), pp. 143–162.

[28] M.X. Goemans, D.P. Williamson, Improved approximation algorithms for maximum cut and satisfiability problems using semidefinite programming, *J. Assoc. Comput. MAch.* 42/6 (1995), pp. 1115–1145.

[29] J. Gu, Optimization algorithms for the satisfiability (SAT) problem, in: D.-Z. Du and J. Sun, eds., *Advances in Optimization and Approximation*, Nonconvex Optimization and its Applications, Vol. 1, Kluwer Academic Publishers, pp.72–154, 1994.

[30] J. Gu, Global optimization Satisfiability (SAT) Problem, *IEEE Trans. on Knowledge and data Engineering*, Vol. 6, No. 3, pp. 361–381, 1994.

[31] J. Gu, Parallel Algorithms for Satisfiability (SAT) Problem, in: P.M. Pardalos et al., eds., *Parallel Processing on Discrete Optimization Problems*, DIMACS Series, American Math. SOciety, Vol. 22, pp. 105–161, 1995.

[32] J. Gu, P.W. Purdom, J. Franko, B. W. Wah, Algorithms for the Satisfiability (SAT) Problem: A Survey, in: D.-Z. Du, J. Gu, P.M. Pardalos, eds., *Satisfiability Problem: Theory and Applications*, DIMACS Series Vol. 35, American Mathematical Society (1998), pp. 19–151.

[33] D. Guesfield, *Algorithms on Strings, Trees, and Sequences: Computer Science and Computational Biology*, Cambridge University Press, 1997.

[34] J. Harant, Some news about the independence number of a graph, *Discussiones Mathematicae Graph Theory*, Vol. 20, pp. 7179, 2000.

[35] J. Harant, A. Pruchnewski, and M. Voigt, On dominating sets and independent sets of graphs, *Combinatorics, Probability and Computing*, Vol. 8, pp. 547553, 1999.

[36] J. Håstad, Clique is hard to approximate within $n^{1-\epsilon}$, in: *Proc. 37th Annual IEEE Symposium on the Foundations of Computer Science (FOCS)* (1996) 627–636.

[37] R. Horst, P.M. Pardalos (Editors), *Handbook of Global Optimization*. Nonconvex Optimization and its Applications Vol. 2, Kluwer Academic Publishers, 1995.

[38] R. Horst, P.M. Pardalos, and N.V. Thoai, *Introduction to Global Optimization*. Kluwer Academic Publishers, Dordrecht, The Netherlands, 2nd edition, 2000.

[39] F.K. Hwang, D.S. Richards, P. Winter, *Steiner Tree Problems*, North-Holland, Amsterdam, 1992.

[40] A.P. Kamath, N.K. Karmarkar, K.G. Ramakrishnan, M.G.C. Resende, Computational experience with an interrior point algorithm on the Satisfiability problem, *Annals of Operations Research* 25 (1990), pp. 43–58.

[41] A.P. Kamath, N.K. Karmarkar, K.G. Ramakrishnan, M.G.C. Resende, A continuous approach to inductive inference, *Mathematical Programming* 57 (1992), pp. 215–238.

[42] S. Khot, Improved inapproximability results for maxclique, chromatic number and approximate graph coloring, in: *Proc. 42nd Annual IEEE Symposium on the Foundations of Computer Science (FOCS)* (2001) 600–609.

[43] B. Khoury, P.M. Pardalos, D. Hearn, Equivalent formulations of the Steiner tree problem in graphs, in: D.-Z. Du, P.M. Pardalos, eds., *Network Optimization Problems*, pp. 53–62, 1993.

[44] D.E. Knuth, The sandwich theorem, *Elec. J. Comb.* **1** (1994).

[45] M. Kojima, N. Megiddo, T. Noma, A. Yoshise, *A Unified Approach to Interior Point Algorithms for Linear Complementarity Problems.* Springer-Verlag, Lecture Notes in Computer Sciences 538 (1991).

[46] M. Kojima, L. Tunçel, Some Fundamental Properties of Successive Convex Relaxation Methods on LCP and Related Problems, *Journal of Global Optimization* Vol 24 (2002), pp. 333–348.

[47] L. Lovaász, A. Schrjiver, Cones of matrices and set functions and 0–1 optimization, *SIAM J. Opt. 1*, pp. 166–190, 1991.

[48] C. Maranas, C. Floudas, P.M. Pardalos, New results in the packing of equal circles in a square, *Discrete Mathematics* 142, pp. 287–293, 1995.

[49] J.E. Mitchell, P.M. Pardalos, M.G.C. Resende, Interior Point Methods for Combinatorial Optimization, in: D.-Z Du and P. Pardalos, eds., *Handbook of Combinatorial Optimization* Vol. 1 (1998), pp. 189-298.

[50] T.S. Motzkin and E.G. Straus, Maxima for graphs and a new proof of a theorem of Turan, *Canadian Journal of Mathematics* 17:4 (1965) 533–540.

[51] K.G. Murty, *Linear complementarity, linear and nonlinear programming.* Heldermann Verlag, Berlin(1988).

[52] Yu.E. Nesterov, Semidefinite relaxation and nonconvex quadratic optimization, *Optimization Methods and Software* Vol.9 (1998) pp. 141–160.

[53] P.M. Pardalos, The Linear Complemementarity Problem, in: S. Gomez and J.P. Hennart, eds., *Advances in Optimization and Numerical Analysis*, Kluwer Academinc Publishers (1994), pp. 39–49.

[54] P.M. Pardalos, Continuous Approaches to Discrete Optimization Problems, in: G. Di Pillo and F. Giannessi, eds., *Nonlinear Optimization and Applications*, Plenum Publishing (1996), pp. 313–328.

[55] P.M. Pardalos, J.B. Rosen, Global optimization approach to the linear complementarity problem, *SIAM J. Scient. Stat. Computing* Vol. 9, No. 2 (1988), pp. 341–353.

[56] P.M. Pardalos, H. Wolkowicz, *Topics on Semidefinite and Interrior-Point Methods*, American Math. Society, 1998.

[57] M.V. Ramana, P.M. Pardalos, Semidefinite Programming, in: T. Terlaky, ed., *Interrior Point Methods of Mathematical Programming*, Kluwer Academic Publishers, 1996, pp. 369–398.

[58] F. Rendl, Semidefinite programming and combinatorial optimization, *Appl. Numer. Math.* 29 (1999), pp. 255–281.

[59] B. Ricceri, S. Simons, eds., *Minimax Theory and Applications*, Kluwer Academic Publishers, 1998.

[60] N.Z. Shor, Dual quadratic estimates in polynomial and Boolean programming, *Annals of Operations Research* 25 (1990), pp. 163–168.

[61] N.Z. Shor and P.I. Stetsyuk, Lagrangian bounds in multiextremal polynomial and discrete optimization problems, *Journal of Global Optimization* 23 (2002), pp. 1-41.

[62] L. Vandenberhe and S. Boyd, *Semidefinite Programming*, SIAM Review, 38 (1996), pp. 49–95.

[63] H. Wolkowicz, Semidefinite Programming, in: P.M. Pardalos, M.G.C. Resende, eds., *Handbook of Applied Optimization*, Oxford University Press, 2002, pp. 40–50.

[64] H. Wolkowicz, Semidefinite and Cone Programming Bibliography/Comments, Available at http://orion.math.uwaterloo.ca/ hwolkowi/henry/software/sdpbibliog.pdf, 2005.

[65] B.Y. Wu, K.-M. Chao, *Spanning Trees and Optimization Problems*, Chapman & Hall/CRC, 2004.

[66] Y. Ye, A .699-approximation algorithm for Max-Bisection, *Math. Programming* 90 (2001), pp. 101–111.

[67] Y. Ye, Approximating quadratic programming with bound and quadratic constraints, *Mathematical Programming*, 84(2):219–226, 1999.

Chapter 3

LOGIC-BASED MODELING

John N Hooker
Graduate School of Industrial Administration
Carnegie Mellon University, Pittsburgh, PA 15213
USA
jh38@andrew.cmu.edu

Abstract   Logic-based modeling can result in decision models that are more natural and easier to debug. The addition of logical constraints to mixed integer programming need not sacrifice computational speed and can even enhance it if the constraints are processed correctly. They should be written or automatically reformulated so as to be as nearly consistent or hyperarc consistent as possible. They should also be provided with a tight continuous relaxation. This chapter shows how to accomplish these goals for a number of logic-based constraints: formulas of propositional logic, cardinality formulas, 0-1 linear inequalities (viewed as logical formulas), cardinality rules, and mixed logical/linear constraints. It does the same for three global constraints that are popular in constraint programming systems: the all-different, element and cumulative constraints.

Clarity is as important as computational tractability when building scientific models. In the broadest sense, models are descriptions or graphic representations of some phenomenon. They are typically written in a formal or quasi-formal language for a dual purpose: partly to permit computation of the mathematical or logical consequences, but equally to elucidate the conceptual structure of the phenomenon by describing it in a precise and limited vocabulary. The classical transportation model, for example, allows fast solution with the transportation simplex method but also displays the problem as a network that is easy to understand.

Optimization modeling has generally emphasized ease of computation more heavily than the clarity and explanatory value of the model. (The transportation model is a happy exception that is strong on both counts.) This is due in part to the fact that optimization, at least in the context of operations research, is often more interested in prescription than description. Practitioners who model a manufacturing plant, for example, typically want a solution that tells them how the plant should be run. Yet a succinct and natural model offers several advantages: it is easier to construct, easier to debug, and more conducive to understanding how the plant works.

This chapter explores the option of enriching mixed integer programming (MILP) models with logic-based constraints, in order to provide more natural and succinct expression of logical conditions. Due to formulation and solution techniques developed over the last several years, a modeling enrichment of this sort need not sacrifice computational tractability and can even enhance it.

One historical reason for the de-emphasis of perspicuous models in operations research has been the enormous influence of linear programming. Even though it uses a very small number of primitive terms, such as linear inequalities and equations, a linear programming model can formulate a remarkably wide range of problems. The linear format almost always allows fast solution, unless the model is truly huge. It also provides such analysis tools as reduced costs, shadow prices and sensitivity ranges. There is therefore a substantial reward for reducing a problem to linear inequalities and equations, even when this obscures the structure of the problem.

When one moves beyond linear models, however, there are less compelling reasons for sacrificing clarity in order to express a problem in a language with a small number of primitives. There is no framework for discrete or discrete/continuous models, for example, that offers the advantages of linear programming. The mathematical programming community has long used MILP for this purpose, but MILP solvers are not nearly as robust as linear programming solvers, as one would expect because MILP formulates NP-hard problems. Relatively small and innocent-looking problems can exceed the capabilities of any existing solver, such as the market sharing problems identified by Williams [35] and studied by Cornuejols and Dawande [16]. Even tractable problems may be soluble only when carefully formulated to obtain a tight linear relaxation or an effective branching scheme.

In addition MILP often forces logical conditions to be expressed in an unnatural way, perhaps using big-M constraints and the like. The formulation may be even less natural if one is to obtain a tight linear relaxation. Current solution technology requires that the traveling salesman problem, for example, be written with exponentially many constraints in order to represent a simple all-different condition. MILP may provide no practical formulation at all for important problem classes, including some resource-constrained scheduling problems.

It is true that MILP has the advantage of a unified solution approach, since a single branch-and-cut solver can be applied to any MILP model one might write. Yet the introduction of logic-based and other higher-level constraints no longer sacrifices this advantage.

The discussion begins in Section 3.1 with a brief description of how solvers can accommodate logic-based constraints: namely, by automatically converting them to MILP constraints, or by designing a solver that integrates MILP and constraint programming. Section 3.2 describes what constitutes a good formulation for MILP and for an integrated solver. The remaining sections describe good formulations for each type of constraint: formulas of propositional logic, cardinality formulas, 0-1 linear inequalities (viewed as logical formulas), cardinality rules, and mixed logical/linear constraints. Section 3.8 briefly discusses three global constraints that are popular in constraint programming systems: the all-different, element and cumulative constraints. They are not purely logical constraints, but they illustrate how logical expressions are a special case of a more general approach to modeling that offers a variety of constraint types. The chapter ends with a summary.

1. Solvers for Logic-Based Constraints

1.1 Two Approaches

There are at least two ways to deal with logic-based constraints in a unified solver.

- One possibility is to provide automatic reformulation of logic-based constraints into MILP constraints, and then to apply a standard MILP solver [26, 28]. The reformulation can be designed to result in a tight linear relaxation. This is a viable approach, although it obscures some of the structure of the problem.

- A second approach is to design a unified solution method for a diversity of constraint types, by integrating MILP with constraint programming (CP). Surveys of the relevant literature on hybrid solvers may be found in [11, 18, 21, 22, 27, 36].

Whether one uses automatic translation or a CP/MILP hybrid approach, constraints must be written or automatically reformulated with the algorithmic implications in mind. A good formulation for MILP is one with a tight linear relaxation. A good formulation for CP is as nearly "consistent" as possible. A consistent constraint set is defined rigorously below, but it is roughly analogous to a convex hull relaxation in MILP. A good formulation for a hybrid approach should be good for both MILP and CP whenever possible, but the strength of a hybrid approach is that it can benefit from a formulation that is good in either sense.

1.2   Structured Groups of Constraints

Very often, the structure of a problem is not adequately exploited unless constraints are considered in groups. A group of constraints can generally be given a MILP translation that is tighter than the combined translations of the individual constraints. The consistency-maintenance algorithms of CP are more effective when applied to a group of constraints whose overall structure can be exploited.

Structured groups can be identified and processed in three ways.

Automatic detection. Design solvers to detect special structure and process it appropriately, as is commonly done for network constraints in MILP solvers. However, since modelers are generally aware of the problem structure, it seems more efficient to obtain this information from them rather than expecting the solver to rediscover it.

Hand coding. Ask modelers to exploit structured constraint groups by hand. They can write a good MILP formulation for a group, or they can write a consistent formulation.

Structural labeling. Let modelers label specially structured constraint groups so that the solver can process them accordingly. The CP community implements this approach with the concept of a *global* constraint, which represents a structured set of more elementary constraints.

As an example of the third approach, the global constraint all-different(a, b, c) requires that a, b and c take distinct values and thus represents three inequations $a \neq b$, $a \neq c$ and $b \neq c$. To take another example, the global constraint cnf$(a \vee b, a \vee \neg b)$ can alert the solver that its arguments are propositional formulas in conjunctive normal form (defined below). This allows the solver to process the constraints with specialized inference algorithms.

2.   Good Formulations

2.1   Tight Relaxations

A good formulation of an MILP model should have a tight linear relaxation. The tightest possible formulation is a *convex hull formulation*, whose continuous relaxation describes the convex hull of the model's feasible set. The convex hull is the intersection of all half planes containing the feasible set. At a minimum it contains inequalities that define all the facets of the convex hull and equations that define the affine hull (the smallest dimensional hyperplane that contains the feasible set). Such a formulation is ideal because it allows one to find an optimal solution by solving the linear programming relaxation.

Since there may be a large number of facet-defining inequalities, it is common in practice to generate *separating* inequalities that are violated by the optimal solution of the current relaxation. This must be done during the solution process, however, since at the modeling stage one does not know what solutions will be obtained from the relaxation. Fortunately, some common constraint types may have a convex hull relaxation that is simple enough to analyze and describe in advance. Note, however, that even when one writes a convex hull formulation of each constraint or each structured subset of constraints, the model as a whole is generally not a convex hull formulation.

It is unclear how to measure the tightness of a relaxation that does not describe the convex hull. In practice, a "tight" relaxation is simply one that provides a relatively good bound for problems that are commonly solved.

2.2 Consistent Formulations

A good formulation for CP is *consistent*, meaning that its constraints explicitly rule out assignments of values to variables that cannot be part of a feasible solution. Note that consistency is not the same as satisfiability, as the term might suggest.

To make the idea more precise, suppose that constraint set S contains variables $x_1, \ldots, x_n$. Each variable x_j has a domain D_j, which is the initial set of values the variable may take (perhaps the reals, integers, etc.) Let a *partial assignment* (known as a *compound label* in the constraints community) specify values for some subset of the variables. Thus a partial assignment has the form

$$(x_{j_1}, \ldots, x_{j_k}) = (v_{j_1}, \ldots, v_{j_k}), \text{ where each } j_\ell \in D_{j_\ell} \qquad (3.1)$$

A partial assignment (3.1) is *redundant* for S if it cannot be extended to a feasible solution of S. That is, every complete assignment

$$(x_1, \ldots, x_n) = (v_1, \ldots, v_n), \text{ where each } j \in D_j$$

that is consistent with (3.1) violates some constraint in S.

By convention, a partial assignment violates a constraint C only if it assigns some value to every variable in C. Thus if $D_1 = D_2 = \mathcal{R}$, the assignment $x_1 = -1$ does not violate the constraint $x_1 + x_2^2 \geq 0$ since x_2 has not been assigned a value. The assignment is redundant, however, since it cannot be extended to a feasible solution of $x_1 + x_2^2 \geq 0$. No value in x_2's domain will work.

A constraint set S is *consistent* if every redundant partial assignment violates some constraint in S. Thus the constraint set $\{x_1 + x_2^2 \geq 0\}$ is not consistent.

It is easy to see that a consistent constraint set S can be solved without backtracking, provided S is satisfiable. First assign x_1 a value $v_1 \in D_1$ that violates no constraints in S (which is possible because S is satisfiable). Now assign x_2 a value $v_2 \in D_2$ such that $(x_1, x_2) = (v_1, v_2)$ violates no constraints in S. Consistency guarantees that such a value exists. Continue in this fashion until a feasible solution is constructed. The key to this greedy algorithm is that it can easily recognize a redundant partial assignment by checking whether it violates some individual constraint.

The greedy algorithm can be viewed as a branching algorithm that branches on variables in the order $x_1, \ldots, x_n$. Consistency ensures that no backtracking is required, whatever the branching order. It is in this sense that a consistent constraint set is analogous to a convex hull formulation. Either formulation allows one to find a feasible solution without backtracking, in the former case by using the greedy algorithm just described, and in the latter case by solving the linear programming relaxation. However, just as a convex hull formulation for each constraint does not ensure a convex hull formulation for the whole problem, an analogous fact holds for consistency. Achieving consistency for each constraint or several subsets of constraints need not achieve consistency for the entire constraint set. Nonetheless the amount of backtracking tends to be less when some form of consistency is achieved for subsets of constraints.

Consistency maintenance is the process of achieving consistency, which is ordinarily done by adding new constraints to S, much as valid inequalities are generated in MILP to approximate a convex hull formulation. One difference between CP and MILP, however, is that they generate different types of constraints. Just as MILP limits itself to linear inequalities, CP ordinarily generates only certain kinds of constraints that are easy to process, primarily *in-domain* constraints that have the form $x_j \in \bar{D}_j$. Here $\bar{D}_j$ is a reduced domain for x_j, obtained by deleting some values from D_j that cannot be part of a feasible solution. Generation of linear inequalities can in principle build a convex hull relaxation, but unfortunately generation of in-domain constraints cannot in general build a consistent constraint set (although it can achieve hyperarc consistency, defined below). Since generating in-domain constraints is equivalent to reducing the domains D_j, consistency maintenance is often described as *domain reduction* .

Hyperarc consistency (also called *generalized arc consistency*) is generally maintained for variables with finite domains. It implies that no assignment to a single variable is redundant. That is, given any variable x_j, no assignment $x_j = v$ for $v \in D_j$ is redundant. Hyperarc consistency is obtained by generating all valid in-domain constraints and therefore achieves maximum domain reduction. It can be viewed as computing the projection of the feasible set onto each variable. A wide variety of "filtering" algorithms has been developed to maintain hyperarc consistency for particular types of constraints.

Bounds consistency is defined when the system maintains *interval domains* $[a_j, b_j]$ for numerical variables x_j. A constraint set S is bounds consistent if neither $x_j = a_j$ nor $x_j = b_j$ is redundant for any j. Thus bounds consistency achieves the narrowest interval domains, just as hyperarc consistency achieves the smallest finite domains. Bounds consistency is generally maintained by interval arithmetic and by specialized algorithms for global constraints with numerical variables.

Both hyperarc and bounds consistency tend to reduce backtracking, because CP systems typically branch by splitting a domain. If the domain is small or narrow, less splitting is required to find a feasible solution. Small and narrow domains also make domain reduction more effective as one descends into the search tree.

2.3 Prime Implications

There is a weaker property than consistency, namely completeness, that can be nearly as helpful when solving the problem. Completeness can be achieved for a constraint set by generating its prime implications, which, roughly speaking, are the strongest constraints that can be inferred from the constraint set. Thus when it is impractical to write a consistent formulation, one may be able to write the prime implications of the constraint set and achieve much the same effect.

Recall that a constraint set S is consistent if any redundant partial assignment for S violates some constraint in S. S is *complete* if any redundant partial assignment for S is *redundant* for some constraint in S. Checking whether a partial assignment is redundant for a constraint is generally harder than checking whether it violates the constraint, but in many cases it is practical nonetheless.

In some contexts a partial assignment is redundant for a constraint only if it violates the constraint. In such cases completeness implies consistency.

Prime implications, roughly speaking, are the strongest constraints that can be inferred from a constraint set. To develop the concept, suppose constraints C and D contain variables in $x = (x_1, \ldots, x_n)$. C *implies* constraint D if all assignments to x that satisfy C also satisfy D. Constraints C and D are *equivalent* if they imply each other.

Let H be some family of constraints, such as logical clauses, cardinality formulas, or 0-1 inequalities (defined in subsequent sections). We assume H is semantically finite, meaning that there are a finite number of nonequivalent constraints in H. For example, the set of 0-1 linear inequalities in a given set of variables is semantically finite, even though there are infinitely many inequalities.

Let an H-*implication* of a constraint set S be a constraint in H that S implies. Constraint C is a *prime H-implication* of S if C is a H-implication of S, and every H-implication of S that implies C is equivalent to C. The following is easy to show.

LEMMA 3.1 *Given a semantically finite constraint set H, every H-implication of a constraint set S is implied by some prime H-implication of S. Thus if $S \subset H$, S is equivalent to the set of its prime H-implications.*

For example, let H be the family of 0-1 linear inequalities in variables x_1, x_2, and let S consist of

$$x_1 + x_2 \geq 1$$
$$x_1 - x_2 \geq 0$$

S has the single prime H-implication $x_1 \geq 1$ (up to equivalence). Any 0-1 inequality implied by S is implied by $x_1 \geq 1$. Since $S \subset H$, S is equivalent to $\{x_1 \geq 1\}$.

Lemma 3.1 may not apply when H is semantically infinite. For instance, let x_1 have domain $[0, 1]$ and H be the set of constraints of the form $x_1 \geq \alpha$ for $\alpha \in [0, 1)$. Then all constraints in H are implied by $S = \{x_1 \geq 1\}$ but none are implied by a prime H-implication of S, since S has no prime H-implications.

The next section states precisely how Lemma 3.1 allows one to recognize redundant partial assignments.

2.4   Recognizing Redundant Partial Assignments

Let a *clause* in variables $x_1, \ldots, x_n$ with domains $D_1, \ldots, D_n$ be a constraint of the form

$$(x_{j_1} \neq v_1) \vee \cdots \vee (x_{j_p} \neq v_p) \tag{3.2}$$

where each $j_i \in \{1, \ldots, n\}$ and each $v_{j_i} \in D_{j_i}$. A clause with zero disjuncts is necessarily false, by convention. A constraint set H *contains all clauses* for a set of variables if every clause in these variables is equivalent to some constraint in H.

LEMMA 3.2 *If a semantically finite constraint set H contains all clauses for the variables in constraint set S, then any partial assignment that is redundant for S is redundant for some prime H-implication of S.*

This is easy to see. If a partial assignment $(x_{j_1}, \ldots, x_{j_p}) = (v_1, \ldots, v_p)$ is redundant for S, then S implies the clause (3.2). By Lemma 3.1, some prime H-implication P of S implies (3.2). This means the partial assignment is redundant for P.

As an example, consider the constraint set S consisting of

$$\begin{align} 2x_1 + 3x_2 &\geq 4 \quad (a) \\ 3x_1 + 2x_2 &\leq 5 \quad (b) \end{align} \tag{3.3}$$

where the domain of each x_j is $\{0, 1, 2\}$. Since the only solutions of (3.3) are $x = (1, 1)$ and $(0, 2)$, there are two redundant partial assignments: $x_1 = 2$ and $x_2 = 0$. Let H consist of all linear inequalities in x_1, x_2, plus clauses in x_1, x_2. The H-prime implications of (3.3), up to equivalence, are (3.3b) and

$$
\begin{aligned}
x_1 + 2x_2 &\geq 3 \\
-x_1 + 2x_2 &\geq 1 \\
2x_1 - x_2 &\leq 1
\end{aligned}
\tag{3.4}
$$

It can be checked that each redundant partial assignment is redundant for at least one (in fact two) of the prime implications. Note that the redundant partial assignment $x_2 = 0$ is not redundant for either of the original inequalities in (3.3); only for the inequalities taken together. Knowledge of the prime implications therefore makes it easier to identify this redundancy.

A constraint set S is *complete* if every redundant partial assignment for S is redundant for some constraint in S. From Lemma 3.2 we have:

COROLLARY 3.3 *If S contains all of its prime H-implications for some semantically finite constraint set H that contains all clauses for the variables in S, then S is complete.*

Thus if inequalities (3.4) are added to those in (3.3), the result is a complete constraint set.

3. Propositional Logic

3.1 Basic Ideas

A *formal logic* is a language in which deductions are made solely on the basis of the form of statements, without regard to their specific meaning. In *propositional logic*, a statement is made up of *atomic propositions* that can be true or false. The form of the statement is given by how the atomic propositions are joined or modified by such logical expressions as *not* ($\neg$), *and* ($\wedge$), and inclusive *or* ($\vee$).

A proposition defines a *propositional function* that maps the truth values of its atomic propositions to the truth value of the whole proposition. For example, the proposition

$$(a \vee \neg b) \wedge (\neg a \vee b)$$

contains atomic propositions a, b and defines a function f given by the following *truth table*:

$$
\begin{array}{cc|c}
a & b & f(a,b) \\
\hline
0 & 0 & 1 \\
0 & 1 & 0 \\
1 & 0 & 0 \\
1 & 1 & 1 \\
\hline
\end{array}
\tag{3.5}
$$

Here 0 and 1 denote *false* and *true* respectively. One can define additional symbols for implication ($\rightarrow$), equivalence ($\equiv$), and so forth. For any pair of formulas A, B

$$
\begin{aligned}
A \rightarrow B &=_{\text{def}} \neg A \vee B \\
A \equiv B &=_{\text{def}} (A \rightarrow B) \wedge (B \rightarrow A)
\end{aligned}
$$

Note that (3.5) is the truth table for equivalence. A *tautology* is a formula whose propositional function is identically true.

A formula of propositional logic can be viewed as a constraint in which the variables are atomic propositions and have domains $\{0, 1\}$, where 0 and 1 signify false and true. The constraint is satisfied when the formula is true. Many complex logical conditions can be naturally rendered as propositional formulas, as for example the following.

Alice will go to the party only if Charles goes.	$a \rightarrow c$
Betty will not go to the party if Alice or Diane goes.	$(a \vee d) \rightarrow \neg b$
Charles will go to the party unless Diane or Edward goes.	$c \rightarrow (\neg d \wedge \neg e)$
Diane will not go to the party unless Charles or Edward goes.	$d \rightarrow (c \vee e)$
Betty and Charles never go to the same party.	$\neg(b \wedge c)$
Betty and Edward are always seen together.	$b \equiv e$
Charles goes to every party that Betty goes to.	$b \rightarrow c$

(3.6)

3.2   Conversion to Clausal Form

It may be useful to convert formulas to clausal form, particularly since the well-known resolution algorithm is designed for clauses, and clauses have an obvious MILP representation.

In propositional logic, clauses take the form of conjunctions of *literals*, which are atomic propositions or their negations. We will refer to a clause of this sort as a *propositional clause*. A propositional formula is in *clausal form*, also known as *conjunctive normal form* (CNF), if it is a conjunction of one or more clauses.

Any propositional formula can be converted to clausal form in at least two ways. The more straightforward conversion requires exponential space in the worst case. It is accomplished by applying some elementary logical equivalences.

De Morgan's laws
$$\neg(F \wedge G) \equiv \neg F \vee \neg G$$
$$\neg(F \vee G) \equiv \neg F \wedge \neg G$$

Distribution laws
$$F \wedge (G \vee H) \equiv (F \wedge G) \vee (F \wedge H)$$
$$F \vee (G \wedge H) \equiv (F \vee G) \wedge (F \vee H)$$

Double negation
$$\neg\neg F \equiv F$$

(Of the two distribution laws, only the second is needed.) For example, the formula

$$(a \wedge \neg b) \vee \neg(a \vee \neg b)$$

may be converted to CNF by first applying De Morgan's law to the second disjunct,

$$(a \wedge \neg b) \vee (\neg a \wedge b)$$

and then applying distribution,

$$(a \vee \neg a) \wedge (a \vee b) \wedge (\neg b \vee \neg a) \wedge (\neg b \vee b)$$

The two tautologous clauses can be dropped. Similarly, the propositions in (3.6) convert to the following clauses.

$$
\begin{aligned}
&\neg a \vee c \\
&\neg a \vee \neg b \\
&\neg b \vee \neg d \\
&\neg c \vee \neg d \\
&\neg c \vee \neg e \\
&c \vee \neg d \vee e \\
&\neg b \vee \neg c \\
&\neg b \vee e \\
&b \vee \neg e \\
&\neg b \vee c
\end{aligned}
\tag{3.7}
$$

The precise conversion algorithm appears in Fig. 3.1.

Exponential growth for this type of conversion can be seen in propositions of the form

$$(a_1 \wedge b_1) \vee \ldots \vee (a_n \wedge b_n) \tag{3.8}$$

```
Let F be the formula to be converted.
Set k = 0 and S = ∅.
Replace all subformulas of the form G ≡ H with (G → H) ∧ (H → G).
Replace all subformulas of the form G ⊃ H with ¬G ∨ H.
Perform Convert(F).
The CNF form is the conjunction of clauses in S.

Function Convert(F)
    If F is a clause then add F to S.
    Else if F has the form ¬¬G then perform Convert(G).
    Else if F has the form G ∧ H then
        perform Convert(G) and Convert(H).
    Else if F has the form ¬(G ∧ H) then perform Convert(¬G ∨ ¬H).
    Else if F has the form ¬(G ∨ H) then perform Convert(¬G ∧ ¬H).
    Else if F has the form G ∨ (H ∧ I) then
        Perform Convert(G ∨ H) and Convert(G ∨ I).
```

Figure 3.1. Conversion of F to CNF without additional variables. A formula of the form $(H \wedge I) \vee G$ is regarded as having the form $G \vee (H \wedge I)$.

The formula translates to the conjunction of 2^n clauses of the form $L_1 \vee \ldots \vee L_n$, where each L_j is a_j or b_j.

By adding new variables, however, conversion to CNF can be accomplished in linear time. The idea is credited to Tseitin [34] but Wilson's more compact version [38] simply replaces a disjunction $F \vee G$ with the conjunction

$$(x_1 \vee x_2) \wedge (\neg x_1 \vee F) \wedge (\neg x_2 \vee G),$$

where x_1, x_2 are new variables and the clauses $\neg x_1 \vee F$ and $\neg x_2 \vee G$ encode implications $x_1 \to F$ and $x_2 \to G$, respectively. For example, formula (3.8) yields the conjunction,

$$(x_1 \vee \ldots \vee x_n) \wedge \bigwedge_{j=1}^{n} (\neg x_j \vee a_j) \wedge (\neg x_j \vee b_j),$$

which is equivalent to (3.8) and grows linearly with the length of (3.8). The precise algorithm appears in Figure 3.2.

3.3   Achieving Consistency

A partial assignment *falsifies* (violates) a propositional formula F when it assigns a truth value to every atomic proposition in F and makes F false. A set S of propositional formulas is *consistent* when every redundant partial assignment for S falsifies a formula in S.

Clause sets are particularly convenient for studying consistency, since a partial assignment is redundant for a clause if and only if it actually falsifies the clause. Lemma 3.2 therefore provides a sufficient condition for consistency.

```
Let F be the formula to be converted.
Set k = 0 and S = ∅.
Replace all subformulas of the form G ≡ H with (G → H) ∧ (H → G).
Replace all subformulas of the form G ⊃ H with ¬G ∨ H.
Perform Convert(F).
The CNF form is the conjunction of clauses in S.

Function Convert(F)
    If F has the form C then add F to S.
    Else if F has the form ¬¬G then perform Convert(G).
    Else if F has the form G ∧ H then
        Perform Convert(G) and Convert(H).
    Else if F has the form ¬(G ∧ H) then perform Convert(¬G ∨ ¬H).
    Else if F has the form ¬(G ∨ H) then perform Convert(¬G ∧ ¬H).
    Else if F has the form C ∨ (G ∧ H) then
        Perform Convert(C ∨ G) and Convert(C ∨ H).
    Else if F has the form C ∨ ¬(G ∨ H) then
        Perform Convert(C ∨ (¬G ∧ ¬H)).
    Else if F has the form C ∨ G ∨ H then
        Add C ∨ x_{k-1} ∨ x_{k-2} to S.
        Perform Convert(¬x_{k-1} ∨ G) and Convert(¬x_{k-2} ∨ H).
        Set k = k - 2.
    Else
        Write F as G ∨ H.
        Add x_{k-1} ∨ x_{k-2} to S.
        Perform Convert(¬x_{k-1} ∨ G) and Convert(¬x_{k-2} ∨ H).
```

Figure 3.2. Linear-time conversion to CNF (adapted from [21]). The letter C represents any clause. It is assumed that F does not contain variables $x_1, x_2, \ldots$.

Let a *prime clausal implication* of a clause set S be a prime H-implication where H is the set of clauses in the variables of S.

COROLLARY 3.4 *Any partial truth assignment that is redundant for clause set S falsifies some prime clausal implication of S. Thus a clause set that contains all of its prime clausal implications is consistent.*

One can achieve consistency for clause set S by applying the *resolution algorithm*, which generates the prime clausal implications of S. If each of the two clauses $C, D \in S$ contains exactly one atomic proposition a that appears in the other clause with opposite sign, then the *resolvent* of C and D is the disjunction of all literals that occur in C or D except a and $\neg a$. For instance, the third clause below is the resolvent of the first two.

$$\frac{\begin{array}{c} a \vee b \vee c \\ \neg a \vee b \vee \neg d \end{array}}{b \vee c \vee \neg d} \tag{3.9}$$

A clause C *absorbs* clause D if all the literals of C appear in D. It is clear that C implies D if and only if C absorbs D. The resolution algorithm is applied to

> **While** S contains clauses C, D with a resolvent R absorbed by no clause in S:
>   Remove from S all clauses absorbed by R.
>   Add R to S.

Figure 3.3. The resolution algorithm applied to clause set S

S by generating all resolvents that are not absorbed by some clause in S, and adding these resolvents to S. It repeats the process until no such resolvents can be generated. The precise algorithm appears in Fig. 3.3.

THEOREM 3.5 (QUINE [31, 32]) *The set S' that results from applying the resolution algorithm to clause set S contains precisely the prime clausal implications of S. In particular, S is infeasible if and only if S' consists of the empty clause.*

Resolution applied to the example problem (3.7) yields the four prime implications

$$
\begin{aligned}
&\neg a \lor c \\
&\neg b \\
&\neg d \\
&\neg e
\end{aligned}
\tag{3.10}
$$

Charles will go to the party if Alice does, and everyone else will stay home. The party will be attended by Charles and Alice, by Charles alone, or by nobody.

The resolution algorithm has exponential worst-case complexity [17] and exponential average-case complexity under weak assumptions [14]; it also tends to blow up in practice. Yet it can be very helpful when applied to small constraint sets. Due to Theorem 3.5 and Corollary 3.4,

COROLLARY 3.6 *The resolution algorithm achieves consistency when applied to a set of propositional clauses.*

For instance, (3.10) can serve as a consistent representation of (3.6) or (3.7). A number of additional theorems that relate resolution to various types of consistency can be found in [21].

Resolution also achieves hyperarc consistency. The domain of an atomic proposition a can be reduced to $\{1\}$ if and only if the singleton clause a is a prime implication, and it can be reduced to $\{0\}$ if and only if $\neg a$ is a prime implication. In the example (3.6), the domains of b, d and e are reduced to $\{0\}$. Unfortunately, the entire resolution algorithm must be carried out to achieve hyperarc consistency, so that one may as well aim for full consistency as well.

3.4   Tight MILP Formulations

The feasible set of a proposition can be regarded as 0-1 points in $\mathcal{R}^n$ as well as points in logical space. This means one can design an MILP representation by defining a polyhedron that contains all and only the 0-1 points that satisfy the proposition. The best MILP representation contains inequalities that define the facets of the convex hull of these 0-1 points, as well as equations that define their affine hull.

For example, the proposition $a \equiv b$ has feasible points $(0,0)$ and $(1,1)$. Their affine hull is defined by the equation $a = b$. The convex hull is the line segment from $(0,0)$ to $(1,1)$ and has two facets defined, for example, by $a \geq 0$ and $a \leq 1$. So we have the convex hull MILP representation

$$a = b$$
$$a, b \in \{0,1\}$$

The inequalities $a \geq 0$ and $a \leq 1$ can be omitted because the linear relaxation of the MILP formulation replaces $a, b \in \{0,1\}$ with $a, b \in [0,1]$.

To take another example, the clause $a \vee b$ has feasible points $(0,1)$, $(1,0)$ and $(1,1)$. The facets of the convex hull of these points are $a + b \geq 1$, $a \leq 1$ and $b \leq 1$. The last two inequalities are again redundant, and we have the convex hull MILP representation

$$a + b \geq 1$$
$$a, b \in \{0,1\}$$

The inequality $a + b \geq 1$ is the only nonelementary facet of the convex hull (i.e., the only facet other than a 0-1 bound on a variable).

One can also derive the inequality $a + b \geq 1$ by observing that since at least one of a, b must be true, the sum of their values must be at least 1. In fact any clause

$$\bigvee_{j \in P} a_j \vee \bigvee_{j \in N} \neg b_j$$

can be given the convex hull formulation

$$\sum_{j \in P} a_j - \sum_{j \in N} (1 - b_j) \geq 1$$

One way to write an MILP formulation of an arbitrary proposition is to convert it to clausal form and write the corresponding inequality for each clause. This might be called the *clausal representation*. Unfortunately, it is not in general a convex hull formulation, even if the clause set is consistent. For instance, the third (and redundant) clause in (3.9) is not facet-defining even though the clause set is consistent. Nonredundant clauses in a consistent set can also fail to be facet defining, as in the following example:

$$a \vee b \qquad\qquad a + b \geq 1$$
$$a \vee c \qquad\qquad a + c \geq 1 \qquad\qquad (3.11)$$
$$b \vee c \qquad\qquad b + c \geq 1$$

None of the inequalities defines a facet, and in fact the inequality $a + b + c \geq 2$ is the only nonelementary facet.

Although converting a proposition to clausal form need not give rise to a convex hull formulation, there are some steps one can take to tighten the MILP formulation.

- Convert the proposition to clausal form without adding variables. That is, use the algorithm in Fig. 3.1 rather than in Fig. 3.2. As Jeroslow pointed out [25], the addition of new variables can result in a weaker relaxation. For example, $(a \wedge b) \vee (a \wedge c)$ can be converted to the clause set on the left below (using the algorithm of Fig. 3.1), or to the set on the right (using the algorithm of Fig. 3.2):

$$
\begin{array}{cc}
\begin{array}{c} a \\ b \vee c \end{array}
&
\begin{array}{c}
x_1 \vee x_2 \\
\neg x_1 \vee a \\
\neg x_1 \vee b \\
\neg x_2 \vee a \\
\neg x_2 \vee c
\end{array}
\end{array}
$$

  The relaxation of the first set fixes $a = 1$, but the relaxation of the second set only requires that $a \in [1/2, 1]$.

- Apply the resolution algorithm to a clause set before converting the clauses to inequalities. It is easily shown that resolvents correspond to valid inequalities that tighten the relaxation. The clausal representation of problem (3.6), for example, is much tighter (in fact, it is a convex hull formulation) if one first applies resolution to generate the consistent formulation (3.10):

$$a \leq c$$
$$b = d = e = 0$$

Resolution can even tighten the relaxation to the point that it is infeasible. Consider for example the clause set on the left below, whose relaxation is shown on the right.

$$
\begin{array}{cc}
\begin{array}{c}
a \vee b \\
\neg a \vee b \\
a \vee \neg b \\
\neg a \vee \neg b
\end{array}
&
\begin{array}{c}
a + b \geq 1 \\
(1 - a) + b \geq 1 \\
a + (1 - b) \geq 1 \\
(1 - a) + (1 - b) \geq 1 \\
a, b \in [0, 1]
\end{array}
\end{array}
$$

The relaxation has the feasible solution $(a, b) = (1/2, 1/2)$. Yet resolution reduces the clause set to the empty clause, whose relaxation $0 \geq 1$ is infeasible. This shows that resolution has greater power for detecting infeasibility than linear relaxation.

- If it is computationally difficult to apply the full resolution algorithm, use a partial algorithm. For instance, one might generate all resolvents having length less than k for some small k. One might also apply *unit resolution*, which is the resolution algorithm modified so that at least one clause of every pair of clauses resolved is a unit clause (i.e., consists of one literal). It can be shown that unit resolution generates precisely the clauses that correspond to rank 1 Chvátal cutting planes [13].

- To obtain further cutting planes, view the clauses as a special case of cardinality formulas, and apply the inference algorithm of Section 3.4 below. This procedure derives the facet-defining inequality $a+b+c \geq 2$ for (3.11), for instance.

Table 3.1 displays the prime implications and a convex hull formulation of some simple propositions. For propositions 1–2 and 4–8, the prime implications correspond to a convex hull formulation. Formula 7 is very common in MILP modeling, because it asks one to linearize the relation $ab = c$, which occurs when condition c is true if and only both a and b are true. One might be tempted to write the MILP model

$$c \geq a + b - 1$$
$$2c \leq a + b$$

The first inequality is facet defining, but the second is not.

The facet-defining inequality for proposition 9 is the smallest example (smallest in terms of the dimension of the space) of a nonclausal facet-defining inequality for set of 0-1 points. The inequality of proposition 10 is the smallest example of a facet-defining inequality with a coefficient not in $\{0, 1, -1\}$.

4. Cardinality Formulas

A cardinality formula states that at least k of a set of literals is true. Cardinality formulas retain some of the properties of propositional logic while making it easier to model problems that require counting.

4.1 Basic Properties

A cardinality formula can be written

$$\{L_1, \ldots, L_m\} \geq k \tag{3.12}$$

Table 3.1.   Prime implications and convex hull formulations of some simple propositions. The set of prime implications of a proposition can serve as a consistent formulation of that proposition.

	Proposition	Prime Implications	Convex Hull Formulation
1.	$a \lor b$	$a \lor b$	$a + b \geq 1$
2.	$a \to b$	$\neg a \lor b$	$a \leq b$
3.	$a \equiv b$	$\neg a \lor b$	$a = b$
		$a \lor \neg b$	
4.	$a \to (b \lor c)$	$\neg a \lor b \lor c$	$a \leq b + c$
5.	$(a \lor b) \to c$	$\neg a \lor b$	$a \leq b$
		$\neg a \lor c$	$a \leq c$
6.	$(a \lor b) \equiv c$	$a \lor b \lor \neg c$	$c \leq a + b$
		$\neg a \lor c$	$a \leq c$
		$\neg b \lor c$	$b \leq c$
7.	$(a \land b) \equiv c$	$\neg a \lor \neg b \lor c$	$c \geq a + b - 1$
		$a \lor \neg c$	$a \geq c$
		$b \lor \neg c$	$b \geq c$
8.	$(a \equiv b) \equiv c$	$a \lor b \lor c$	$a + b + c \geq 1$
		$a \lor \neg b \lor \neg c$	$b + c \leq a + 1$
		$\neg a \lor b \lor \neg c$	$a + c \leq b + 1$
		$\neg a \lor \neg b \lor c$	$a + b \leq c + 1$
9.	$(a \lor b) \land$	$a \lor b$	$a + b + c \geq 2$
	$(a \lor c) \land$	$a \lor c$	
	$(b \lor c)$	$b \lor c$	
10.	$(a \lor b \lor c) \land$	$a \lor b \lor c$	$a + b + c + 2d \geq 3$
	$(a \lor d) \land$	$a \lor d$	
	$(b \lor d) \land$	$b \lor d$	
	$(c \lor d)$	$c \lor d$	

for $m \geq 0$ and $k \geq 1$, where the L_js are literals containing distinct variables. The formula asserts that at least k of the literals are true. To assert that at most k literals are true, one can write

$$\{\neg L_1, \ldots, \neg L_m\} \geq m - k$$

By convention a cardinality formula (3.12) with $m < k$ is necessarily false. A cardinality formula (3.12) is a clause when $k = 1$. Thus the clauses of cardinality logic are precisely the cardinality formulas that are propositional clauses.

The propositional representation of (3.12) requires $\binom{m}{k-1}$ clauses:

$$\bigvee_{j \in J} L_j, \quad \text{all } J \subset \{1, \ldots, m\} \text{ with } |J| = m - k + 1 \qquad (3.13)$$

It is useful to write cardinality formula (3.12) in the form $A \geq k$, where A is a set of literals.

LEMMA 3.7 ([12]) *$A \geq k$ implies $B \geq \ell$ if and only if $|A \setminus B| \leq k - \ell$.*

For example, $\{a, \neg b, c, d, e\} \geq 4$ implies $\{a, \neg b, c, \neg d\} \geq 2$.

A modeling example might go as follows. A firm wishes to hire one or more employees from a pool of four applicants (Alice, Betty, Charles, Diane), including one manager. Only one of the applicants (Betty) does not belong to a minority group. The firm must observe the following constraints.

At least two women must be hired.	$\{a, b, d\} \geq 2$	(i)
At least two minority applicants must be hired.	$\{a, c, d\} \geq 2$	(ii)
Only Alice and Betty are qualified to be managers.	$\{a, b\} \geq 1$	(iii) (3.14)
Alice won't work for the firm unless it hires Charles.	$\{\neg a, c\} \geq 1$	(iv)

4.2   Redundant Partial Assignments

Consistency can be achieved for a single cardinality formula simply by generating all of the propositional clauses it implies, using (3.13). However, there is no need to do so, since it is easy to check whether a partial assignment is redundant for a given cardinality formula (3.12): it is redundant if it falsifies more than $m - k$ literals in $\{L_1, \ldots, L_m\}$.

One can achieve consistency for a set S of cardinality formulas by generating all the propositional clauses S implies and applying the standard resolution algorithm to these clauses. However, this could result in a large number of clauses.

A more efficient way to recognize redundant partial assignments is to achieve completeness by generating *prime cardinality implications* of S; that is, prime H-implications where H is the set of cardinality formulas. Since H contains all clauses in the variables of S, we have from Lemma 3.2 and Corollary 3.3:

COROLLARY 3.8 *Any partial truth assignment that is redundant for a set S of cardinality formulas is redundant for some prime cardinality implication of S. Thus S is complete if it contains all of its prime cardinality implications.*

Prime cardinality implications can be generated with a procedure one might call *cardinality resolution* [19], which is a specialization of the 0-1 resolution procedure described in the next section. Cardinality resolution generates two kinds of resolvents: classical resolvents and diagonal sums.

A cardinality formula set S has a *classical resolvent R* if

(a) there are propositional clauses C_1, C_2 with resolvent R such that each C_i is implied by a cardinality formula in S, and

(b) no cardinality formula in S implies R.

Thus either clause below the line is a classical resolvent of the set S of formulas above the line:

$$\frac{\{a, b, c\} \geq 2 \\ \{\neg a, c, d\} \geq 2}{\begin{array}{c} \{b, d\} \geq 1 \\ \{c\} \geq 1 \end{array}} \qquad (3.15)$$

The clause $\{b, d\} \geq 1$, for example, is the resolvent of $\{a, b\} \geq 1$ and $\{\neg a, d\} \geq 1$, each of which is implied by a formula in S.

The cardinality formula set S has a *diagonal sum* $\{L_1, \ldots, L_m\} \geq k + 1$ if

(a) there are cardinality formulas $A_j \geq k$ for $j = 1, \ldots, m$ such that each $A_j \geq k$ is implied by some $C_j \in S$,

(b) each $A_j \subset \{L_1, \ldots, L_m\}$,

(c) $L_j \notin A_j$ for each j, and

(d) no cardinality formula in S implies $\{L_1, \ldots, L_m\} \geq k + 1$.

Consider for example the following set S of cardinality formulas:

$$\begin{array}{c} \{a, b, e\} \geq 2 \\ \{b, c, e\} \geq 2 \\ \{a, c, d\} \geq 1 \end{array} \qquad (3.16)$$

> **While** S has a classical resolvent or diagonal sum R
> that is implied by no cardinality formula in S:
>   Remove from S all cardinality formulas implied by R.
>   Add R to S.

Figure 3.4. The cardinality resolution algorithm applied to cardinality formula set S

A diagonal sum $\{a, b, c, d\} \geq 2$ of S can be derived from the following formulas $A_j \geq k$:

$$
\begin{aligned}
\{ \quad b, c, d \} &\geq 1 \\
\{ a, \quad c, d \} &\geq 1 \\
\{ a, b, \quad d \} &\geq 1 \\
\{ a, b, c \quad \} &\geq 1
\end{aligned}
\tag{3.17}
$$

Note that each formula in (3.17) is implied by (at least) one of the formulas in S.

The cardinality resolution algorithm appears in Fig. 3.4. The following theorem is a special case of the 0-1 resolution theorem stated in the next section.

THEOREM 3.9 ([19, 20]) *The set that results from applying the cardinality resolution algorithm to cardinality formula set S contains precisely the prime cardinality implications of S. Completeness can be achieved for S by applying the cardinality resolution algorithm, even if the diagonal summation step is omitted and only classical resolvents are generated.*

The theorem can be illustrated with example (3.14). Formulas (ii) and (iv) yield the classical resolvent

$$
\{c\} \geq 1
\tag{3.18}
$$

Each of the following diagonal sums can be derived from (i) and (3.18):

$$
\begin{aligned}
\{a, b, c\} &\geq 2 \\
\{b, c, d\} &\geq 2
\end{aligned}
\tag{3.19}
$$

Now the formulas (i),(ii) and the two formulas in (3.19) yield the diagonal sum

$$
\{a, b, c, d\} \geq 3
\tag{3.20}
$$

The prime implications of (3.14) consist of the two formulas (3.18) and (3.20). They say that the firm must hire at least three people, and it must hire Charles. These prime implications also allow one to recognize all redundant partial assignments for (3.14). For instance, the partial assignment $c = 0$ is redundant for (and in fact violates) the prime implication (3.18) but is not redundant for any of the original constraints (3.14).

Since diagonal summation is not needed to achieve completeness, it does not allow one to recognize any additional redundant partial assignments. For instance, all partial assignments that are redundant for $\{a, b, c, d\} \geq 2$ are already redundant for the clauses in (3.16). Yet generation of diagonal sums can accelerate the process of achieving completeness, and it can reduce the number of constraints one must check to determine whether a partial assignment is redundant.

A possibly faster way to recognize *some* redundant partial assignments is to generate a set of 0-1 linear inequalities that are known to imply resolvents. A cardinality formula

$$\{a_1, \ldots, a_m, \neg b_1, \ldots, \neg b_n\} \geq k \qquad (3.21)$$

is easily written as a 0-1 inequality:

$$\sum_{j=1}^{m} a_j - \sum_{j=1}^{n} (1 - b_j) \geq k \qquad (3.22)$$

It is convenient to say that a 0-1 inequality implies (3.21) if it implies (3.22).

THEOREM 3.10 (BARTH [8]) *All classical resolvents of two cardinality formulas are implied by the sum of the two corresponding 0-1 inequalities.*

An alternative proof of the theorem appears in [21]. As an example, the sum of the inequalities that correspond to the cardinality formulas in (3.15) appears below the line:

$$a + b + c \geq 2$$
$$(1 - a) + c + d \geq 2 \qquad (3.23)$$
$$\overline{b + 2c + d \geq 3}$$

Both classical resolvents in (3.15) are implied by this sum. The sum (3.23) also helps one to recognize redundant partial assignments. For example, $c = 0$ is redundant for $b + 2c + d \geq 3$ but not for either of the formulas in (3.15).

It is easy to check whether a partial assignment

$$(x_{j_1}, \ldots, x_{j_p}) = (v_1, \ldots, v_p)$$

is redundant for a 0-1 inequality $\alpha x \geq \beta$. It is redundant if and only if $\alpha \bar{x} < \beta$, where

$$\bar{x}_j = \begin{cases} v_j & \text{if } j \in \{j_1, \ldots, j_p\} \\ 1 & \text{if } j \notin \{j_1, \ldots, j_p\} \text{ and } \alpha_j > 0 \\ 0 & \text{otherwise} \end{cases}$$

Theorem 3.10 is paralleled by a similar result for diagonal sums, but it does not assist in the recognition of redundant partial assignments.

THEOREM 3.11 (BARTH [8]) *Any diagonal sum $A \geq k + 1$ of a cardinality formula set S is implied by the sum of $|A|$ or fewer cardinality formulas in S.*

Again an alternate proof appears in [21]. For example, the diagonal sum $\{a, b, c, d\} \geq 2$ derived from the set S in (3.16) is implied by the sum of the three formulas in S, which is

$$2a + 2b + 2c + d + 2e \geq 5$$

Theorems 3.10 and 3.11 do not in general produce prime implications, because the sums they describe imply only cardinality formulas that appear in the first round of resolution.

4.3 Tight MILP Relaxations

The single 0-1 inequality (3.22) is a convex hull formulation of the cardinality formula (3.21) [40]. A set of cardinality formulas can be relaxed by writing the inequality (3.22) for each formula, but the relaxation is likely to be weak. Tighter relaxations can be obtained by applying the cardinality resolution algorithm, in whole or in part, and writing the resolvents as 0-1 inequalities. Thus the example (3.14) has the MILP formulation

$$a + b + c + d \geq 3$$
$$c = 1$$
$$a, b, c, d \in \{0, 1\}$$

This happens to be a convex hull formulation.

Generating sums as indicated in Theorems 3.10 and 3.11 does not strengthen the MILP formulation, even though it strengthens the logical formulation by excluding more redundant partial assignments.

5. 0-1 Linear Inequalities

One can regard 0-1 linear inequalities as logical propositions as well as numerical constraints. A 0-1 linear inequality is an inequality of the form $ax \geq \alpha$ where each $x_j \in \{0, 1\}$. We will suppose that α and each coefficient a_j are integers.

There is a large literature (surveyed in [29, 39]) that investigates how to tighten a 0-1 formulation by adding cutting planes. It is also possible to write a good 0-1 formulation for logical purposes. In particular there is a resolution procedure that yields all prime 0-1 implications of a set of 0-1 inequalities. From Lemma 3.2 we have:

COROLLARY 3.12 *Any partial assignment that is redundant for a set S of 0-1 linear inequalities is redundant for some prime 0-1 implication of S. Thus S is complete if it contains all of its prime 0-1 implications, up to equivalence.*

The 0-1 resolution procedure requires that one be able to check whether one 0-1 linear inequality implies another, but unfortunately this is an NP-complete problem. Checking whether $ax \geq \alpha$ implies $bx \geq \beta$ is equivalent to checking whether the minimum of bx subject to $ax \geq \alpha, x \in \{0,1\}^n$ is at least β. The latter is the general 0-1 knapsack problem, which is NP-complete. Due to this difficulty, 0-1 resolution is most likely to be useful when applied to a special class of 0-1 inequalities, such as cardinality clauses.

The 0-1 formulas that are clauses are precisely those that represent propositional clauses. We will refer to these as *clausal inequalities*. Two clausal inequalities resolve in a manner analogous to the corresponding propositional clauses. A set S of 0-1 linear inequalities has a *classical resolvent* R if R is a clausal inequality and

(a) there are clausal inequalities C_1, C_2 with resolvent R such that C_1 is implied by an inequality in S, C_2 is implied by an inequality in S, and

(b) no inequality in S implies R.

A 0-1 inequality $ax \geq \alpha$ implies a clause $x_{j_1}^{v_1} \vee \cdots \vee x_{j_k}^{v_k}$ if and only if the partial assignment $(x_{j_1}, \ldots, x_{j_k}) = (1 - v_1, \ldots, 1 - v_k)$ is redundant for $ax \geq \alpha$, which can be checked as described earlier.

To take an example, either clausal inequality below the line is a classical resolvent of the inequalities above the line:

$$x_1 + 2x_2 + x_3 \geq 2$$
$$x_1 - 2x_2 + x_3 \geq 0$$
$$\overline{}$$
$$x_1 \geq 1$$
$$x_3 \geq 1$$

To define diagonal sums, it is convenient to write a 0-1 inequality $ax \geq \alpha$ in the form $ax \geq \delta + n(a)$, where $n(a)$ is the sum of the negative components of a and δ is the *degree* of the inequality. Thus clausal inequalities have degree 1. A set S of 0-1 inequalities has a *diagonal sum* $ax \geq \delta + 1 + n(a)$ if there is a nonempty $J \subset \{1, \ldots, n\}$ such that

(a) $a_j = 0$ for $j \notin J$,

(b) there are 0-1 inequalities $a^j x \geq \delta + n(a^j)$ for $j \in J$ such that each $a^j x \geq \delta + n(a^j)$ is implied by some inequality in S,

(c) each $a^i x \geq \delta + n(a^i)$ satisfies

$$a_j^i = \begin{cases} a_j - 1 & \text{if } j = i \text{ and } a_j > 0 \\ a_j + 1 & \text{if } j = i \text{ and } a_j < 0 \\ a_j & \text{otherwise} \end{cases}$$

(d) no inequality in S implies $ax \geq \delta + 1 + n(a)$.

> **While** S has a classical resolvent or diagonal sum R
> that is implied by no inequality in S:
>   Remove from S all inequalities implied by R.
>   Add R to S.

Figure 3.5. The 0-1 resolution algorithm applied to set S of 0-1 inequalities

Consider for example the set S consisting of

$$x_1 + 5x_2 - 3x_3 + x_4 \geq 4 - 3$$
$$2x_1 + 4x_2 - 3x_3 + x_4 \geq 4 - 3$$
$$2x_1 + 5x_2 - 2x_3 + x_4 \geq 4 - 2 \tag{3.24}$$
$$2x_1 + 5x_2 - 3x_3 \qquad \geq 4 - 3$$

Note that the inequalities are identical except that the diagonal element in each is reduced by one in absolute value. Since each inequality in (3.24) is implied by a member of S (namely, itself), S has the diagonal sum

$$2x_1 + 5x_2 - 3x_3 + x_4 \geq 5 - 3$$

in which the degree is increased by one.

The 0-1 resolution algorithm appears in Fig. 3.5. The completeness theorem for 0-1 resolution must be stated carefully, since there are infinitely many 0-1 linear inequalities in a given set of variables $x = (x_1, \ldots, x_n)$. Let a finite set H be a *monotone* set of 0-1 linear inequalities if H contains all propositional clauses and given any inequality $ax \geq \delta + n(a)$ in H, H contains all 0-1 inequalities $a'x \geq \delta' + n(a')$ such that $|a'_j| \leq |a_j|$ for all j and $0 \leq \delta' \leq \delta$.

THEOREM 3.13 ([20]) *If H is a monotone set of 0-1 linear inequalities, the set that results from applying the 0-1 resolution algorithm to a set S of 0-1 linear inequalities contains precisely the prime H-implications of S, up to equivalence.*

Again, diagonal summation does not allow one to recognize additional redundant assignments, although it can reduce the number of inequalities one must examine to check for redundancy.

6.   Cardinality Rules

A cardinality rule is a generalized form of cardinality clause that provides more expressive power. A cardinality rule has the form

$$\{L_1, \ldots, L_m\} \geq k \rightarrow \{M_1, \ldots, M_n\} \geq \ell \tag{3.25}$$

where literals L_i and M_j contain distinct variables. The rule states that if at least k of the literals L_i are true, then at least ℓ of the literals M_j are true. It is assumed that $m \geq k \geq 0$, $n \geq 0$, and $\ell \geq 1$.

Procedure Facet $(\{a_1,\ldots,a_m\} \geq k \to \{b_1,\ldots,b_n\} \geq \ell)$
If $m > k = 1$ **then**
    For all $i \in \{1,\ldots,m\}$ perform **Facet** $(a_i \to \{b_1,\ldots,b_n\} \geq \ell)$.
Else if $n = \ell > 1$ **then**
    For all $j \in \{1,\ldots,n\}$ perform **Facet** $(\{a_1,\ldots,a_m\} \geq k \to b_j)$.
Else
    Generate the facet-defining inequality
        $\ell(a_1 + \cdots + a_m) - (m - k + 1)(b_1 + \ldots + b_n) \leq \ell(k-1)$.
    If $m > k$ and $n > 0$ **then**
        For all $\{i_1,\ldots,i_{m-1}\} \subset \{1,\ldots,m\}$,
            perform **Facet** $(\{a_{i_1},\ldots,a_{i_{m-1}}\} \geq k \to \{b_1,\ldots,b_n\} \geq \ell)$.
    If $\ell > 1$ and $m > 0$ **then**
        For all $\{j_1,\ldots,j_{n-1}\} \subset \{1,\ldots,n\}$,
            Perform **Facet** $(\{a_1,\ldots,m\} \geq k \to \{b_{j_1},\ldots,b_{j_{n-1}}\} \geq \ell - 1)$.

Figure 3.6.   An algorithm, adapted from [40], for generating a convex hull formulation of the cardinality rule (3.26). It is assumed that $a_i, b_j \in \{0,1\}$ is part of the formulation. The cardinality clause $\{a_i\} \geq 1$ is abbreviated a_i. The procedure is activated by calling it with (3.26) as the argument.

No specialized inference method has been developed for cardinality rules, but a convex hull MILP formulation for an individual cardinality rule is known. Without loss of generality, assume all literals in (3.25) are positive:

$$\{a_1,\ldots,a_m\} \geq k \to \{b_1,\ldots,b_n\} \geq \ell \qquad (3.26)$$

Any negated literal $\neg a_i$ is represented in the convex hull formulation by $1 - a_i$ rather than a_i.

THEOREM 3.14 (YAN, HOOKER [40]) *The algorithm of Fig. 3.6 generates a convex hull formulation for the cardinality rule (3.26).*

This result has been generalized in [4].

The following example is presented in [40]. Let a_i state that a plant is built at site i for $i = 1, 2$, and let b_j state that product j is made for $j = 1, 2, 3$. Plant construction must observe the following constraints.

If at least 2 plants are built, $\{a_1, a_2, a_3\} \geq 2 \to \{b_1, b_2, b_3\} \geq 2$
at least 2 new products
should be made.
At most 1 new product should $\{\neg a_1, \neg a_2\} \geq 1 \to \{\neg b_1, \neg b_2, \neg b_3\} \geq 2$
be made, unless plants
are built at both sites
1 and 2.

$$\qquad\qquad\qquad\qquad\qquad\qquad\qquad\qquad\qquad\qquad\qquad\qquad (3.27)$$

The convex hull formulation for the first rule is

$$
\begin{aligned}
2(a_1 + a_2 + a_3) - 2(b_1 + b_2 + b_3) &\le 2 \\
2(a_1 + a_2) - b_1 - b_2 - b_3 &\le 2 \\
2(a_1 + a_3) - b_1 - b_2 - b_3 &\le 2 \\
2(a_2 + a_3) - b_1 - b_2 - b_3 &\le 2 \\
a_1 + a_2 + a_3 - 2(b_1 + b_2) &\le 1 \\
a_1 + a_2 + a_3 - 2(b_1 + b_3) &\le 1 \\
a_1 + a_2 + a_3 - 2(b_2 + b_3) &\le 1 \\
a_1 + a_2 - b_1 - b_2 &\le 1 \\
a_1 + a_2 - b_1 - b_3 &\le 1 \\
a_1 + a_2 - b_2 - b_3 &\le 1 \\
a_1 + a_3 - b_1 - b_2 &\le 1 \\
a_1 + a_3 - b_1 - b_3 &\le 1 \\
a_1 + a_3 - b_2 - b_3 &\le 1 \\
a_2 + a_3 - b_1 - b_2 &\le 1 \\
a_2 + a_3 - b_1 - b_3 &\le 1 \\
a_2 + a_3 - b_2 - b_3 &\le 1 \\
a_i, b_j &\in \{0, 1\}
\end{aligned}
$$

The convex hull formulation for the second rule in (3.27) is

$$
\begin{aligned}
2(1 - a_1) - (1 - b_1) - (1 - b_2) - (1 - b_3) &\le 0 \\
2(1 - a_2) - (1 - b_1) - (1 - b_2) - (1 - b_3) &\le 0 \\
(1 - a_1) - (1 - b_1) - (1 - b_2) &\le 0 \\
(1 - a_1) - (1 - b_1) - (1 - b_3) &\le 0 \\
(1 - a_1) - (1 - b_2) - (1 - b_3) &\le 0 \\
(1 - a_2) - (1 - b_1) - (1 - b_2) &\le 0 \\
(1 - a_2) - (1 - b_1) - (1 - b_3) &\le 0 \\
(1 - a_2) - (1 - b_2) - (1 - b_3) &\le 0 \\
a_i, b_j &\in \{0, 1\}
\end{aligned}
$$

7.   Mixing Logical and Continuous Variables

Very often logical and continuous variables must be combined in a constraint. A constraint involving continuous variables, for example, may be enforced only when a certain logical expression is true. One versatile syntax for expressing such constraints is a *conditional constraint*:

$$
g(a) \Rightarrow Ax \ge b \tag{3.28}
$$

where $g(a)$ is a formula of propositional logic containing atomic propositions $a = (a_1, \ldots, a_n)$. Formula (3.28) says that $Ax \ge b$ is enforced if $g(a)$ is true.

It is straightforward for a solver to implement conditional constraints. The constraint in the consequent of (3.28) is posted whenever the antecedent $g(a)$ becomes true in the course of branching and constraint propagation. The rule is "fired" when the currently fixed atomic propositions a_j are enough to ensure that $g(a)$ is true; that is, when every extension of the current partial assignment makes $g(a)$ true. It is clearly advantageous to put the antecedent in clausal form, since this allows one to check whether every extension makes $g(a)$ true by checking whether the partial assignment already satisfies every clause. For example, the partial assignment $a_1 = 1$ necessarily satisfies the clause set $\{a_1 \vee a_2, a_1 \vee a_3\}$ because it makes a literal in both clauses true, whereas $a_2 = 1$ does not necessarily satisfy the clause set.

Relaxations can also be generated for conditional constraints, as will be seen below.

7.1   Modeling with Conditional Constraints

To illustrate the use of conditional constraints, consider a fixed charge problem in which an activity incurs no cost when its level $x = 0$ and a fixed cost f plus variable cost cx when $x > 0$. This is normally formulated in MILP with the help of a big-M constraint:

$$z = fa + cx$$
$$x \leq Ma$$
$$x \geq 0, a \in \{0, 1\}$$

where M is the maximum level of the activity and z is the total cost. Binary variable a is 1 when the activity level is positive and zero otherwise. A conditional formulation could be written

$$a \Rightarrow z \geq f + cx$$
$$\neg a \Rightarrow x = 0 \qquad\qquad (3.29)$$
$$x, z \geq 0$$

To take another situation, suppose it costs f_1 to build plant A only, f_2 to build plant B only, and f_3 to build both plants, where $f_3 \neq f_1 + f_2$. The capacities of plants A and B are M_a and M_b, respectively, and total production is x. One might write the total cost z in an MILP by defining a 0-1 variable a that takes the value 1 when plant A is built, and similarly for b. A third 0-1 variable c is 1 when $a = b = 1$. Thus $c = ab$, an expression that must be linearized. The model becomes

$$z = af_1 + bf_2 + c(f_3 - f_1 - f_2)$$
$$c \geq a + b - 1$$
$$a \geq c$$
$$b \geq c$$
$$x \leq aM_a + bM_b$$
$$a, b, c \in \{0, 1\}$$

With a little thought one might realize that the cost constraint simplifies if a is set to 1 when *only* plant A is built, and similarly for $b = 1$, so that $a + b \leq 1$.

Also $a = b = 0$ when $c = 1$, which can be written with the constraints $a + c \leq 1$ and $b + c \leq 1$. These three inequalities can be combined into a single inequality $a + b + c \leq 1$. This simplifies the model but slightly complicates the capacity constraint.

$$
\begin{aligned}
&z = af_1 + bf_2 + cf_3 \\
&a + b + c \leq 1 \\
&x \leq aM_a + bM_b + c(M_a + M_b) \\
&a, b, c \in \{0, 1\}
\end{aligned}
\tag{3.30}
$$

A logic-based approach permits one to write the model as it is originally conceived:

$$
\begin{aligned}
&(\neg a \wedge \neg b) \Rightarrow (x = z = 0) \\
&(a \wedge \neg b) \Rightarrow (x \leq M_a, z = f_a) \\
&(\neg a \wedge b) \Rightarrow (x \leq M_b, z = f_b) \\
&(a \wedge b) \Rightarrow (x \leq M_a + M_b, z = f_c)
\end{aligned}
\tag{3.31}
$$

7.2   Relaxations for Conditionals

MILP formulations have the advantage that a continuous relaxation is readily available from the model itself, simply by dropping integrality constraints. However, conditional constraints can trigger the generation of comparable or even better relaxations.

To obtain a relaxation, however, one must indicate which groups of conditionals have the property that the disjunction of their antecedents is a tautology (a chore that could be done automatically if desired). A natural way to indicate this is to use an *inequality-or* global constraint. The disjunction of the antecedents of the two conditionals (3.29), for example, is the tautology $a \vee \neg a$. The two conditionals can therefore be written

$$
\text{inequality-or}\left\{ \begin{pmatrix} a \\ z \geq f + cx \end{pmatrix}, \begin{pmatrix} \neg a \\ x = 0 \end{pmatrix} \right\}
\tag{3.32}
$$

Similarly, the conditionals (3.31) can be written

$$
\text{inequality-or}\left\{ \begin{pmatrix} \neg a \wedge \neg b \\ x = z = 0 \end{pmatrix}, \begin{pmatrix} a \wedge \neg b \\ x \leq M_a \\ z = f_a \end{pmatrix}, \begin{pmatrix} \neg a \wedge b \\ x \leq M_b \\ z = f_b \end{pmatrix}, \begin{pmatrix} \neg a \wedge \neg b \\ x \leq M_a + M_b \\ z = f_c \end{pmatrix} \right\}
\tag{3.33}
$$

In general inequality-or has the form

$$
\text{inequality-or}\left\{ \begin{pmatrix} g_1(a) \\ A^1 x \geq b^1 \end{pmatrix}, \dots, \begin{pmatrix} g_k(a) \\ A^K x \geq b^K \end{pmatrix} \right\}
\tag{3.34}
$$

Constraint (3.34) implies a disjunction of linear systems

$$
\bigvee_{k=1}^{K} A^k x \geq b^k
\tag{3.35}
$$

that can be relaxed in two ways. The big-M relaxation is

$$A^k x \geq b^k - M^k(1 - y_k), \quad k = 1, \ldots, K$$

$$\sum_{k=1}^{K} y_k = 1 \tag{3.36}$$

$$y_k \geq 0, \quad k = 1, \ldots, K$$

where M_i^k is a valid upper bound on $A^k x - b^k$. The smallest possible value for M_i^k, which yields the tightest relaxation, is

$$M_i^k = b_i^k - \min_{k' \neq k} \left\{ \min_x \left\{ A_i^k x \mid A_i^{k'} x \leq b_i^{k'} \right\} \right\}$$

If some $M_i^k = \infty$, then x must be bounded to prevent this. Beaumont [9] showed that the relaxation (3.34) simplifies for a disjunction of inequalities

$$\bigvee_{k=1}^{K} a^k x \geq b_k$$

where $0 \leq x \leq m$. The relaxation becomes a single inequality

$$\left(\sum_{k=1}^{K} \frac{a^k}{M_k} \right) x \geq \sum_{k=1}^{K} \frac{b_k}{M_k} - K + 1 \tag{3.37}$$

where

$$M_k = b_k - \min_{k' \neq k} \left\{ \min_x \left\{ a_i^k x \mid a_i^{k'} x \leq b_{k'}, \ 0 \leq x \leq m \right\} \right\} \tag{3.38}$$

A second relaxation for (3.35) is the disjunctive convex hull relaxation of Balas [2, 3]:

$$A^k x^k \geq b^k y_k, \quad k = 1, \ldots, K$$

$$x = \sum_{k=1}^{K} x^k$$

$$\sum_{k=1}^{K} y_k = 1 \tag{3.39}$$

$$y_k \geq 0, \quad k = 1, \ldots, K$$

The disjunctive convex hull relaxation provides the tightest linear relaxation but requires additional variables $x^1, \ldots, x^K$.

For example, the disjunction in (3.32) can be written

$$(-cx + z \geq f) \vee (-x \geq 0) \tag{3.40}$$

where $0 \leq x \leq M$ (it turns out that no bound on z is needed). To apply Beaumont's big-M relaxation (3.37) we obtain from (3.38) that $M_1 = f$ and $M_2 = M$. Thus the big-M relaxation is

$$z \geq \left(c + \frac{f}{M}\right) x \tag{3.41}$$

The disjunctive convex hull relaxation (3.39) of (3.40) simplifies to

$$z \geq fy + cx, \quad 0 \leq y \leq 1 \tag{3.42}$$

This is equivalent to the big-M relaxation (3.41), which is the projection of (3.42) onto x, z. For this particular model, the big-M relaxation is as tight as the disjunctive convex hull relaxation and is simpler.

Sometimes the disjunctive convex hull relaxation is simpler than the big-M relaxation. Consider for example a problem in which one can install either of two machines, or neither. Machine A has cost 50 and capacity 5, while machine B has cost 80 and capacity 10. If z is the total fixed cost and x the output, the model is

$$\text{inequality-or} \left\{ \begin{pmatrix} \neg a \wedge \neg b \\ x = 0 \end{pmatrix}, \begin{pmatrix} a \\ z \geq 50 \\ 0 \leq x \leq 5 \end{pmatrix}, \begin{pmatrix} b \\ z \geq 80 \\ 0 \leq x \leq 10 \end{pmatrix} \right\}$$

The big-M relaxation is

$$\begin{aligned}
x &\leq 10(y_1 + y_2) \\
x &\leq 5(2 - y_1) \\
x &\leq 5(1 + y_2) \\
z &\geq 50y_1 \\
z &\geq 80y_2 \\
y_1 + y_2 &\leq 1 \\
x, y_1, y_2 &\geq 0
\end{aligned} \tag{3.43}$$

which projects onto x, z as follows:

$$\begin{aligned}
z &\geq (40/13)x \\
z &\geq 16x - 80 \\
0 &\leq x \leq 10
\end{aligned}$$

The disjunctive convex hull relaxation is simpler than (3.43):

$$\begin{aligned}
x &\leq 5y_1 + 10y_2 \\
z &\geq 50y_1 + 80y_2 \\
y_1 + y_2 &\leq 1 \\
x, y_1, y_2 &\geq 0
\end{aligned}$$

It is also tighter, since it projects onto x, z as follows:

$$z \geq 8x$$
$$0 \leq x \leq 10$$

Typically, however, the disjunctive convex hull relaxation is more complicated than the big-M relaxation.

Finally, the disjunctive convex hull relaxation of the plant building constraint (3.33) is precisely the linear relaxation of (3.30).

8.   Additional Global Constraints

The CP community has developed a large lexicon of global constraints. Three particularly useful ones are briefly examined here: *all-different*, *element* and *cumulative*. All have been studied with respect to both relaxations and consistency maintenance. A number of hybrid modeling examples using these and other global constraints may be found in [21].

8.1   The All-different Constraint

The all-different constraint is ubiquitous in scheduling applications. It is written

$$\text{all-different}\{y_1, \ldots, y_k\} \tag{3.44}$$

where $y_1, \ldots, y_k$ are variables with finite domains. The constraint states that $y_1, \ldots, y_k$ take distinct values.

The all-different constraint permits an elegant formulation of the traveling salesman problem. The object is to find the cheapest way to visit each of n cities exactly once and return home. Let y_i be the ith city in the tour, and let c_{jk} be the cost of traveling from city j to city k. The problem can be written

$$\min \quad \sum_{i=1}^{n} c_{y_i y_{i+1}} \tag{3.45}$$
$$\text{subject to} \quad \text{all-different}\{y_1, \ldots, y_n\}$$

where y_{n+1} is identified with y_1. (It may be more practical to write the traveling salesman problem with the *cycle* constraint [21], since it can be relaxed using well-known cutting planes for the problem.)

The best-known domain reduction algorithm, that of Régin [33], achieves hyperarc consistency. Let G be a bipartite graph whose vertices correspond to the variables $y_1, \ldots, y_k$ and to the elements $v_1, \ldots, v_n$ in the union of the variable domains $D_1, \ldots, D_k$. G contains an edge (y_j, v_k) whenever $v_k \in D_j$. Hyperarc consistency is obtained as follows. First find a maximum cardinality bipartite matching on G. For each vertex that is not covered by the matching, mark all edges that are part of an alternating path that starts at that vertex. By a theorem of Berge [10], these edges belong to at least one, but not every, maximum cardinality matching. For the same reason, mark every edge that belongs to some alternating cycle. (An alternating path or cycle is one in which every second edge belongs to the matching.) Now every unmarked edge in the matching belongs to every maximum cardinality matching, by Berge's theorem. One can therefore delete all unmarked edges from G that are not part of the matching. The reduced domains D_j contain values v_k for which (x_j, v_k) is a remaining edge.

For example, consider the constraint all-different$(y_1, \ldots, y_5)$ in which the initial domains are as shown on the left below.

$$
\begin{aligned}
D_1 &= \{1\} && \rightarrow \{1\} \\
D_2 &= \{2, 3, 5\} && \rightarrow \{2, 3\} \\
D_3 &= \{1, 2, 3, 5\} && \rightarrow \{2, 3\} \\
D_4 &= \{1, 5\} && \rightarrow \{5\} \\
D_5 &= \{1, 2, 3, 4, 5, 6\} && \rightarrow \{4, 6\}
\end{aligned}
$$

The algorithm reduces the domains to those shown on the right.

The convex hull relaxation for all-different is known for the case when the variables have the same domains, even if it is rather weak and exponentially long [21, 37]. Let each domain D_j be a set $\{t_1, \ldots, t_n\}$ of numerical values, with $t_1 \leq \ldots \leq t_n$. A convex hull relaxation of (3.44) can be written as follows:

$$
\sum_{j=1}^{n} x_j = \sum_{j=1}^{n} t_j
$$

$$
\sum_{j=|J|+1}^{n} t_j \geq \sum_{j \in J} x_j \geq \sum_{j=1}^{|J|} t_j, \quad \text{all } J \in \{1, \ldots, n\} \text{ with } |J| < n \tag{3.46}
$$

For example, consider all-different$\{y_1, y_2, y_3\}$ with domain $\{15, 20, 25\}$. The convex hull relaxation is

$$
\begin{aligned}
y_1 + y_2 + y_3 &= 60 \\
45 \geq y_1 + y_2 &\geq 35 \\
45 \geq y_1 + y_3 &\geq 35 \\
45 \geq y_2 + y_3 &\geq 35 \\
25 \geq y_1, y_2, y_3 &\geq 15
\end{aligned}
$$

8.2   The Element Constraint

The element constraint selects a value to assign to a variable. It can be written

$$
\text{element}(y, (v_1, \ldots, v_n), z) \tag{3.47}
$$

If the value of y is fixed to i, the constraint fixes z to v_i. If the domain D_y of y is larger than a singleton, the constraint essentially defines a domain for z:

$$z \in \{v_i \mid i \in D_y\} \tag{3.48}$$

The current domain D_z of z can also be used to reduce the domain of y:

$$y \in \{i \mid v_i \in D_z\} \tag{3.49}$$

If the values v_i are numerical, there is an obvious convex hull relaxation for (3.47):

$$\min\{v_i \mid i \in D_y\} \le z \le \max\{v_i \mid i \in D_y\} \tag{3.50}$$

The element constraint is particularly useful for implementing variable indices, an example of which may be found in the objective function of the traveling salesman problem (3.45). A constant with a variable index, such as v_y, may be replaced by the variable z and the constraint (3.47).

An extension of the element constraint implements variable indices for variables:

$$\text{element}(y, (x_1, \ldots, x_n), z) \tag{3.51}$$

If y is fixed to i, the constraint posts a new constraint $z = x_i$. Otherwise the constraint is essentially a disjunction:

$$\bigvee_{i \in D_y} (z = x_i)$$

Hyperarc consistency can be achieved for (3.51) as follows [21]. Let D_y, D_{x_i}, D_z be the current domains, and let $\bar{D}_y$, $\bar{D}_{x_i}$ and $\bar{D}_z$ be the reduced domains. Then

$$\bar{D}_z = D_z \cap \bigcup_{i \in D_y} D_{x_i}$$

$$\bar{D}_y = D_y \cap \{i \mid D_z \cap D_{x_i} \ne \emptyset\} \tag{3.52}$$

$$\bar{D}_{x_i} = \begin{cases} \bar{D}_z & \text{if } \bar{D}_y = \{i\} \\ D_{x_j} & \text{otherwise} \end{cases}$$

Disjunctive relaxations may be derived for (3.51) using the principles of the previous section. It is proved in [21] that (3.51) has the following convex hull relaxation when all the variables x_i have the same upper bound m_0:

$$\sum_{i \in D_y} x_i \le z \le \sum_{i \in D_y} x_i - (|D_y| - 1)m_0 \tag{3.53}$$

When each $x_i \leq m_i$ one can write the relaxation

$$\left(\sum_{i \in D_y} \frac{1}{m_i} \right) z \leq \sum_{i \in D_y} \frac{x_i}{m_i} + |D_y| - 1$$

$$\left(\sum_{i \in D_y} \frac{1}{m_i} \right) z \geq \sum_{i \in D_y} \frac{x_i}{m_i} - |D_y| + 1$$

(3.54)

and use the relaxation (3.53) as well by setting $m_0 = \max_i \{m_i\}$.

An expression of the form x_y, where x is a variable, can be implemented by replacing it with z and the constraint (3.51).

8.3   The Cumulative Constraint

The *cumulative* constraint [1] is used for resource-constrained scheduling problems. Jobs must be scheduled so that the total resource consumption at any moment is within capacity.

The constraint may be written

$$\text{cumulative}(t, d, r, L) \qquad (3.55)$$

where $t = (t_1, \ldots, t_n)$ is a vector of start times for jobs $1, \ldots, n$, $d = (d_1, \ldots, d_n)$ is a vector of job durations, $r = (r_1, \ldots, r_n)$ a vector of resource consumption rates, and scalar L the amount of available resources. The domain of each t_j is given as $[a_j, b_j]$, which defines an earliest start time a_j and a latest start time b_j for job j. The cumulative constraint requires that

$$\sum_{\substack{j \\ t_j \leq t < t_j + d_j}} r_j \leq L, \quad \text{all } t$$

and $a \leq t \leq b$.

An important special case is the L-machine scheduling problem, in which each resource requirement r_j is 1, representing one machine. The cumulative constraint requires that jobs be scheduled so that no more than L machines are in use at any time.

The cumulative constraint can be approximated in an MILP formulation if one discretizes time, but this can result in a large number of variables. There is also an MILP formulation that uses continuous time variables, but it has a weaker relaxation and is difficult to solve [23].

A number of domain reduction procedures have been developed for the cumulative constraint (e.g., [1, 5, 6, 7, 15, 30]), although none of them are guaranteed to achieve hyperarc consistency. They are generally based on edge-finding ideas.

One simple relaxation of the cumulative constraint provides a bound on the minimum makespan T, which can be no less than the "area" of the jobs (duration times resource requirement) divided by L:

$$T \geq \sum_j d_j r_j / L \qquad (3.56)$$

Suppose for example that there are five jobs and $a = (0,0,0,0,0)$, $d = (2,4,3,3,3)$, $r = (4,2,3,3,3)$, $L = 6$. The relaxation (3.56) yields a bound $T \geq 7\frac{1}{6}$, which may be compared with the minimum makespan of $T = 8$.

Another relaxation may be obtained by analyzing the problem in two parts: the "lower" problem, in which the upper bounds $b_j = \infty$, and the "upper" problem, in which the lower bounds $a_j = -\infty$. Relaxations for the upper and lower problem can then be combined to obtain a relaxation for the entire problem. Only the upper problem is studied here, but the lower problem is closely parallel.

Facet-defining inequalities exist when there are subsets of jobs with the same release time, duration, and resource consumption rate.

THEOREM 3.15 (HOOKER AND YAN [24]) *Suppose jobs* $j_1, \ldots, j_k$ *satisfy* $a_{j_i} = a_0$, $d_{j_i} = d_0$ *and* $r_{j_i} = r_0$ *for* $i = 1, \ldots, k$. *Let* $Q = \lfloor L/r_0 \rfloor$ *and* $P = \lceil k/Q \rceil - 1$. *Then the following defines a facet of the convex hull of the upper problem.*

$$t_{j_1} + \cdots + t_{j_k} \geq (P+1)a_0 + \tfrac{1}{2}P[2k - (P+1)Q]d_0 \qquad (3.57)$$

provided $P > 0$. *Furthermore, bounds of the form* $t_j \geq a_j$ *define facets.*

The following valid inequalities are in general non-facet-defining but exist in all problems.

THEOREM 3.16 (HOOKER AND YAN [24]) *Renumber any subset of jobs* $j_1, \ldots, j_k$ *using indices* $q = 1, \ldots, k$ *so that the products* $r_q d_q$ *occur in nondecreasing order. The following is a valid inequality for the upper problem:*

$$t_{j_1} + \cdots + t_{j_k} \geq \sum_{q=1}^{k} \left((k - q + \tfrac{1}{2})\frac{r_q}{L} - \tfrac{1}{2} \right) d_q \qquad (3.58)$$

Theorems 3.15 and 3.16 can be used to obtain a bound on minimum tardiness. Supposing in the above example that $b = (2,3,4,5,6)$, we have the following relaxation of the minimum tardiness problem, where z_j is the tardiness of job j. The second constraint is facet-defining and derives from Theorem 3.15, and the third constraint is from Theorem 3.16.

$$\min \quad \sum_{j=1}^{5} z_j$$

$$\text{subject to} \quad z_j \geq t_j + d_j - b_j, \quad j = 1, \ldots, 5$$
$$t_3 + t_4 + t_5 \geq 3$$
$$t_1 + t_2 + t_3 + t_4 + t_5 \geq 9\tfrac{11}{12}$$
$$t_j \geq 0, \quad \text{all } j$$

The optimal value 4.91 of this relaxation provides a valid bound on the minimum tardiness, which in this case is 6.

9. Conclusion

Table 3.2 lists the logic-based constraints discussed above and briefly indicates how one might write or reformulate the constraint to achieve consistency or completeness, as well as how one might write a tight MILP formulation. Consistency and completeness are important for CP-based or CP/MILP hybrid solvers, and they can also help generate tighter MILP formulations. The MILP reformulation is obviously essential if one intends to use an MILP, and it also provides a tight relaxation for hybrid solvers. The situation is summarized in Table 3.3.

The consistency/completeness methods are designed to be applied to groups of constraints and can be implemented either by hand or by the solver. The MILP reformulation would be carried out automatically.

As mentioned at the outset, there are advantages to applying a unified solver directly to logic-based constraints, rather than sending them to MILP solver in inequality form. The advantages are not only computational, but such constraints as all-different, element and cumulative are difficult to formulate in an MILP framework.

The discussion here has focused on obtaining consistency/completeness and a tight relaxation before solution starts. Similar methods can be used in the course of solution. CP and hybrid solvers, for example, may restore hyperarc consistency or some approximation of it whenever domains are reduced by branching or filtering. This is regularly done for all-different, element and cumulative constraints. Resolution could be repeatedly applied to clause sets and to antecedents of conditional constraints in order to maintain full consistency. Cardinality and 0-1 resolution may be too costly to apply repeatedly. It may also be advantageous to update linear relaxations.

Several types of logics other than those discussed here can be given MILP formulations. They include probabilistic logic, belief logics, many-valued logics, modal logics, default and nonmonotonic logics, and if one permits infinite-

Table 3.2. Catalog of logic-based constraints

Type of constraint and examples	To achieve consistency or completeness	To obtain a tight MILP formulation
Formula of propositional logic $a \vee \neg b \vee c$ $a \equiv (b \rightarrow c)$	Convert to clausal form using the algorithm of Fig. 3.1 or 3.2 and apply resolution to achieve full consistency, or apply a limited form of resolution.	Convert to clausal form without adding variables (using the algorithm of Fig. 3.1) and write clauses as 0-1 inequalities. If desired apply cardinality resolution.
Cardinality clauses $\{a, \neg b, c\} \geq 2$	Apply cardinality resolution to achieve completeness. (Diagonal sums are not needed for completeness but can accelerate checking for redundant assignments.)	Apply cardinality resolution and convert the cardinality clauses to 0-1 inequalities.
0-1 linear inequalities, viewed as logical propositions $3x_1 - 2x_2 + 4x_3 \geq 5$	Apply 0-1 resolution to achieve completeness (best for small or specially structured inequality sets).	Generate cutting planes.
Cardinality rules $\{a, \neg b, c\} \geq 2$ $\rightarrow \{d, e\} \geq 1$		Generate a convex hull formulation as described in Fig. 3.6.
Conditional constraints $(b \vee \neg c) \rightarrow (Ax \geq a)$	Consistency unnecessary, but put every antecedent into clausal form.	Use inequality-or global constraints to identify disjunctions of linear systems and relax using a convex hull relaxation (3.39) or a big-M relaxation given by (3.36) or (3.37).
All-different all-different$\{y_1, \ldots, y_n\}$	Use Régin's matching algorithm to achieve hyperarc consistency.	Use the convex hull formulation (3.46) if it is not too large.
Element element$(y, (v_1, \ldots, v_n), z)$	Use formulas (3.48) and (3.49) to achieve hyperarc consistency	Use the range (3.50).
Extended element element$(y, (x_1, \ldots, x_n), z)$	Use formulas (3.52) to achieve hyperarc consistency.	Use relaxations (3.53) and (3.54).
Cumulative cumulative(t, d, r, L)		Use relaxations (3.57) and (3.58).

Table 3.3. Advantages of consistent and tight formulations

Type of formulation	Advantage for CP solver	Advantage for hybrid CP/MILP solver	Advantage for MILP solver
Consistent or complete logic-based formulation	Reduces backtracking.	Reduces backtracking.	May help generate tight MILP formulation.
Tight MILP formulation		May provide tight relaxation for better bounds.	Provides tight relaxation for better bounds.

dimensional integer programming, predicate logic. All of these are discussed in [12]. Many of the resulting MILP formulations provide practical means of reasoning in these various logics. The formulations are omitted here, however, because it is unclear whether it is practical or useful to incorporate them with other types of constraints in a general-purpose model. This remains an issue for further research.

References

[1] Aggoun, A., and N. Beldiceanu, Extending CHIP in order to solve complex scheduling and placement problems, *Mathematical and Computer Modelling* **17** (1993) 57–73.

[2] Balas, E., Disjunctive programming: Cutting planes from logical conditions, in O. L. Mangasarian, R. R. Meyer, and S. M. Robinson, eds., *Nonlinear Programming 2*, Academic Press (New York, 1975), 279-312.

[3] Balas, E., Disjunctive programming, *Annals Discrete Mathematics* **5** (1979): 3-51.

[4] E. Balas, A. Bockmayr, N. Pisaruk and L. Wolsey, On unions and dominants of polytopes, manuscript, http://www.loria.fr/~bockmayr/ dp2002-8.pdf, 2002.

[5] Baptiste, P.; C. Le Pape, Edge-finding constraint propagation algorithms for disjunctive and cumulative scheduling, *Proceedings, UK Planning and Scheduling Special Interest Group 1996* (PLANSIG96), Liverpool, 1996.

[6] Baptiste, P., and C. Le Pape, Constraint propagation and decomposition techniques for highly disjunctive and highly cumulative project scheduling problems. *Principles and Practice of Constraint Programming (CP 97)*, Springer-Verlag (Berlin, 1997) 375–89.

[7] Baptiste, P.; C. Le Pape, Constraint propagation and decomposition techniques for highly disjunctive and highly cumulative project scheduling problems, *Constraints* **5** (2000) 119–39.

[8] Barth, P., *Logic-Based 0-1 Constraint Solving in Constraint Logic Programming*, Kluwer (Dordrecht, 1995).

[9] Beaumont, N., An algorithm for disjunctive programs, *European Journal of Operational Research* **48** (1990): 362-371.

[10] Berge, C., *Graphes et hypergraphes*, Dunod (Paris, 1970).

[11] Bockmayr, A., and J. N. Hooker, Constraint programming, in K. Aardal, G. Nemhauser and R. Weismantel, eds., *Handbook of Discrete Optimization*, North-Holland, to appear.

[12] Chandru, V. and J. N. Hooker, *Optimization Methods for Logical Inference*, Wiley (New York, 1999).

[13] Chvátal, V., Edmonds polytopes and a hierarchy of combinatorial problems, *Discrete Mathematics* **4** (1973): 305-337.

[14] Chvátal, V., and E. Szemeredi, Many hard examples for resolution, *Journal of the ACM* **35** (1988) 759-768.

[15] Caseau, Y., and F. Laburthe, Cumulative scheduling with task intervals, *Proceedings of the Joint International Conference and Symposium on Logic Programming*, 1996.

[16] Cornuejols, G., and M. Dawande, A class of hard small 0-1 programs, *INFORMS Journal on Computing* **11** (1999) 205-210.

[17] Haken, A., The intractability of resolution, *Theoretical Computer Science* **39** (1985): 297-308.

[18] S. Heipcke, *Combined Modelling and Problem Solving in Mathematical Programming and Constraint Programming*, PhD thesis, University of Buckingham, 1999.

[19] Hooker, J. N., Generalized resolution and cutting planes, *Annals of Operations Research* **12** (1988): 217-239.

[20] Hooker, J. N., Generalized resolution for 0-1 linear inequalities, *Annals of Mathematics and Artificial Intelligence* **6** (1992): 271-286.

[21] Hooker, J. N., *Logic-Based Methods for Optimization: Combining Optimization and Constraint Satisfaction*, Wiley (New York, 2000).

[22] Hooker, J. N., Logic, optimization and constraint programming, *INFORMS Journal on Computing* **14** (2002) 295-321.

[23] Hooker, J. N., Logic-based Benders decomposition for planning and scheduling, manuscript, 2003.

[24] Hooker, J. N., and Hong Yan, A relaxation of the cumulative constraint, *Principles and Practice of Constraint Programming* (CP02), Lecture Notes in Computer Science **2470**, Springer (Berlin, 2002) 686-690.

[25] Jeroslow, R. E., *Logic-Based Decision Support: Mixed Integer Model Formulation, Annals of Discrete Mathematics* **40**. North-Holland (Amsterdam, 1989).

[26] McKinnon, K. I. M., and H. P. Williams, Constructing integer programming models by the predicate calculus, *Annals of Operations Research* **21** (1989) 227-245.

[27] Milano, M., ed., *Constraint and Integer Programming: Toward a Unified Methodology*, Kluwer (2003).

[28] Mitra, G., C. Lucas, S. Moody, and E. Hadjiconstantinou, Tools for reformulating logical forms into zero-one mixed integer programs, *European Journal of Operational Research* **72** (1994) 262-276.

[29] Nemhauser, G. L., and L. A. Wolsey, *Integer and Combinatorial Optimization*, Wiley (1999).

[30] Nuijten, W. P. M., *Time and Resource Constrained Scheduling: A Constraint Satisfaction Approach*, PhD Thesis, Eindhoven University of Technology, 1994.

[31] Quine, W. V., The problem of simplifying truth functions, *American Mathematical Monthly* **59** (1952): 521-531.

[32] Quine, W. V., A way to simplify truth functions, *American Mathematical Monthly* **62** (1955): 627-631.

[33] Régin, J.-C., A filtering algorithm for constraints of difference in CSPs, *Proceedings, National Conference on Artificial Intelligence* (1994): 362-367.

[34] Tseitin, G. S., On the complexity of derivations in the propositional calculus, in A. O. Slisenko, ed., *Structures in Constructive Mathematics and Mathematical Logic, Part II* (translated from Russian, 1968), 115-125.

[35] Williams, H. P. *Model Building in Mathematical Programming*, 4th ed., Wiley (Chichester, 1999).

[36] Williams, H. P., and J. M. Wilson, Connections between integer linear programming and constraint logic programming–An overview and introduction to the cluster of articles, *INFORMS Journal on Computing* **10** (1998) 261-264.

[37] Williams, H.P., and Hong Yan, Representations of the all different predicate of constraint satisfaction in integer programming, *INFORMS Journal on Computing* **13** (2001) 96-103.

[38] Wilson, J. M., Compact normal forms in propositional logic and integer programming formulations, *Computers and Operations Research* **17** (1990): 309-314.

[39] Wolsey, L. A., *Integer Programming*, Wiley (1998).

[40] Yan, H., and J. N. Hooker, Tight representation of logical constraints as cardinality rules, *Mathematical Programming* **85** (1999): 363-377.

Chapter 4

MODELLING FOR FEASIBILITY - THE CASE OF MUTUALLY ORTHOGONAL LATIN SQUARES PROBLEM

Gautam Appa[1], Dimitris Magos[2], Ioannis Mourtos[3] and Leonidas Pitsoulis[4]

[1] *Department of Operational Research*
London School of Economics, London
United Kingdom
g.appa@lse.ac.uk

[2] *Department of Informatics*
Technological Educational Institute of Athens
12210 Athens, Greece
dmagos@teiath.gr

[3] *Department of Economics*
University of Patras
26500 Rion, Patras, Greece
imourtos@upatras.gr

[4] *Department of Mathematical and Physical Sciences*
Aristotle University of Thessaloniki
54124 Thessaloniki, Greece
pitsouli@gen.auth.gr

Abstract   In this chapter we present various equivalent formulations or models for the Mutually Orthogonal Latin Squares (MOLS) problem and its generalization. The most interesting feature of the problem is that for some parameters the problem may be infeasible. Our evaluation of different formulations is geared to tackling this feasibility problem. Starting from a Constraint Programming (CP) formulation which emanates naturally from the problem definition, we develop several Integer Programming (IP) formulations. We also discuss a hybrid CP-IP approach in both modelling and algorithmic terms. A non-linear programming formulation and an interesting modelling approach based on the intersection of matroids are also considered.

Keywords: Integer Programming, Constraint Programming, Matroids, Mutually Orthogonal Latin Squares, Feasibility Problems.

1. Introduction

Various combinatorial optimization problems can be formulated in several different ways as integer programs (IP). For example, there are at least eight different formulations of the travelling salesman problem [30]. The quadratic assignment problem [9], vehicle routing [24], uncapacitated lot sizing [7], shared fixed costs [32], production planning [8, 23] and uncapacitated facility location [11] are some further cases of combinatorial optimization problems with multiple formulations. These reflect the efforts of various researchers to view the problem from a different perspective, in order to gain more information that will lead to a better solution technique.

In the last decade, constraint programming (CP) has evolved as an alternative method for dealing with certain combinatorial optimization problems, traditionally examined only as integer programs. The ability of CP to incorporate, and also to efficiently handle different types of non-linear constraints implies a broader declarative and modelling capacity. In other words, CP allows for more natural problem formulations. Examples of its success in tackling practical problems are now in the public domain. One such application in scheduling a basketball conference appears in [29], while a list of further applications is included in [20]. Research has also been conducted on providing empirical evidence on the competitiveness of hybrid CP-IP methods for specific problems (e.g. see [2, 27]).

Given multiple formulations for an optimization problem, the question of evaluation criteria arises. Since our goal is to solve the problem at hand, a natural criterion would be the efficacy of the best solution method available for the formulation. For example, if an IP problem can be reformulated to yield a totally unimodular constraint matrix, thus guaranteeing an integer solution to its linear programming (LP) relaxation, the resulting formulation is clearly the most beneficial. Examples of this kind can be found in [11] and [32].

In general IP problems, finding all the constraints defining the convex hull of integer solutions is equivalent to obtaining a linear program whose extreme points are all integral ([17]). There have been several advances in obtaining the convex hull of IP problems, examples including matching ([14]) and uncapacitated lot sizing ([7]). A more recent strand of research is devoted to obtaining the convex hull of the vectors satisfying CP predicates, like cardinality rules (see [34] and also [6] for a generalisation) and the *all_different* constraint (in [33]). When the convex hull is described by an exponential number of inequalities, this knowledge remains theoretically interesting but computationally almost useless, unless a polynomial separation algorithm for these inequalities is available. Well-known examples of separation algorithms appear in [18] for the subtour elimination constraints of the TSP and in [19] for the inequalities defining the matching polytope. Another example related to our work is the separation of the inequalities defining the convex hull of the *all different* constraint ([4]). In general though, the search for all the constraints of the convex hull might be both hopeless and possibly unnecessary in most interesting IP problems. A related fact is that the vast majority of the known constraints of the convex hull might never become binding during the search for an optimal solution.

From an optimization perspective, the generally accepted criterion for comparing different IP formulations is the *tightness* of their LP relaxations. More formally let F_i, $i = 1, 2$, represent two different IP formulations of a maximization problem and O_i represent the objective function value of their LP relaxations. Formulation F_1 is better than F_2 if $O_2 \geq O_1$ because the *duality gap* (also called *integrality* gap) - i.e. the gap between the IP solution and the LP solution - is smaller for F_1. But this is probably only a practical guide. Conceptually at least, it is possible to form a situation where F_1 is better than F_2 for objectives O_i but not for another set of objectives O_i'. A more realistic comparison is possible if the feasible region defined by F_1 is smaller than the one defined by F_2, i.e. F_1 defines a tighter relaxation (see [30] for the TSP and also [28] for other examples).

In theory, optimization problems and feasibility problems are equivalent in the sense that one can be transformed into the other so that the computational complexity of both remains the same. But in practice a genuine feasibility problem is different. The main thrust of this chapter is to highlight this difference by illustrating model formulations of the Mutually Orthogonal Latin Squares problem (MOLS) - a combinatorial problem which can be naturally thought of as a feasibility problem.

What do we mean by a genuine feasibility problem? We can loosely define it as a problem for which the existence of a feasible solution for some value of a defining parameter is in doubt. Hence, the only question of interest is whether there exists a feasible solution to our model for a given set of input values. The MOLS problem discussed in this chapter is precisely of this type, since it refers to the existence of a set of k MOLS.

What distinguishes a genuine feasibility problem from an optimization problem in algorithmic terms? There are at least two important considerations. First recall that in a normal IP problem the duality gap is an important comparator for different formulations. In contrast, the actual objective function value of the LP relaxation of the IP is of no interest in a feasibility problem because all we are interested in is *any* feasible solution to the IP. Hence, the concept of duality gap, or tightness of linear relaxations, is totally irrelevant in comparing different formulations of a feasibility problem. Secondly, it is possible that there is no feasible solution, therefore the formulation has to be good in its capacity to explore all possible combinations of permissible integer values of its variables and show explicitly that none of them satisfy all the constraints.

At its best, the LP relaxation of a good formulation should be infeasible for all infeasible instances of the problem.

In the rest of this chapter, we look at different formulations of the MOLS problem to show that multiple answers are possible. After defining the MOLS problem in the next section, various formulations of the problem are presented in Section 4.3. Several pure and mixed IP formulations, a CP formulation, a hybrid CP-IP formulation and two combinatorial approaches, one with permutations and another involving matroids are provided. Finally, Section 4.4 discusses the comparative advantages of each model, partly with respect to computational experience obtained for small feasible and infeasible instances of the MOLS problem.

2.   Definitions and notation

In this section we will present the definition of a Latin square as well as the notion of orthogonality, and state the problems which are of interest. The definitions and the notation follow the lines of [13, 16].

DEFINITION 4.1 *A **partial Latin square (PLS)** of order n is an $n \times n$ square matrix, where each of its cells is either empty or contains one of the values $0, .., (n-1)$ that appears exactly once in the corresponding row and column. A **Latin square (LS)** is a partial Latin square without empty cells.*

If a PLS can be extended to a LS, then it is called *completable* or *incomplete Latin square*. Given a partial Latin square, the problem of filling the maximum number of empty cells is called the **partial Latin square extension (PLSE)** problem, which has been studied in [15, 16, 22], while in [10] it has been shown that deciding whether or not a partial Latin square is completable is NP-complete.

DEFINITION 4.2 *Two Latin squares of order n are called **orthogonal (OLS)** if and only if each of the n^2 ordered pairs $(0,0), ..., (n-1, n-1)$ appears exactly once in the two squares.*

A pair of OLS of order 4 appears in Table 4.1. Definition 4.2 is extended to sets of $k > 2$ Latin squares, which are called **mutually orthogonal (MOLS)** if they are pairwise orthogonal. There can be at most $n-1$ MOLS of order n (see [25]). In this chapter we are concerned mainly with the following problem.

DEFINITION 4.3 *the **kMOLS problem:** Given n and k, construct k mutually orthogonal latin squares of order n.*

Following the definition of the kMOLS problem, given some order n, constructing a single Latin square is the 1MOLS problem, while finding an OLS pair defines the 2MOLS problem. Note that the notion of orthogonality directly

0	1	2	3
1	0	3	2
2	3	0	1
3	2	1	0

0	1	2	3
2	3	0	1
3	2	1	0
1	0	3	2

Table 4.1. A pair of OLS of order 4

applies to partial Latin squares also, therefore we could state the following generalization of the kMOLS and the PLSE problems.

DEFINITION 4.4 *the kMOPLS problem: Given k mutually orthogonal partial Latin squares of order n, fill the maximum number of emtpy cells.*

The kMOLS problem occurs as a special case of the kMOPLS when the PLSs have only empty cells, while the PLSE problem is similarly obtained for $k = 1$. Further related problems, such as deciding whether or not a LS has one or more orthogonal mates, can also be expressed as special cases of the kMOPLS problem. A concept related to orthogonality is that of a *transversal*.

DEFINITION 4.5 *A transversal of a Latin square is a set of n cells, each in a different row and column, which contain pairwise different values.*

As an example, consider the bordered cells of the second square in Table 4.1. It is easy to prove that a Latin square has an orthogonal mate if and only if it can be decomposed into n disjoint transversals [25]. Similarly, a set of $k - 1$ MOLS is extendible to a set of k MOLS if an only if the $k - 1$ MOLS have n disjoint transversals in common.

The kMOLS problem exhibits inherent symmetries, which will appear in the models presented in this chapter. A single Latin square embodies three entities, conventionally corresponding to the n-sets of rows, columns and values. Evidently, any of these sets can be permuted without violating the fact that each value must exist exactly once in each row and column. Thus, two Latin squares are called *isotopic* if one can be obtained from the other by permuting its rows, columns and values. Extending this concept to MOLS, we call two sets of k MOLS isotopic if one can be derived from the other by applying certain permutations to the rows, columns, and values of each of the k squares. A different form of symmetry is obtained if we observe that the roles of rows, columns and values are interchangeable. For example, by interchanging the roles of rows and columns of a Latin square L, we derive its transpose L^T, which is also a Latin square. Since there exist 3! permutations of the three entities, each Latin square may be used to obtain up to 5 other Latin squares L'. Any such square L' is said to be *conjugate* to L.

We introduce some notation to facilitate our presentation hereafter. Consider first the 1MOLS problem and assume that I denotes the row set, J the column set and Q the set of values that appear in L. One can now refer to the cell (i,j) that contains value q, where $i \in I$, $j \in J$ and $q \in Q$. Based on the above definition of conjugacy, one can also refer to cell (i,q) containing j, or cell (j,q) containing value i and so on.

For a set of k MOLS, we refer to each of the k Latin squares as $L_1, L_2, \ldots, L_k$. The number of sets required is now $k+2$, i.e. k sets for the values of each square and further 2 sets for rows and columns. The role of these $k+2$ sets is again conventional, therefore interchangeable. Formally, let $K = \{1, \ldots, k\}$, $\hat{K} = K \cup \{k+1, k+2\}$ and define sets M_i, $i \in \hat{K}$, each of cardinality n. Alternatively, denote the row and column sets as I and J, respectively, and also sets M_i, $i \in K$ of cardinality n. For $S \subseteq K$ (or $S \subseteq \hat{K}$), assume $S = \{i_1, \ldots, i_s\}$ and let $M^S = M_{i_1} \times \cdots \times M_{i_s}$. Also let $m^S \in M^S$, where $m^S = (m^{i_1}, \ldots, m^{i_s})$. A k-tuple is denoted as $(m^1, m^2, \ldots, m^k)$.

3.   Formulations of the kMOLS problem

This section exhibits various modelling approaches to the kMOLS problem. Despite the number of models illustrated, it is expected that further discrete models could be obtained. The motivation for developing the formulations of this section arises from broader research efforts, which are mainly concentrated on answering feasibility questions that remain open in the MOLS literature and also to assess the potential of integrating CP with IP. For clarity of presentation, each of the following subsections discusses first a model for the 1MOLS or 2MOLS problem and then provides, or implicitly indicates, its generalisation to the kMOLS problem.

The first modelling approach proposed is that of CP (Section 4.3.1). Several IP formulations are presented in Sections 4.3.2-4.3.4, while hybrid CP-IP models are discussed in Section 4.3.5. Section 4.3.6 presents a nonlinear combinatorial formulation using permutations while Section 4.3.7 introduces a formulation arising from matroid intersection.

3.1   CP formulation

The CP formulation is illustrated first because it arises more naturally from the problem's description. For the 2MOLS case, the two squares are denoted as L_p, $p \in K$ where $K = \{1, 2\}$. Let $X_{ij}^p, \in \{0, 1, ..., n-1\}$ be the variables denoting the value assigned to the cell (i,j) in square p. For each square, an *all_different* predicate on the n cells of every row and column ensures that the squares are Latin. To express the orthogonality condition we define the variables $Z_{ij} = X_{ij}^1 + n \cdot X_{ij}^2$, for $i, j = 0, 1, \ldots, n-1$. There are n^2 possible values for Z_{ij}, i.e. $Z_{ij} \in \{0, 1, \ldots, n^2-1\}$, which have a $1-1$ correspondence with all n^2 ordered pairs (i, j), for $i, j = 0, 1, \ldots, n-1$. The two squares are orthogonal iff all Z_{ij}s are pairwise different. The CP formulation is as follows.

$$all\_different\{X_{ij}^p : i \in I\}, \; \forall j \in J, \, p \in K$$
$$all\_different\{X_{ij}^p : j \in J\}, \; \forall i \in I, \, p \in K$$
$$all\_different\{Z_{ij} : i \in I, j \in J\} \tag{4.1}$$
$$Z_{ij} = X_{ij}^1 + n \cdot X_{ij}^2, \; \forall i \in I, j \in J$$
$$X_{ij}^p, \in \{0, 1, ..., n-1\}, \; Z_{ij} \in \{0, 1, ..., n^2 - 1\}, \; \forall i \in I, j \in J, p \in K$$

Generalising this model for kMOLS, let again variables $X_{ij}^p, p \in K$, denote the contents of the Latin square L_p. Note that $K = \{1, ..., k\}$ in this case. Let also variables Z_{ij}^{qr} with $q, r = 1, ..., k; q < r$, enforce the orthogonality between any of the $\binom{k}{2}$ pairs (q, r) of Latin squares, by imposing that $Z_{ij}^{rq} = X_{ij}^r + n \cdot X_{ij}^q$. Variables X_{ij}^p and Z_{ij}^{rq} have domains of cardinality n and n^2, respectively. *All_different* constraints are required to ensure that *(a)* cells in each row and column of all squares contain pairwise different values and *(b)* all ordered pairs of values appear exactly once in any pair of Latin squares (i.e. all Z_{ij}^{qr} variables are pairwise different, for a certain pair (L_q, L_r) of Latin squares). The formulation follows.

$$all\_different\{X_{ij}^p : i \in I\}, \; \forall j \in J, \, p \in K$$
$$all\_different\{X_{ij}^p : j \in J\}, \; \forall i \in I, \, p \in K$$
$$all\_different\{Z_{ij}^{qr} : i \in I, j \in J\}, \; \forall q, r \in K, \, q < r \tag{4.2}$$
$$Z_{ij}^{qr} = X_{ij}^q + n \cdot X_{ij}^r, \; \forall i \in I, j \in J, q, r \in K, q < r$$
$$X_{ij}^p, \in \{0, 1, ..., n-1\}, \; Z_{ij}^{qr} \in \{0, 1, ..., n^2 - 1\},$$
$$\forall i \in I, j \in J, q, r \in K, \; q < r$$

The above model contains $2kn + \frac{k(k-1)}{2}$ *all_different* constraints, $\frac{k(k-1)}{2}n^2$ equality constraints and $kn^2 + \frac{k(k-1)}{2}n^2$ variables. More concisely, the model includes $O(n^2)$ constraints and $O(n^2)$ variables.

3.2 IP formulation with 2 indices

As in the CP formulation, the most apparent direction for obtaining an IP model is to consider a set of 2-index n-ary variables for each Latin square. The problem lies in the fact that the *all_different* predicate is a nonlinear constraint. For example, the constraint *all_different*$\{X_{ij}^p : i \in I\}$ of (4.1) is equivalent to $\frac{n(n-1)}{2}$ "not-equal" constraints of the type $X_{i_1 j}^p \neq X_{i_2 j}^p$, where $i_1 \neq i_2$. As shown below, "linearising" this constraint requires either an exponential number of linear constraints or a polynomial, but still large, number of auxiliary variables.

We recall first the convex hull of integer vectors satisfying an *all_different* predicate, as analysed in [33]. To illustrate this result, let $N = \{0, 1, ..., n-1\}$ and consider the constraint *all_different*$\{y_i : i \in N\}$, where $y_i \in \{0, 1, ..., h - 1\}$, $h \geq n$. Hence, we are examining an *all_different* predicate on n variables, whose common domain contains h values. The convex hull of all non-negative integer n-vectors is the polytope $P_{all\_diff} = conv\{y \in \mathbb{Z}^n : all\_different\{y_i : i \in N\}\}$, which is defined by the following facet-defining inequalities [33].

$$\sum_{S \subseteq N : |S| = l} \{y_t : t \in S\} \geq \frac{l(l-1)}{2}, \forall S \subseteq N, \forall l \in \{1, ..., n\} \quad (4.3)$$

$$\sum_{S \subseteq N : |S| = l} \{y_t : t \in S\} \leq \frac{l(2h - l - 1)}{2}, \forall S \subseteq N, \forall l \in \{1, ..., n\} \quad (4.4)$$

The number of these inequalities is $2(2^n - 1)$, i.e. $O(2^n)$. Hence, obtaining an IP model for the 2MOLS case implies the replacement of each predicate in (4.1) by the associated inequalities (4.3) and (4.4). The *all_different* predicate with the largest cardinality is the one imposed on the n^2 Z_{ij} variables. Since the number of inequalities required to represent this constraint is $O(2^{n^2})$, our IP model requires $O(n^2)$ variables and $O(2^{n^2})$ constraints. As an example, we list all constraints of this IP model for $n = 2$. Note that $h = n$ in this case.

$$
\begin{aligned}
X_{00}^p + X_{01}^p &= 1, \text{ for } p = 1, 2 \\
X_{00}^p + X_{10}^p &= 1, \text{ for } p = 1, 2 \\
X_{01}^p + X_{11}^p &= 1, \text{ for } p = 1, 2 \\
X_{10}^p + X_{11}^p &= 1, \text{ for } p = 1, 2 \\
Z_{ij} &= X_{ij}^1 + n \cdot X_{ij}^2, \text{ for } i, j = 0, 1 \\
1 &\leq Z_{00} + Z_{01} \leq 5 \\
1 &\leq Z_{00} + Z_{10} \leq 5 \\
1 &\leq Z_{00} + Z_{11} \leq 5 \\
1 &\leq Z_{01} + Z_{10} \leq 5 \\
1 &\leq Z_{01} + Z_{11} \leq 5 \\
1 &\leq Z_{10} + Z_{11} \leq 5
\end{aligned}
\qquad (4.5)
$$

$$3 \leq Z_{00} + Z_{01} + Z_{10} \leq 6$$
$$3 \leq Z_{00} + Z_{01} + Z_{11} \leq 6$$
$$3 \leq Z_{00} + Z_{10} + Z_{11} \leq 6$$
$$3 \leq Z_{01} + Z_{10} + Z_{11} \leq 6$$
$$Z_{00} + Z_{01} + Z_{10} + Z_{11} = 6$$
$$X_{ij}^p \in \{0,1\}, \text{ for } p = 1,2, i,j = 0,1$$
$$Z_{ij} \in \{0,1,2,3\}, \text{ for } i,j = 0,1$$

Notice that a number of inequalities, e.g. $0 \leq Z_{ij} \leq 3$, are not illustrated in (4.5) because they are implied by the domain of the integer variables X_{ij}^p. To reduce this enormous size of constraints, one can include in the model only a polynomial number of the inequalities (4.3) and (4.4), thus obtaining a relaxation of the original model. For example, if one includes inequalities (4.3) and (4.4) only for $S = N$, each *all_different* predicate is replaced by the single equality:

$$\sum \{y_t : t \in N\} = \frac{n(n-1)}{2}$$

In that case, our relaxed model for 2MOLS would contain only $n^2 + 4n + 1$ equality constraints. The inequalities omitted from the original formulation can always be added when violated by the current LP solution, via the separation algorithm described in [4]. The general case of kMOLS is treated in a similar manner.

An equivalent model is obtained if we consider the binary "$\neq$" constraints. The standard way of dealing with the nonlinear constraint $x \neq y$, where $x, y \in \{0, ..., n-1\}$ is to replace it with the inequalities

$$1 + x - y \leq nz \qquad (4.6)$$
$$1 - x + y \leq n(1 - z)$$

where z is a binary variable. It follows that each *all_different* predicate of cardinality n can be modelled via $O(n^2)$ binary variables and $O(n^2)$ inequalities analogous to (4.6). Again, the dominant constraint, in terms of size, is the one imposed on the Z_{ij} variables. It is not difficult to see that the IP model for the 2MOLS case requires $O(n^2)$ n-ary variables, $O(n^4)$ binary variables and $O(n^4)$ constraints.

3.3 IP formulation with 3 indices

Let us shift our focus towards IP models that are based on $0 - 1$ (binary) variables instead of n-ary ones. A simple observation is that a one-index n-ary variable X_i, $i = 1, ..., n$ can be converted to a binary variable with two indices (e.g. x_{ij}), where the additional index can be regarded as indexing the domain of the initial variable. The equivalence is simple to formalise: $x_{ij} = 1$ if $X_i = j$ and $x_{ij} = 0$ otherwise.

We initiate our discussion by considering the 1MOLS problem, i.e. a single Latin square. The n-ary variable X_{ij} of the previous section will be replaced by the binary variable x_{ijq}, where $q \in Q$ and $|Q| = n$. Since each value q must be contained exactly once in each row i, exactly one of the variables $\{x_{ijq} : j \in J\}$ will be one. This implies the constraint $\sum \{x_{ijq} : j \in J\} = 1$, for each $i \in I$ and $q \in Q$. Two more constraints of this type can be derived by taking into account that the roles of the three ground sets are interchangeable. The formulation of the 1MOLS as a $0 - 1$ IP model follows (see also [15]).

$$\sum \{x_{ijq} : i \in I\} = 1, \forall j \in J, \, q \in Q$$

$$\sum \{x_{ijq} : j \in J\} = 1, \forall i \in I, \, q \in Q \qquad (4.7)$$

$$\sum \{x_{ijq} : q \in Q\} = 1, \forall i \in I, \, j \in J$$

$$x_{ijq} \in \{0, 1\} \, \forall i \in I, \, j \in J, \, q \in Q$$

The above model is a special case of the **set-partitioning problem** ([5]) and also represents the **planar 3-index assignment problem** ([15, 26]).

Proceeding with the 2MOLS case, two variables with 3 indices must be defined, one for each Latin square L_t, $t = 1, 2$. Hence, define variables x_{ijq}^t for $t \in K$, where $K = \{1, 2\}$. Evidently, the constraints (4.7) must be repeated in order to ensure that each of the squares L_1, L_2 remains Latin. In order to establish the orthogonality condition, we need the binary equivalent of variable Z_{ij} of Section 4.3.1. It is easy to see that the number of additional indices is two instead of one, exactly because the cardinality of the domain of variable Z_{ij} is n^2. Thus, we also define the binary variable z_{ijqr}, $(q, r) \in Q^2$, which must be 1 if and only if both x_{ijq}^1 and x_{ijr}^2 become 1. In other words, variable z_{ijqr} must become one if and only if pair (q, r) appears in cell (i, j) in the two squares. This is accomplishable by the following set of linear constraints.

$$x_{ijq}^1 + x_{ijr}^2 \leq z_{ijqr} + 1$$

$$x_{ijq}^1 \geq z_{ijqr} \qquad (4.8)$$

$$x_{ijr}^2 \geq z_{ijqr}$$

The important feature is that only variables x_{ijq}^t, $t = 1, 2$, must be integer. Variables z_{ijqr} can be allowed to take values in the $[0, 1]$ interval, since they will be forced to values 0 or 1 once variables x_{ijq}^i obtain integer values. Furthermore, exactly one of the variables $\{z_{ijqr} : i \in I, j \in J\}$ must be one for each pair (q, r) to appear exactly once (orthogonality condition). This implies a constraint of the type $\sum \{z_{ijqr} : i \in I, j \in J\} = 1$ for each $(q, r) \in Q^2$. Notice that five more constraints of this type exist, again by taking into account the interchangeability of the indices. Nevertheless, including only one is sufficient for the validity of our model.

$$\sum \{x_{ijq}^t : i \in I\} = 1, \; \forall j \in J, \; q \in Q, \; t \in K$$

$$\sum \{x_{ijq}^t : j \in J\} = 1, \; \forall i \in I, \; q \in Q, \; t \in K$$

$$\sum \{x_{ijq}^t : q \in Q\} = 1, \; \forall i \in I, \; j \in J, \; t \in K$$

$$x_{ijq}^1 \geq z_{ijqr}, \; \forall i \in I, \; j \in J, \; (q,r) \in Q^2$$

$$x_{ijr}^2 \geq z_{ijqr}, \; \forall i \in I, \; j \in J, \; (q,r) \in Q^2 \qquad (4.9)$$

$$x_{ijq}^1 + x_{ijr}^2 \leq z_{ijqr} + 1, \; \forall i \in I, \; j \in J, \; (q,r) \in Q^2$$

$$\sum \{z_{ijqr} : i \in I, \; j \in J\} = 1, \; \forall (q,r) \in Q^2$$

$$x_{ijq}^t \in \{0,1\}, \; \forall i \in I, \; j \in J, \; q \in Q, \; t \in K$$

$$0 \leq z_{ijqr} \leq 1, \; \forall i \in I, \; j \in J, \; (q,r) \in Q^2$$

This model is defined on $2n^3$ binary and n^4 real variables, while it includes $7n^2 + 3n^4$ constraints, i.e. the model size is $O(n^4)$ with respect to both variables and constraints. Its generalisation to the kMOLS problem is slightly more involved but emanates directly from the corresponding CP model (4.2).

3.4 IP formulation with 4 indices

Recall the 3-index formulation for the 1MOLS problem (4.7) and notice that its symmetric structure reveals exactly the symmetry of our problem, i.e. the fact that the roles of the sets are purely conventional. This is also reflected by the fact that this model is simply derived by summing over 1 out of 3 available indices, which arise from the 3 entities/sets involved in the problem (rows, columns, values). As discussed in Section 4.2, the 2MOLS problem involves 4 sets, which can be naturally mapped to a 4-index variable $x_{ijm^1m^2}$, where m^1, m^2 stand for the values in the two squares and i, j represent rows and columns, respectively. Variable $x_{ijm^1m^2}$ equals 1 if pair (m^1, m^2) appears in cell (i, j) and equals 0 otherwise. The constraint enforcing the orthogonality condition is clearly $\sum \{x_{ijm^1m^2} : i \in I, j \in J\} = 1$. To emphasise the interchangeability of the 4 sets, we will re-write this variable as $x_{m^1m^2m^3m^4}$, where any two of the four sets M_i, $i \in \{1, ..., 4\}$ can be thought of as representing the rows or columns. The orthogonality constraint is similarly re-written, with five more constraint types obtained by interchanging the roles of the sets. The following model appeared first in [12].

$$\sum \{x_{m^1 m^2 m^3 m^4} : m^1 \in M_1, m^2 \in M_2\} = 1, \forall m^3 \in M_3, \; m^4 \in M_4$$

$$\sum \{x_{m^1 m^2 m^3 m^4} : m^1 \in M_1, m^3 \in M_3\} = 1, \forall m^2 \in M_2, \; m^4 \in M_4$$

$$\sum \{x_{m^1 m^2 m^3 m^4} : m^1 \in M_1, m^4 \in M_4\} = 1, \forall m^2 \in M_2, \; m^3 \in M_3$$

$$\sum \{x_{m^1 m^2 m^3 m^4} : m^2 \in M_2, m^3 \in M_3\} = 1, \forall m^1 \in M_1, \; m^4 \in M_4 \quad (4.10)$$

$$\sum \{x_{m^1 m^2 m^3 m^4} : m^2 \in M_2, m^4 \in M_4\} = 1, \forall m^1 \in M_1, \; m^3 \in M_3$$

$$\sum \{x_{m^1 m^2 m^3 m^4} : m^3 \in M_3, m^4 \in M_4\} = 1, \forall m^1 \in M_1, \; m^2 \in M_2$$

$$x_{m^1 m^2 m^3 m^4} \in \{0, 1\}, \forall m^1 \in M_1, m^2 \in M_2, \; m^3 \in M_3, \; m^4 \in M_4 \quad (4.11)$$

Observe that the IP formulation is obtained by summing over all possible subsets of 2 out of 4 indices. Under the notation introduced in Section 4.3.2, $\hat{K} = \{1, ..., 4\}$ and the above model is simply written as follows.

$$\sum \{x_{m^K} : m^{\hat{K} \setminus \{i_1, i_2\}} \in M^{\hat{K} \setminus \{i_1, i_2\}}\} = 1,$$
$$\forall m^{\{i_1, i_2\}} \in M^{\{i_1, i_2\}}, \; \forall (i_1, i_2) \in \hat{K}^2, \; i_1 < i_2 \quad (4.12)$$

$$x_{m^{\hat{K}}} \in \{0, 1\}, \; \forall m^{\hat{K}} \in M^{\hat{K}}$$

It is not difficult to deduce that the above formulation remains identical for the kMOLS case as long as set $\hat{K}$ is defined accordingly, i.e. $\hat{K} = \{1, ..., k, k + 1, k + 2\}$. This formulation represents also the **multidimensional planar assignment problem** (see [1]).

Let A_n^k denote the $(0, 1)$ matrix of the constraints (4.12). The matrix A_n^k has n^k columns and $\binom{k}{2} \cdot n^2$ rows, i.e. n^2 constraints for each of the $\binom{k}{2}$ distinct $(i_1, i_2) \in \hat{K}^2$. The convex hull of the integer points satisfying constraints (4.12) is the k-**dimensional planar assignment polytope**, denoted as P_I^k. Formally, $P_I^k = conv\{x \in \{0, 1\}^{n^k} : A_n^k x = 1\}$. Integer vectors of P_I^k have a $1 - 1$ correspondence to sets of k MOLS. Again, P_I^k defines a special case of the set-partitioning polytope. The set-packing relaxation of P_I^k, denoted as $\tilde{P}_I^k$, is defined exactly as P_I^k but with '=' replaced by '$\leq$'. Formally, $\tilde{P}_I^k = conv\{x \in \{0, 1\}^{n^k} : A_n^k x \leq 1\}$. The integer vectors of $\tilde{P}_I^k$ have a $1 - 1$ correspondence to sets of k incomplete MOLS, i.e. MOLS where not all cells have been assigned values.

3.5   CP-IP formulations

The notion of a hybrid formulation, in the sense that constraints are expressed via both CP and IP, constitutes a rather recent topic. A relevant discussion appears in [20]. We illustrate this modelling approach for the case of 2MOLS.

Assume that, we wish to formulate the problem of checking whether a certain Latin square L_1 has an orthogonal mate L_2. Latin square L_2 should be decomposable into n disjoint transversals (recall Definition 4.5). Supposing that L_1 is given, each transversal should include n specific cells of L_2, such that all the corresponding cells of L_1 contain the same value. Moreover, all n cells of each transversal of L_2 should contain each value $q \in Q$ exactly once. Define $T_{r_0} = \{(i_0, j_0) \in I \times J : \text{cell } (i_0, j_0) \text{ of } L_1 \text{ contains value } r_0\}$, for all $r_0 \in R$, where R denotes the set of values for square L_1. For example, for the first Latin square of Table 4.1, $T_0 = \{(0,0),(1,1),(2,2),(3,3)\}$. The IP model is essentially model (4.7) with the additional constraints ensuring that square L_2 is orthogonal to L_1. It is not difficult to see that these constraints are of the form $\sum\{x_{i_0 j_0 q} : (i_0, j_0) \in T_{k_0}\} = 1$ for each q_0 and for each r.

$$\sum\{x_{ijq} : i \in I\} = 1, \forall j \in J,\ q \in Q$$
$$\sum\{x_{ijq} : j \in J\} = 1, \forall i \in I,\ q \in Q$$
$$\sum\{x_{ijq} : q \in Q\} = 1, \forall i \in I,\ j \in J \qquad (4.13)$$
$$\sum\{x_{i_0 j_0 q} : (i_0, j_0) \in T_{r_0}\} = 1,\ \forall q \in Q,\ r_0 \in R$$
$$x_{ijq} \in \{0,1\},\ \forall i \in I,\ j \in J,\ q \in Q$$

To complete the model for the 2MOLS case, we need to include the first 2 constraints of the CP model (4.1) in order for square L_1 to be Latin. In order to connect the CP and IP components, we simply have to include the rule $X_{i_0 j_0} = r_0 \Longrightarrow (i_0, j_0) \in T_{r_0}$. The overall model is illustrated below.

$$all\_different\{X_{ij} : i \in I\},\ \forall j \in J$$
$$all\_different\{X_{ij} : j \in J\},\ \forall i \in I$$
$$X_{i_0 j_0} = r_0 \Longrightarrow (i_0, j_0) \in T_{r_0}$$
$$\sum\{x_{ijq} : i \in I\} = 1,\ \forall j \in J,\ q \in Q$$
$$\sum\{x_{ijq} : j \in J\} = 1,\ \forall i \in I,\ q \in Q \qquad (4.14)$$
$$\sum\{x_{ijq} : q \in Q\} = 1,\ \forall i \in I,\ j \in J$$
$$\sum\{x_{i_0 j_0 q} : (i_0, j_0) \in T_{r_0}\} = 1,\ \forall q \in Q,\ r_0 \in R$$
$$x_{ijq} \in \{0,1\},\ \forall i \in I,\ j \in J,\ q \in Q$$
$$X_{ij} \in \{0, ..., n-1\},\ \forall i \in I,\ j \in J$$

The above model is implicitly used in [3] in the context of an LP-based proof for the infeasibility of 2MOLS for $n = 6$.

We provide an alternative CP-IP formulation, only to illustrate the modelling strength of this hybrid approach. Recall first model (4.9), which involves a rather large number of auxiliary variables, i.e. the n^4 z_{ijqr} variables. It is easy to see that these variables could be replaced by their 2-index counterparts of the CP model (4.1). Hence, assume variables $z_{ij} \in \{0, ..., n^2 - 1\}$, which must evidently obtain pairwise different values. Since variables x_{ijq}^t are now binary rather than n-ary, we must reform the constraint of (4.1) relating variables z_{ij} with variables x_{ijq}^t. To achieve this, it is sufficient to observe that variable X_{ij} of model (4.1) takes value q if and only if variable x_{ijq}^1 becomes 1. Since exactly one of the variables $\{x_{ijq}^1 : q \in Q\}$ will be 1, it is valid to write that $X_{ij} = \sum\{q \cdot x_{ijq}^1 : q \in Q\}$ and similarly $Y_{ij} = \sum\{q \cdot x_{ijq}^2 : q \in Q\}$. Hence, the constraint involving variables z_{ij} and x_{ijq}^t is now written as $z_{ij} = \sum\{q \cdot x_{ijq}^1 : q \in Q\} + n \cdot \sum\{r \cdot x_{ijr}^2 : r \in Q\}$. The hybrid model is as follows.

$$\sum\{x_{ijq}^t : i \in I\} = 1, \forall j \in J,\ q \in Q,\ t \in K$$

$$\sum\{x_{ijq}^t : j \in J\} = 1, \forall i \in I,\ q \in Q,\ t \in K$$

$$\sum\{x_{ijq}^t : q \in Q\} = 1, \forall i \in I,\ j \in J,\ t \in K$$

$$z_{ij} = \sum\{q \cdot x_{ijq}^1 : q \in Q\} + n \cdot \sum\{r \cdot x_{ijr}^2 : r \in Q\},$$
$$\forall i \in I,\ j \in J \qquad (4.15)$$

$$all\_different\{z_{ij} : i \in I,\ j \in J\}$$
$$x_{ijq}^t \in \{0, 1\}, \forall i \in I,\ j \in J,\ q \in Q,\ t \in K$$
$$z_{ij} \in \{0, ..., n^2 - 1\},\ \forall i \in I,\ j \in J$$

Notice that the above model can become a pure IP model by replacing the single *all_different* constraint with inequalities (4.3) and (4.4). Equivalently, only a small subset of those inequalities can be included, the rest being added on demand via a separation routine. This observation is also valid for model (4.14) that contains two sets of *all_different* constraints.

For the kMOLS problem to be modelled via both CP and IP, it is sufficient to decompose the problem in such a way that CP handles a subset of the k Latin squares and IP deals with the rest. Clearly, the exact models can be of various forms, a fact indicative of the flexibility of the CP-IP modelling approach.

3.6   A nonlinear formulation via permutation matrices

Let D denote a $n \times n$ matrix with $0 - 1$ entries. Matrix D is a permutation matrix of order n if it can be derived from the identity matrix of order n by applying row and column permutations. More formally, a $0 - 1$ matrix D is a permutation matrix if and only if:

$$De = e$$
$$e^T D = e^T$$

where $e \in \mathbb{R}^n$ denotes the vector of all 1's. Let $X^1, ..., X^n$ be permutation matrices of order n and consider a matrix L defined as follows.

$$L = 1 \cdot X^1 + 2 \cdot X^2 + ... + n \cdot X^n = \sum_{q=1}^{n} q \cdot X^t$$

Then L is a Latin square if and only if

$$X^1 + X^2 + ... + X^n = \sum_{q=1}^{n} X^q = f_n$$

where f_n denotes the $n \times n$ matrix of all 1s. Essentially, each permutation matrix defines a collection of n row-wise and column-wise disjoint cells in the $n \times n$ matrix. To construct a Latin square, one needs n such collections that are also disjoint in the sense that no two collections have any cells in common.

Let $Q = \{1, ..., n\}$. The formulation of the 1MOLS problem using permutation matrices is the following.

$$
\begin{aligned}
X^q e &= e, \ \forall q \in Q \\
e^T X^q &= e^T, \ \forall q \in Q \\
\sum \{X^q : q \in Q\} &= f_n \\
X^q &\in \mathbb{R}^{n \times n}, \ \forall q \in Q
\end{aligned}
\tag{4.16}
$$

It is not difficult to observe that formulation (4.7) can be obtained from (4.16) if variable x_{ijq} in (4.7) denotes the entry of cell (i, j) in matrix X^q.

Proceeding to the 2MOLS case, we will essentially describe a model analogous to (4.9). We consider two sets of $0 - 1$ matrices, namely X^q and Y^q, $q \in Q$ and include constraints (4.16) for each square. To express the orthogonality condition, we must include the constraint

$$e^T (X^q \circ Y^r) e = 1 \tag{4.17}$$

for all pairs (q, r). Note that symbol $\circ$ denotes the *Hadamard* product of two matrices. The model for 2MOLS is the following.

$$X^q e = e, \ \forall q \in Q$$
$$e^T X^q = e^T, \ \forall q \in Q$$
$$\sum \{X^q : q \in Q\} = f_n$$
$$Y^q e = e, \ \forall q \in Q$$
$$e^T Y^q = e^T, \ \forall q \in Q$$
$$\sum \{Y^q : q \in Q\} = f_n$$
$$e^T (X^q \circ Y^r) e = 1, \ \forall (q, r) \in Q^2$$
$$X^q, Y^q \in \mathbb{R}^{n \times n}, \ \forall q \in Q$$

Clearly, constraint (4.17) states that each pair of values (q, r) must appear exactly once. This constraint can also be written in the following nonlinear form.

$$\sum_{i=1}^{n} \sum_{j=1}^{n} x_{ijq} \cdot y_{ijr} = 1, \ \forall (q, r) \in Q^2$$

The above constraint can be linearised via inequalities (4.8).

3.7   Matroid Intersection Formulation

Let us first recall a number of definitions concerning independence systems and matroids ([35]).

DEFINITION 4.6 *Given a finite set E and some family of subsets $\mathcal{I} \subset 2^E$, the set system $\mathcal{M} = (E, \mathcal{I})$ is called a **matroid** if the following axioms hold:*

(I1.) $\emptyset \in \mathcal{I}$.

(I2.) If $X \in \mathcal{I}$ and $Y \subset X \ \Rightarrow Y \in \mathcal{I}$.

(I3.) If $X, Y \in \mathcal{I}$ and $|X| = |Y| + 1 \ \Rightarrow \ \exists \, x \in X/Y$ such that $Y \cup \{x\} \in \mathcal{I}$.

*If only (I1) and (I2) are true then the system is called an **independent system**.*

Axiom (I2) will be called the **subinclussiveness** property, while axiom (I3) the **accessibility** property. The elements of $\mathcal{I}$ are called **independent**, while those not in $\mathcal{I}$ are called **dependent**. Maximal independent sets are called **bases** and minimal dependent sets are called **circuits**. For some $X \in E$ the **rank** of X is defined as

$$r(X) = \max\{|Y| : Y \subseteq X, Y \in \mathcal{I}\}.$$

An independent system is characterised by its **oracle**, meaning a mechanism that checks whether a given set is independent. It is normally implied that the oracle procedure is of polynomial complexity with respect to the size of the set.

In what follows we will formulate the kMOLS problem as a problem of finding a maximum cardinality base in an independent system which is the intersection of partition matroids. We will start by describing the 1MOLS case in order to clarify the tools which we will use, and generalize further to the kMOLS case as we did in the previous sections. Assume the sets I, J and Q as defined in Section 4.3.3 and consider the tripartite hypergraph $H(V, E)$ where:

- $V := I \cup J \cup Q$

- $E := I \times J \times Q$

Now consider the following three partitions of the hyperedge set:

$$\Pi_{IJ} = \bigcup_{\forall i \in I, j \in J} P(i, j)$$

where $P(i, j) := \{e \in E : e \text{ is incident to vertices } i \in I, j \in J\}$,

$$\Pi_{IQ} = \bigcup_{\forall i \in I, q \in Q} P(i, q)$$

where $P(i, q) := \{e \in E : e \text{ is incident to vertices } i \in I, q \in Q\}$,

$$\Pi_{JQ} = \bigcup_{\forall j \in J, q \in Q} P(j, q)$$

where $P(j, q) := \{e \in E : e \text{ is incident to vertices } j \in J, q \in Q\}$. The cardinality of each partition is n^2, while the cardinality of each set $P(\cdot, \cdot)$ is n. Each set $P(\cdot, \cdot)$ contains those hyperedges that are simultaneously incident to a pair of edges, each from a different vertex set. We define for each partition the following families of subsets of E:

$$
\begin{aligned}
\mathcal{I}_{IJ} &= \{X \subseteq E : |X \cap P(i, j)| \leq 1, i \in I, j \in J\}, \\
\mathcal{I}_{IQ} &= \{X \subseteq E : |X \cap P(i, k)| \leq 1, i \in I, q \in Q\}, \\
\mathcal{I}_{JQ} &= \{X \subseteq E : |X \cap P(j, k)| \leq 1, j \in J, q \in Q\}.
\end{aligned}
$$

It is easy to verify that each of the set systems $(E, \mathcal{I}_{IJ})$, $(E, \mathcal{I}_{IQ})$ and $(E, \mathcal{I}_{JQ})$ is a partition matroid. The intersection of these matroids will be denoted by $M(E, \mathcal{I}_{IJ} \cap \mathcal{I}_{IQ} \cap \mathcal{I}_{JQ}) = M(E, \mathcal{I})$ and it will not be a matroid in general. It is also easy to check the following:

REMARK 1 *Any independent set $X \in \mathcal{I}$ corresponds to an PLS of order n.*

To show the above it is enough to consider, as usual, that the first two indices of each hyperedge $e = (i, j, k) \in X$ denote the corresponding row and column of the array, while the third index denotes the value appearing in cell (i, j). Then, for each hyperedge $e = (ijk) \in X$, the independence relationship $\mathcal{I}_{IJ}$ ensures that the pair (i, j) appears once among all hyperedges in X, and therefore we can form an (possibly incomplete) array. In addition, the independence relationships $\mathcal{I}_{IQ}$ and $\mathcal{I}_{JQ}$ ensure that value q appears nowhere else in row i and column j of the incomplete array. This partial Latin square though is not guaranteed to be completable. Let the family of bases of $M(E, \mathcal{I})$ be $\mathcal{B}(\mathcal{I})$. We can then state the following.

REMARK 2 *Every $B \in \mathcal{B}(\mathcal{I})$ with $|B| = n^2 = r(E)$ corresponds to a Latin square of order n.*

This follows easily, since the independent set having cardinality n^2 implies that the corresponding array is complete.

Thus the framework we have set up provides a formulation of Latin squares in terms of bases of an independence system, defined as an intersection of three partition matroids. We know that the bases in the family $\mathcal{B}(M)$ are not all of the same cardinality, since the underlying system is not a matroid but an intersection of matroids. The subsets of every base with cardinality n^2, i.e. a base which corresponds to a Latin square, are completable partial Latin squares. Every base with cardinality strictly smaller than n contains a PLS that is not completable. The question of distinguishing which PLS's are completable is important and has been studied in the polyhedral combinatorics framework in [15]. This can be transalated in our framework as whether a given independent set is contained in a base with cardinality n^2, or equivalently whether a given independent set can be extended to a basis of size n^2. We can state the PLSE problem with respect to this matroid formulation as follows.

DEFINITION 4.7 *(PLSE problem) Given the independence system $M(E, \mathcal{I})$ and some $X \in \mathcal{I}$ find some Y of maximum cardinality such that $X \cup Y \in \mathcal{I}$.*

Having formulated the PLSE problem as a maximization problem in 3-matroid intersection, the results of [21] imply that, assuming a proper weighting function on E, a natural greedy algorithm is a 3-approximation algorithm for the PLSE problem (a t-approximation algorithm is a polynomial time algorithm which finds a solution within a factor of t from the optimum). This result has been proved with linear programming techniques in [22].

There are two ways to formulate the kMOLS problem in the matroid framework. Since we already have characterized Latin squares as maximum cardinality bases in the intersection of three matroids, we could search for those that are pairwise orthogonal within the family of bases. Given any two distinct bases B_1, B_2 with $|B_1| = |B_2| = n^2$ in $M(E, \mathcal{I})$, or equivalently two Latin squares of order n, observe that their elements $e_1 = (i_1, j_1, k) \in B_1$ and $e_2 = (i_2, j_2, l) \in B_2$ contain all n^2 combinations of any two vertices belonging to different vertex partitions of the hypergraph. Since all the combinations of any two vertices are contained in both bases, it follows that every ordered pair (i, j) in B_1 appears also in B_2. Therefore, there exists an ordering of their elements having the following form:

$$B_1 = \{\{1, 1, k_1\}, \{1, 2, k_2\}, \ldots \{1, n, k_n\}, \ldots, \{n, n, k_{n^2}\}\}$$
$$B_2 = \{\{1, 1, l_1\}, \{1, 2, l_2\}, \ldots \{1, n, l_n\}, \ldots, \{n, n, l_{n^2}\}\}$$

By definition, the Latin squares associated with these two bases are orthogonal iff the set of pairs $\{(k_i, l_i), i = 1, ..., n^2\}$ contain all n^2 ordered pairs of values. Clearly, the n pairs $(1, 1), (2, 2), \ldots, (n, n)$ must also appear, hence the following necessary condition can be deduced.

REMARK 3 *For any two bases B_1, B_2 of size n^2, if the corresponding pairs of latin squares are orthogonal then $|B_1 \cap B_2| = n$.*

It is easy to see that the above remark does not provide a sufficient condition. Actually it is a restatement of the following known lemma in the theory of Latin squares.

LEMMA 4.8 *Any set of MOLS is **equivalent** to a set of MOLS where each square has the first row in **natural order** and one of the squares has its first column also in natural order.*

Here, by natural order we mean the not permutted $\{1, 2, \ldots, n\}$. In order to formulate the kMOLS problem as the intersection of matroids, we generalize the 1MOLS matroid formulation to include the presence of more than one value sets, say $Q_1, Q_2, \ldots, Q_k$, while the row set I and column set J remain as previously. Since the role of those sets is interchangeable as mentioned in Section 4.3.2, let $K := \{1, \ldots, k, k + 1, k + 2\}$ and define the $(k + 2)$-partite hypergraph $H(V, E)$ as follows:

- $V := \bigcup_{i=1}^{k+2} V_i$, where $|V_i| = n$ and $V_i \cap V_j = \emptyset$, for all $i, j \in K$.

- $E := V_1 \times \cdots \times V_{k+2}$.

We can assume that $Q_i = V_i, i = 1, \ldots, k$ while $V_{k+1} = I$ and $V_{k+2} = J$. Any two hyperedges $e = (v_{i_1}, v_{i_2}, \ldots, v_{i_{k+2}})$ and $f = (v_{j_1}, v_{j_2}, \ldots, v_{j_{k+2}})$ are said to be *disjoint* iff $v_{i_t} \neq v_{j_t}, t = 1, \ldots, k + 2$. A hyperedge $e =$

$(v_{i_1}, \ldots, v_{i_{k+2}})$ is said to be *incident* to some vertex $v \in V_t$ iff $v_{i_t} = v$. For each pair of vertex sets say $\{V_i, V_j\}$ define the following partition of E

$$\Pi_{V_i V_j} := \bigcup_{\forall v_i \in V_i, v_j \in V_j} P(v_i, v_j),$$

where

$$P(v_i, v_j) := \{e \in E : e \text{ is incident to } v_i \text{ and } v_j\}.$$

Hence, each subset of the partition contains those edges that are simultaneously incident to a pair of vertices, each from a different vertex set. A set of edges is independent iff it contains no two edges that are incident to the same vertex, that is the independence relationship is defined as

$$\mathcal{I}_{V_i V_j} := \{X \subseteq E : |X \cap P(v_i, v_j)| \leq 1, \text{ for all } v_i \in V_i, v_j \in V_j\}$$

It is not a difficult task to prove that the system $M_{V_i V_j} = (E, \mathcal{I}_{V_i V_j})$ is a partition matroid. The intersection of all $\binom{k+2}{2}$ partition matroids, which correspond to all possible vertex set pairs $\{V_i, V_j\}$ defines the independence system which describes the kMOLS problem. Let this system be $M(E, \mathcal{I})$ where $\mathcal{I} = \bigcap_{\forall \{i,j\} \in K} \mathcal{I}_{V_i V_j}$. Each independent set in this system defines a set of k partial MOLS. Evidently, only sets of cardinality n^2 represent a solution to the kMOLS problem.

DEFINITION 4.9 (kMOLS) *Given the independence system $M(E, \mathcal{I})$ by its independence oracle find a set $X \in \mathcal{I}$ such that $|X| = n^2$.*

Alternatively we could also formulate the maximization version of the problem, which results in the kMOPLS problem.

DEFINITION 4.10 (kMOPLS) *Given the independence system $M(E, \mathcal{I})$ by its independence oracle and some independent set X find some $Y \subset E$ of maximum cardinality such that $X \cup Y \in \mathcal{I}$.*

The independence oracle mentioned in Definitions 4.9 and 4.10 is based on a simple labelling algorithm. This algorithm proceeds in an additive fashion, starting from the empty set and adding one element of the set (i.e. an edge) at a time. It is not difficult to show that this algorithm runs in $\mathcal{O}(\binom{k}{2}|I|)$ steps.

4.   Discussion

Having presented a wide variety of models allows us to better investigate the comparative advantages of each model, since multiple criteria are on hand for comparisons among them. Our first observation is that the CP model (Section

4.3.1) has clearly the smallest size, if size is defined only with respect to the number of variables and the number of constraints. If domain cardinalities are taken into account, however, the difference becomes less significant since the CP approach requires n-ary instead of binary variables. It is clear, however, that the CP representation is more compact than any other and easier to comprehend. Also, as the number of MOLS, i.e. parameter k, grows, the size of the CP model grows in constant terms, i.e. it still requires $O(n^2)$ variables and constraints. The same is true for the IP models of Sections 4.3.2 and 4.3.3, whereas the IP model of Section 4.3.4 grows exponentially with respect to k.

To illustrate that there is no easy or clear cut answer when comparing two models, we compare in more detail models (4.9) and (4.10). It is clear that the first model is larger in terms of both variables and constraints. Therefore, the associated LP relaxation will take longer to solve, a fact also supported by computational experience. Moreover, constraints like (4.8) are known to produce very *weak* LP-relaxations (see [28]). The only advantage of model (4.9) is that it uses only n^3 binary variables against n^4 of the second model. Although this could imply a less difficult to solve IP, observe that every integer feasible solution to the first model would require $2n^2$ integer variables to be set to 1, in contrast to the n^2 variables required to be 1 by the second model.

A more significant feature of model (4.10) is its simplicity and symmetry. In other words, the structure of this model reflects more precisely the inherent symmetry of the OLS problem, since it does not distinguish among the 4 ground sets. Through this observation we may introduce another criterion for comparing our models: the extent to which a given model allows for the problem's structure to be efficiently extracted. In the case of IP models, extracting problem's structure usually implies the identification of strong valid inequalities. Families of strong valid inequalities for the MOLS problem have been obtained (see [1]) only via model (4.10). These families arise from structures of the underlying *intersection graph* (see [31] for definitions) such as cliques, odd-holes, antiwebs and wheels. Incorporating these inequalities as cutting planes within a *Branch & Cut* algorithm accelerates significantly the solution process ([2]).

As already discussed in Section 4.3.1, a feasibility problem might include infeasible instances. Such instances constitute critical benchmarks for comparing different solution methods, in the sense that the entire solution space must be explored. In other words, infeasible instances ask for a *proof of infeasibility*, which might require some kind of enumeration in order to be established. It is well known that the 2MOLS problem is infeasible only for $n = 2, 6$, while the simplest open question for the 3MOLS problem refers to the existence of a solution for $n = 10$. Using the valid inequalities that arise from model (4.10), we can directly establish infeasibility for $n = 2$ without even solving an LP. In particular, we can obtain a valid inequality, arising from an antiweb of the intersection graph, which states that the sum of all variables in model (4.10) for $n = 2$ must be less than or equal to 2. Given that exactly 4 variables must be 1 in any feasible solution for $n = 2$, this inequality provides a proof of infeasibility. In contrast, the known inequalities for the 2-index IP model described in Section 4.3.2, i.e. the inequalities arising from the convex hull of *all_different* predicates, are not sufficient to provide such a proof. In other words, if we solve model (4.5) as an LP, the outcome will still be a feasible fractional solution. Clearly, infeasibility can be proved via all IP models if *Branch & Bound* is applied, i.e. through enumeration.

Concerning the case $n = 6$, an LP-based proof of infeasibility is illustrated in [3]. This proof uses known results about the classification of Latin squares of order 6 into 12 equivalence classes and therefore requires the solution of only 12 LPs of the type obtained from the LP relaxation of (4.13). Establishing infeasibility through a single LP, after adding violated inequalities, has not been possible for any IP model. Using valid inequalities for model (4.10) improves significantly over simple *Branch & Bound* in terms of reducing both the solution time and the number of nodes in the search tree (see [2] for details). Nevertheless, the most competitive method, in terms of solution time, for enumerating the entire solution space for $n = 6$ is CP. Exactly because of the capacity of CP to quickly explore alternative vectors of values, model (4.1) appears as the most appropriate for dealing with the kMOLS problem for small values of the parameters k, n. For slightly larger orders, the *Branch & Cut* algorithm behaves better although it has to solve larger LPs. The reason is its ability to either prune infeasibe subproblems earlier during the search or because the cutting planes force the LP to an integer extreme point, therefore terminating the search.

However, as discussed in [2], the approach that prevails as n grows, at least for the 2MOLS case, is a hybrid algorithm that integrates the strengths of both CP and IP. Analogous results have been obtained for small orders of the 3MOLS problem. Hence, it appears that utilising two different representations of the problem, in the form of CP and IP models, increases the potential for solving large problem instances. Moreover, large values of k and n imply enormous model sizes, therefore imposing the need for an appropriate decomposition. In this context, more than one modelling approaches might be incorporated. The recent discussion about CP-IP integration (see for example [20]) indicates multiple possibilities for combining the two methods. Based also on the computational experience obtained so far, combining the CP model (4.2) with one of the IP models exhibited in this chapter, or using a CP-IP model, appear as the most prominent directions of future research.

References

[1] Appa G., Magos D., Mourtos I., Jensen J. (2002): Polyhedral approaches to assignment problems. To appear in *Discrete Optimization*.

[2] Appa G., Mourtos I., Magos D. (2002): Integrating Constraint and Integer Programming for the Orthogonal Latin Squares Problem. In van Hentenryck P. (ed.), Principles and Practice of Constraint Programming (CP2002), *Lecture Notes in Computer Science* **2470**, Springer-Verlag, 17-32.

[3] Appa G., Magos D. Mourtos I. (2004): An LP-based proof for the non-existence of a pair of Orthogonal Latin Squares of order 6. *Operations Research Letters* July 2004, 336-344.

[4] Appa G., Magos D., Mourtos I. (2004): LP relaxations of multiple all_different predicates. In *Proceedings of CPAIOR 2004, Lecture Notes in Computer Science*, Springer-Verlag (forthcoming).

[5] Balas E., Padberg M. (1976): Set partitioning: a survey. *SIAM Review* **18**, 710-760.

[6] Balas E., Bockmayr A., Pisaruk N., Wolsey L. (2004): On unions and dominants of polytopes. *Mathematical Programming* **99**, 223-239.

[7] Barany I., Van Roy T.J., Wolsey L.A. (1984): Uncapacitated lot sizing: the convex hull of solutions. *Mathematical Programming Study* **22**, 32-43.

[8] Beale E.M.L., Morton G. (1958): Solution of a purchase-storage Programme: Part I. *OR Quarterly* **9** (3), 174-187.

[9] Burkard R.E., Cela E., Pardalos P.M., Pitsoulis L.S.(1998): The quadratic assignment problem. *Handbook of Combinatorial Optimization* **2**, Kluwer Academic Publishers, 241-337.

[10] Colbourn C.J. (1984): The complexity of completing partial Latin squares. *Discrete Applied Mathematics* **8**, 25-30.

[11] Cornuejols G., Nemhauser G.L., Wolsey L.A. (1990): The uncapacitated facility location problem. In Mirchandani P.B., Francis R.L. (eds.): *Discrete Location Theory*, John Wiley, New York, 119-171.

[12] Dantzig G.B. (1963): *Linear programming and extensions*. Princeton University Press.

[13] Dénes J., Keedwell A.D. (1991): *Latin Squares: New developments in the Theory and Applications*. North-Holland.

[14] Edmonds, J. (1965): Maximum matching and a polyhedron with 0-1 vertices. *Journal of Research of the National Bureau of Standards* **69B**, 125-130.

[15] Euler R., Burkard R. E., Grommes R. (1986): On Latin squares and the facial structure of related polytopes. *Discrete Mathematics* **62**, 155-181.

[16] Gomes C.P., Regis R.G., Shmoys D.B. (2004): An improved approximation algorithm for the partial Latin square extension problem. *Operations Research Letters* **32**, 479-484.

[17] Gomory R.E. (1960): Faces of an Integer Polyhedron. *Proceedings of the National Academy of Science* **57**, 16-18.

[18] Gomory R.E., Hu T.C. (1961): Multi-terminal network flows. *SIAM Journal* **9**, 551-570.

[19] Grötschel M., Holland O. (1985): Solving matching problems with linear programming. *Mathematical Programming* **33**, 243-259.

[20] Hooker J.N. (2000): *Logic-Based Methods for Optimization: Combining Optimization and Constraint Satisfaction.* J.Wiley(NY).

[21] Korte B., Hausmann D. (1978): An analysis of the greedy algorithm for independence systems. *Annals of Discrete Mathematics* **2**, 65-74.

[22] Kumar S.R., Russell A., Sundaram R. (1999): Approximating Latin square extensions. *Algorithmica* **24**, 128-138.

[23] Land A.H. (1958): Solution of a purchase-storage Programme: Part II. *OR Quarterly* **9** (3), 187-197.

[24] Gordeau J.F., Laporte G. (2004): Modelling and optimization of vehicle routing and arc routing problems. *this volume*

[25] Laywine C.F., Mullen G.L. (1998): *Discrete Mathematics using latin squares.* J.Wiley & Sons.

[26] Magos D., Miliotis P. (1994): An algorithm for the planar three-index assignment problem. *European Journal of Operational Research* **77** 141-153.

[27] Milano M., van Hoeve W.J. (2002): Reduced cost-based ranking for generating promising subproblems. In van Hentenryck P. (ed.), Principles and Practice of Constraint Programming (CP2002), *Lecture Notes in Computer Science* **2470**, Springer-Verlag, 1-16.

[28] Nemhauser G.L., Wolsey L.A. (1988): *Integer and Combinatorial Optimization.* J.Wiley.

[29] Nemhauser G.L., Trick M.A. (1998): Scheduling a major colleg basketball conference. *Operations Research* **46**, 1-8.

[30] Orman A.J., Williams H.P. (2003): A survey of different integer programming formulations for the travelling salesman problem. Working Paper, London School of Economics.

[31] Padberg M.W. (1973): On the facial structure of set packing polyhedra. *Mathematical Programming* **5**,199-215.

[32] Rhys J. (1970): A selection problem of shared fixed costs and network flows. *Management Science* **17**, 200-207.

[33] Williams H.P., Yan H. (2001): Representations of the all-different Predicate of Constraint Satisfaction in Integer Programming. *INFORMS Journal on Computing* **13**, 96-103.

[34] Yan H., Hooker J.N. (1999): Tight representation of logical constraints as cardinality rules. *Mathematical Programming* **85**, 363-377.

[35] Welsh, D.J.A. (1976): *Matroid Theory*. Academic Press.

Chapter 5

NETWORK MODELLING

Douglas R. Shier
Department of Mathematical Sciences
Clemson University, Clemson, SC 29634-0975
USA
shierd@clemson.edu

Abstract   Networks form the backbone of many familiar, and economically critical, activities in our modern life, as exemplified by electrical grid systems, computer systems, telecommunication systems, and transportation systems. Network models, and associated optimization algorithms, are important in ensuring the smooth and efficient operation of such systems. In this chapter, we wish to introduce the reader to a variety of less apparent network models. Such models are drawn from application areas such as genomics, sports, artificial intelligence, and decision analysis as well as transportation and communication. Our objective is to illustrate how network modelling can be useful in formulating problems occurring in these diverse areas — problems that at first blush seem quite remote from the standard applications of networks.

Keywords: assignment problems, Euler paths, maximum flows, networks, shortest paths

1.   Introduction

Our modern, technological society presents ample evidence of networks in daily life. Telephone conversations are carried over various communication links (e.g., copper and fiber optic cable) with routing via intermediate switching centers. Airline networks are used to plan trips between specified origin and destination cities, with possible (in fact likely) stops at hub cities or other intermediate airports along the way. Other familiar networks, such as road networks, distribution networks, and of course the Internet, are regularly encountered in the day-to-day operations of modern life.

In all such cases, there are identifiable nodes (cities, airports, street intersections) connected by various arcs (communication links, flights, roads). Important quantitative information is often assigned to nodes (supplies) and edges (cost, length, capacity).

The purpose of this article is to enlarge our appreciation of networks by identifying less obvious manifestations of networks in a diverse set of applications. In this way, we extend the excellent compilations of network applications found in [1, 2]. Our intent is to identify network structures lurking beneath the surface of various discrete problems arising in engineering, computer science, biology, telecommunications, and decision analysis. Our concern will be mainly in formulating such problems, rather than developing specific algorithms for their solution. The reader is referred to [1, 19] for the treatment of efficient algorithms and data structures for solving standard network problems.

To begin, we establish a small amount of terminology in which to phrase our collection of problems. Let $G = (N, A)$ denote a (directed) *graph* with node set N and arc set $A \subseteq N \times N$. If arc $e = (x, y) \in A$, then x and y are *adjacent* nodes, and arc e is *incident* with both these nodes. The *indegree* of node x is the number of arcs entering (incident to) x, while the *outdegree* of node x is the number of arcs leaving (incident from) x. For any $X \subseteq N$, we define $\Gamma(X) = \{y \in N : (x, y) \in A \text{ for some } x \in X\}$, the set of nodes adjacent from X.

A graph $G = (N, A)$ is *bipartite* if its node set N can be partitioned into disjoint nonempty subsets N_1 and N_2 such that each arc $(x, y) \in A$ satisfies $x \in N_1$ and $y \in N_2$.

A *path* P from node v_0 to node v_k is a sequence $v_0 \rightarrow v_1 \rightarrow \cdots \rightarrow v_k$ of adjacent nodes, where all $(v_i, v_{i+1}) \in A$. The *cardinality* of path P is equal to k, the number of arcs in the path. Generally we will assume that all paths are *simple*: i.e., the nodes v_i are distinct.

A *cycle* is a (simple) v_0-v_k path, $k > 1$, in which nodes v_0 and v_k are the same. An *acyclic* graph contains no cycles.

A *network* G consists of a graph (N, A), together with an assignment of quantitative measures (such as cost or length) to its arcs. On occasion, there might also be supplies associated with the nodes of G. The *cost* (or length) of a path is just the sum of its constituent arc costs.

2.   Transit Networks

A transit system that services a metropolitan area is composed of a set of transit *stops* and a set of *routes* that service these stops. In addition, there are published *schedules* that indicate for each route the departure times for each node on the route. A very small example illustrating these concepts is shown by the network in Figure 5.1, containing six stops $s_1, \ldots, s_6$ and four routes $R_1, \ldots, R_4$. As seen in the accompanying Table 5.1, route R_1 is a "local" route involving stops s_1, s_3, s_4, s_6, whereas route R_2 is an "express" route stopping at s_1, s_3, s_6. In addition, there are "connector" routes R_3 and R_4 that allow transfers to routes R_1 and R_2 at the transfer points s_3 and s_4, respectively. Portions of the schedules for these four routes are also displayed in Table 5.1. Given such data, a typical problem encountered in this scenario is that of determining an "optimal" itinerary — for example, an itinerary that leaves after a designated time from a specified origin and arrives as soon as possible at a specified destination.

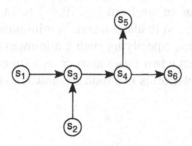

Figure 5.1. A transit system G with 6 stops

Table 5.1. A system of routes and stops

route	stops			
R_1	s_1	s_3	s_4	s_6
	10:00	10:10	10:25	10:40
	10:45	10:55	11:10	11:25
R_2	s_1	s_3	s_6	
	10:08	10:17	10:31	
R_3	s_2	s_3		
	10:29	10:38		
	10:38	10:48		
R_4	s_4	s_5		
	10:43	10:54		
	11:28	11:39		

The point to be emphasized here is that the network G of nodes and arcs shown in Figure 5.1 is not a particularly useful network representation for solving such problems. A much more satisfactory representation both for conceptualization and for computation is a *time-expanded network* $\widetilde{G}$. Each node of this network $\widetilde{G}$ is in fact a pair (i, t) representing stop i at time t. An arc of $\widetilde{G}$ joining node (i, a) to node (j, b) indicates that a vehicle leaving stop i at time a arrives next at stop j at time b. In this way schedule information is explicitly embedded within the network model. Transfers from one route to another at stop i can be easily accommodated by adding arcs from node (i, a) to node (i, b). We illustrate this construction process, using the data in Table 5.1, yielding the time-expanded network $\widetilde{G}$ shown in Figure 5.2. The first run of route R_1 then corresponds to the path $(s_1, 10{:}00) \rightarrow (s_3, 10{:}10) \rightarrow (s_4, 10{:}25) \rightarrow (s_6, 10{:}40)$. The network $\widetilde{G}$ also contains the transfer arcs $(s_4, 10{:}25) \rightarrow (s_4, 10{:}43)$ and $(s_4, 11{:}10) \rightarrow (s_4, 11{:}28)$ between routes R_1 and R_4 as well as the transfer arc $(s_3, 10{:}38) \rightarrow (s_3, 10{:}55)$ between routes R_3 and R_1. Notice that an arc from $(s_3, 10{:}48)$ to $(s_3, 10{:}55)$ has not been included, to reflect the need to allow a certain minimum time to accomplish a transfer between vehicles. Specifying such a minimum transfer time might be necessary in a bus system when a user must cross a street to board the next bus, or to transfer between platforms in a train or subway system.

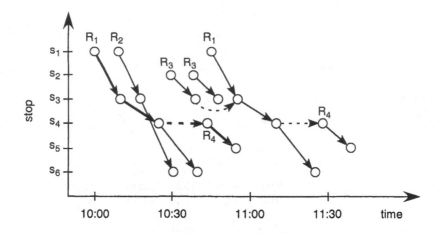

Figure 5.2.   The time-expanded network $\widetilde{G}$

To illustrate routings in this time-expanded network, consider a transit user who wishes to depart at 10:00 from stop s_1 and arrive as soon thereafter at stop s_5. The optimal path in this instance consists of $(s_1, 10{:}00) \rightarrow (s_3, 10{:}10) \rightarrow (s_4, 10{:}25) \rightarrow (s_4, 10{:}43) \rightarrow (s_5, 10{:}54)$. This particular routing has been highlighted in bold in Figure 5.2. That is, the user boards route R_1 at the origin stop s_1 at 10:00, transfers at stop s_4 to route R_4, and then alights at the destination s_5 at 10:54.

An advantage of this network representation is that the time-expanded network $\widetilde{G}$ is necessarily acyclic, since all arcs proceed forward in time. This means that highly efficient shortest path algorithms for acyclic networks [1, 19] can be used to find optimal routings for such networks. Another advantage of this approach is that it can address the *overtake problem* that on occasion occurs in actual transit systems. Namely, it might be better not to board a slower (local) bus that arrives first at a stop, waiting instead for a faster (express) bus that arrives at that stop somewhat later. Indeed, a normal graph representation (such as in Figure 5.1) must allow for a violation of the *optimality principle* of dynamic programming [6], which states here that any subpath of an optimal path should be optimal between its endpoints. In our example, the best routing from s_1 to s_6, leaving after 10:00, utilizes route R_2: $s_1 \rightarrow s_3 \rightarrow s_6$. However, the subpath $s_1 \rightarrow s_3$ of R_2, leaving s_1 at 10:08 and arriving s_3 at 10:17, is not itself an optimal routing. Rather, it is optimal to follow instead route R_1, leaving s_1 at 10:00 and arriving s_3 at 10:10. This problem is obviated in the time-expanded network $\widetilde{G}$ since the destination stop s_3 has been replicated in time. Thus the path $(s_1, 10{:}00) \rightarrow (s_3, 10{:}10)$ in $\widetilde{G}$ is optimal for arriving at stop s_3, whereas the path $(s_1, 10{:}08) \rightarrow (s_3, 10{:}17) \rightarrow (s_6, 10{:}31)$ in $\widetilde{G}$ is optimal for arriving at stop s_6.

The concept of a time-expanded network has origins that reach back to the seminal work of Ford and Fulkerson [9]. Its use in transit systems , in conjunction with computational algorithms, is fully explored in the report [10]. This modelling technique has also proved useful in other contexts, such as multi-period inventory problems, airline fleet scheduling, and crew allocation for long-haul freight transportation.

3. Amplifier Location

A problem encountered in constructing a fiber optic connection between two distant stations is that of determining where to place a fixed number of amplifiers between these stations. There are only certain candidate locations for these intermediate amplifiers, and one wants to select the best locations for the amplifiers.

To be more specific, suppose that the fiber optic link between stations S and T is represented by the interval $[s, t]$, where station S is located at position s and station T is located at position t. Candidate locations for the amplifiers are at the n positions $x_1 < x_2 < \cdots < x_n$. It is required to select exactly r positions for the amplifiers from $\{x_1, x_2, \ldots, x_n\}$ and this is to be done in an "optimal" manner.

To explain our chosen measure of optimality, suppose that positions $x_{j_1} < x_{j_2} < \cdots < x_{j_r}$ have been selected for the amplifiers. Then it is desirable to make the successive distances $z_i = x_{j_i} - x_{j_{i-1}}$ as equal as possible. (For notational convenience, we set $x_{j_0} = s$ and $x_{j_{r+1}} = t$.) Our objective is to select the x_{j_i} in order to minimize the variance of the distances z_i. Notice that $\sum_{i=1}^{r+1} z_i = t - s$ so that the mean $E(Z)$ of the z_i is fixed. Since $\mathrm{var}(Z) = E(Z^2) - [E(Z)]^2$ it suffices to minimize $E(Z^2) = \frac{1}{r+1} \sum_{i=1}^{r+1} z_i^2$ or simply $\sum_{i=1}^{r+1} z_i^2$. As an illustration, Table 5.2 lists $n = 10$ candidate locations x_i, where $s = 0$ and $t = 7.5$ are the locations of the two stations. If $r = 3$ amplifiers are to be chosen, then one possible selection consists of x_4, x_6, x_9. The distances defined by this particular selection are thus $\{x_4 - x_0, x_6 - x_4, x_9 - x_6, x_{11} - x_9\} = \{2.4, 2.2, 1.7, 1.2\}$. We associate the objective function value $f(x_4, x_6, x_9) = (2.4)^2 + (2.2)^2 + (1.7)^2 + (1.2)^2 = 14.93$ with this selection. Our aim is then to minimize $f(x_{j_1}, x_{j_2}, x_{j_3})$ by appropriate choice of the indices j_1, j_2, j_3.

Table 5.2.   Possible locations for amplifiers

s	1	2	3	4	5	6	7	8	9	10	t
0.0	1.1	1.7	1.9	2.4	3.7	4.6	5.1	5.5	6.3	6.6	7.5

To cast the general situation as a network problem, let the set of nodes be $N = \{0, 1, 2, \ldots, n, n + 1\}$ and define arcs (i, j) for all $0 \le i < j \le n + 1$. That is to say, $G = (N, A)$ is the complete directed graph on $n+2$ nodes. Also we define the *cost* of arc (i, j) by $c_{ij} = (x_j - x_i)^2$. It is important to notice that there is a one-to-one correspondence between (i) paths of cardinality $r+1$ from 0 to $n+1$ in G, and (ii) selections of r locations for the amplifiers. For example, the path $0 \to 4 \to 6 \to 9 \to 11$ in G corresponds to selecting the amplifier locations $x_4 = 2.4$, $x_6 = 4.6$, $x_9 = 6.3$. Moreover the cost of this path $c_{0,4} + c_{4,6} + c_{6,9} + c_{9,11} = 5.76 + 4.84 + 2.89 + 1.44 = 14.93$ is exactly $f(x_4, x_6, x_9)$. Consequently, a best placement of r amplifiers can be determined by solving a certain type of shortest path problem on G: namely, finding a shortest path in G from node 0 to node $n + 1$ which is comprised of exactly $r + 1$ arcs. Since G is *acyclic*, this problem can be solved efficiently, using for instance an appropriate FIFO implementation of a label-correcting shortest path algorithm [1, p. 142]. (This problem can also be solved using a dynamic programming approach.) In our example, it turns out that the path $0 \to 3 \to 5 \to 8 \to 11$ is a shortest path with $r + 1 = 4$ arcs and $f(x_3, x_5, x_8) = 14.09$. Thus the optimal placement of amplifiers is at locations $x_3 = 1.9$, $x_5 = 3.7$, $x_8 = 5.5$ within the interval $[0, 7.5]$.

4.   Site Selection

A communications company leases an orbiting satellite and has the option to build several ground-based receiving stations selected from n candidate locations. The cost of constructing a receiving station at location i is given by c_i. However, if stations are built at locations i and j, then the company can earn the revenue r_{ij}. The objective of the company is to determine the number and placement of ground-based stations to maximize net profit.

One possible representation of this problem employs the complete undirected network G_U with node set $N = \{1, 2, \ldots, n\}$ being the candidate stations and undirected edges joining each distinct pair of stations. Associated with each node i is the cost c_i and with each edge (i, j) is the revenue r_{ij}. The issue then is to determine the optimum number of constructed stations k as well as the selected stations $i_1, i_2, \ldots, i_k$.

To illustrate, Table 5.3 shows data for a small problem involving $n = 4$ sites. If all four sites are selected, then the resulting revenue is $(4.2 + 2.2 + 7.8 + 2.1 + 3.3 + 8.9) - (6 + 10 + 5 + 4) = 3.5$. Alternatively, if sites $\{1, 3, 4\}$ are selected then net profit increases to $(2.2 + 7.8 + 8.9) - (6 + 5 + 4) = 3.9$. Presumably, finding the best selection would involve trying all possible subsets of sites — that is, evaluating the net profit for all complete subnetworks of G_U. This approach quickly becomes infeasible.

Table 5.3. Revenues r_{ij} and costs c_i

			j		
i	c_i	1	2	3	4
1	6		4.2	2.2	7.8
2	10			2.1	3.3
3	5				8.9
4	4		r_{ij}		

Rather than modelling this problem with respect to the undirected network G_U, it is advantageous to formulate the problem using a directed network that captures the logical relations between activities. Here the driving logical constraint is that the revenue r_{ij} associated with the pair $\{i, j\}$ can only be realized if both of the activities i and j, with resulting costs c_i and c_j, are also included. Thus we define a directed flow network G whose nodes include both the revenue nodes $\{i, j\}$, abbreviated to ij, as well as the cost nodes (stations) i. Arcs join ij to i, and also ij to j, indicating that the selection of ij requires the concomitant selection of both i and j. In addition, a source node s and sink node t are added, together with arcs from s to each node ij and arcs from each node i to t. It is then natural to associate a capacity r_{ij} with each of the former arcs, and a capacity c_i with each of the latter arcs. Finally, a capacity of ∞ is associated with each arc from ij to i, or ij to j. Also let r_{tot} be the sum of the capacities of all arcs emanating from s: that is, $r_{tot} = \sum_{i,j} r_{ij}$. The bipartite flow network G for our example is illustrated in Figure 5.3, with R indicating the revenue nodes and C the cost nodes.

Rather than concentrating on a maximum flow from s to t in this capacitated network, we consider any potential minimum capacity cut $[X, \overline{X}]$ separating s

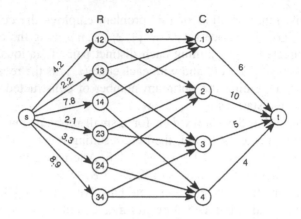

Figure 5.3.   Bipartite flow network

and t. Suppose that $X = \{s\} \cup U \cup V$, where $U \subseteq R$ and $V \subseteq C$. Since arcs joining nodes of R to nodes of C have infinite capacity, the set $\Gamma(U)$ of cost nodes associated with U must be included in X, so that $\Gamma(U) \subseteq V$. Indeed, it would only increase the cost, and decrease the total revenue, to include cost nodes in V that are not required by U, so we can assume $V = \Gamma(U)$.

The capacity $\mathrm{cap}[X, \overline{X}]$ of this cut is then given by the sum of the capacities of the arcs leading from s to $R - U$ plus the capacities of the arcs from $\Gamma(U)$ to t:

$$\mathrm{cap}[X, \overline{X}] = \sum_{ij \notin U} r_{ij} + \sum_{i \in \Gamma(U)} c_i = r_{tot} - \sum_{ij \in U} r_{ij} + \sum_{i \in \Gamma(U)} c_i$$

$$= r_{tot} - \left(\sum_{ij \in U} r_{ij} - \sum_{i \in \Gamma(U)} c_i \right).$$

In other words, the capacity of this cut is a constant minus the net revenue associated with the selection of revenue nodes U. Consequently, to maximize net revenue we can equivalently minimize the cut capacity $\mathrm{cap}[X, \overline{X}]$ in the flow network G. Such a minimum cut $[X, \overline{X}]$ can be found as a byproduct of applying any standard maximum flow algorithm [1, 19]. The optimal set of stations to construct is then found as the set $V = X \cap C$ of cost nodes associated with this minimum cut.

Returning to our motivating example, we have $r_{tot} = 28.5$ and the maximum flow value in G is found to be 24.6. Thus the minimum cut capacity is also 24.6, obtained using $X = \{s, 13, 14, 34, 1, 3, 4\}$. From the equation above, the maximum net revenue possible is $28.5 - 24.6 = 3.9$. The selection of sites $V = X \cap C = \{1, 3, 4\}$, which achieves this net revenue, is then optimal.

In summary, the site selection problem can be productively formulated by constructing a flow network G associated with the logical constraints and finding a minimum capacity cut in G. This construction, originally conceived by Rhys [15] and Balinski [4], can be generalized to include other forms of logical constraints, resulting in a type of "maximum closure" problem [14] in a network (not necessarily bipartite). Applications to open-pit mining, as well as algorithmic aspects, are discussed in [12].

5. Team Elimination in Sports

Towards the end of a sports season, it is of interest to determine when a team has been "mathematically eliminated" from winning its league championship. Suppose that there are n teams comprising a league and that each team plays a total of m games. Table 5.4 shows the current rankings of $n = 4$ teams at some point during the season; each team plays a total of $m = 20$ games. It is apparent that Team 4 is currently eliminated from winning (or being tied for) the championship: even if it won all the remaining games, it would achieve a record of 10 wins and 10 losses, which is insufficient to outrank either Team 1 or Team 2.

Table 5.4. Current team rankings

Team	wins	losses	remaining
1	11	4	5
2	11	5	4
3	10	8	2
4	7	10	3

On the other hand, Team 3 could (optimistically) achieve a 12–8 record, and conceivably it is still in contention for the title. If however Teams 1 and 2 play one another three (or more) times during the remainder of the season, then one of those teams will win at least two more games, giving it a record of at least 13–7, thus eliminating Team 3. This illustrates the importance of the actual schedule of games that remain, rather than simply the number of games remaining to be played.

For concreteness suppose that the schedule of games to be played is as given in Table 5.5. Under these circumstances can Team 3 be mathematically eliminated from winning the title? This will depend on the results of the games being played among the remaining teams (1, 2, 4). In order for Team 3 (with a possible final record of 12–8) not to be eliminated, Team 1 can win at most 1 more game. Similarly, Team 2 can win at most 1 more game and Team 4 at most 5 more games. These values provide an upper bound on the number

Table 5.5.   Games remaining to be played

Team	1	2	3	4
1		3	1	1
2	3		0	1
3	1	0		1
4	1	1	1	

of remaining games to be won by Teams 1, 2, and 4. In addition, the games played (according to the schedule) among these teams must be distributed in a way that does not violate these upper bounds.

To visualize this more clearly, construct a bipartite flow network G with nodes partitioned into sets R (remaining game nodes) and T (team nodes). A game node ij is joined to both Team i and Team j nodes by arcs of infinite capacity. In addition, G contains a source node s and a sink node t. Node s is joined to each game node ij by an arc with capacity equal to the number of games remaining between Teams i and j. Each team node i is joined to t with a capacity equal to the maximum number of permitted wins by Team i. This construction is illustrated in Figure 5.4 for our particular example.

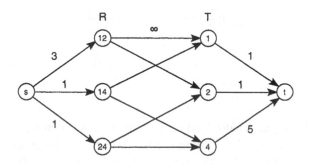

Figure 5.4.   Bipartite flow network associated with Team 3

Suppose now that a maximum s-t flow is found in network G. If this flow *saturates* all the source arcs (has flow equal to capacity), then we have identified a realization of wins for the remaining games in which Team 3 can be in first (or tied for first) place at the end of the season. However, if a maximum flow does not saturate all the source arcs, then Team 3 is mathematically eliminated. In our particular example, we claim that the maximum flow value is no more than 4, which is less than the total capacity 5 of all source arcs. As a result, Team 3 is mathematically eliminated. To verify that the maximum flow value cannot exceed 4, consider the cut $[X, \overline{X}]$ with $X = \{s, 12, 1, 2\}$; thus $[X, \overline{X}] = \{(s, 14), (s, 24), (1, t), (2, t)\}$. The maximum flow value cannot exceed the capacity of this cut, namely $\text{cap}[X, \overline{X}] = 4$.

Notice that this particular cut $[X, \overline{X}]$, which in fact is a cut of minimum capacity in G, provides proof that Team 3 can now be eliminated. Indeed, the nodes $U = X \cap R = \{12\}$ identify a "critical" set of games yet to be played by the teams $V = X \cap T = \{1, 2\} = \Gamma(U)$. These two teams play 3 remaining games, exceeding the total number of allowable wins (2) permitted by Teams 1 and 2.

In a more general setting, we wish to know whether Team k can be eliminated, based on the current records (wins and losses) as well as the remaining schedule. Suppose that Team k has L_k losses, so that it can possibly end up with $m - L_k$ wins at the end of the season. We can assume that $m - L_k \geq W_i$ holds for all i, where W_i is the current number of wins for Team i, since otherwise Team k is automatically eliminated.

Set up a bipartite flow network G with source s and sink t as before, using the remaining games not involving Team k. The capacity for each arc (s, ij) is the number of games left to be played between Teams i and j. The capacity for each arc (i, t) is the maximum permitted number of additional wins $m - L_k - W_i \geq 0$. Arcs joining game nodes to team nodes have infinite capacity.

If the maximum s-t flow value in network G is less than N, the total number of remaining games (excluding Team k), then we conclude that Team k can be mathematically eliminated. The key idea here is that a minimum capacity cut $[X, \overline{X}]$ in G provides a critical set of games $U = X \cap R$ to be played among the teams $V = X \cap T = \Gamma(U)$. Since the maximum flow value is less than N, then so is the minimum cut capacity:

$$\text{cap}[X, \overline{X}] = (\# \text{ of games not in } U) + (\# \text{ of allowable wins for } V) < N.$$

This provides the contradiction

$$(\# \text{ of allowable wins for } V) < (\# \text{ of games played in } U).$$

This type of team elimination problem was apparently first investigated by Schwartz [18] in the context of baseball, where the top finisher(s) of the conference can progress to the playoffs. It should be clear that a similar analysis can be applied to other sports. This problem was subsequently studied by other authors [13, 16].

6. Reasoning in Artificial Intelligence

A number of problems in formal reasoning can be expressed as *simple temporal problems*, in which constraints are placed on the starting and ending points of certain events as well as between the starting and ending points of specified pairs of events. Questions that arise in this context are whether the stated system is in fact consistent, and if so what are the possible times at which events can occur (taking into account all constraints).

To illustrate these concepts, consider the following problem [8] in reasoning about temporal events. John leaves home between 7:10 and 7:20 in the morning, and takes his own vehicle to work. The journey to work takes between 30 and 40 minutes. Fred arrives at work between 8:00 and 8:10, having taken a carpool that typically takes between 40 and 50 minutes. John arrives at work between 10 and 20 minutes after Fred leaves home. We first determine whether the information presented here is in fact *consistent* — i.e., whether there are temporal assignments to the various events that satisfy all stated conditions.

To this end, define x_1 as John's departure time from home and x_2 as John's arrival time at work. Similarly, x_3 is Fred's departure time from home and x_4 is Fred's arrival time at work. To establish an appropriate starting time, let $x_0 = 7{:}00$ a.m. be an initial reference time. Then the temporal constraints of the problem can be expressed as

$$10 \leq x_1 - x_0 \leq 20, \quad 30 \leq x_2 - x_1 \leq 40, \quad 60 \leq x_4 - x_0 \leq 70,$$

$$40 \leq x_4 - x_3 \leq 50, \quad 10 \leq x_2 - x_3 \leq 20.$$

Each of the above constraints has the form $a_{ij} \leq x_j - x_i \leq b_{ij}$, or equivalently $x_j - x_i \leq b_{ij}$, $x_i - x_j \leq -a_{ij}$. Consequently, we will assume henceforth that all constraints are given in the form

$$x_j - x_i \leq c_{ij}. \tag{5.1}$$

Define the *constraint graph* $G = (N, A)$ associated with these constraints having a node for each variable x_i and having an arc (i, j) with cost c_{ij} for each constraint of the form (5.1). The constraint graph derived from our example is shown in Figure 5.5.

It is important to notice that the linear inequalities (5.1) are related to the optimality conditions for shortest path distances in G with respect to the costs (lengths) c_{ij}. More precisely, if d_j denotes the shortest path distance from node r to node j in G, then these distances satisfy the well-known optimality conditions [1]

$$d_j \leq d_i + c_{ij}, \text{ for all } (i, j) \in A, \tag{5.2}$$

as well as the normalization requirement $d_r = 0$.

Now suppose that the constraint graph G contains a cycle K, denoted $i_0 \rightarrow i_1 \rightarrow \cdots \rightarrow i_k \rightarrow i_0$. By adding up the inequalities in (5.1) corresponding to the arcs $(i_0, i_1), (i_1, i_2), \ldots, (i_k, i_0)$ we obtain the inequality $0 \leq c(K)$, where $c(K)$ is the cost of cycle K with respect to the arc costs c_{ij}. This shows that

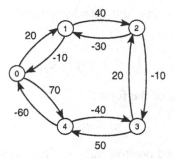

Figure 5.5. A constraint graph

if the system (5.1) is consistent, then all cycles in G must have nonnegative cost. Conversely, suppose that all cycles in G have nonnegative cost. Then we are assured [19] that shortest path distances in G are well defined. In particular, if x_j is the shortest path distance from node $r = 0$ to node j in G, then $x_1, x_2, \ldots, x_n$ is a solution to (5.1).

We conclude that our set of temporal constraints is consistent precisely when the constraint graph G contains no negative cycle. In our example, all cycles of cardinality 2 have positive cost, as well as the cycles $0 \rightarrow 1 \rightarrow 2 \rightarrow 3 \rightarrow 4 \rightarrow 0$ (with cost 40) and $0 \rightarrow 4 \rightarrow 3 \rightarrow 2 \rightarrow 1 \rightarrow 0$ (with cost 10). As a result, the stated conditions in the example are consistent.

In addition, using the shortest path distances from $r = 0$ in G, we can place tight bounds on the possible realization values for each variable x_i. In our example, the shortest path distance from node 0 to node 3 is 30, via the shortest path $0 \rightarrow 4 \rightarrow 3$, showing that $x_3 = 30$ is a feasible value. Similarly the negative of the shortest path distances from each node to node $r = 0$ also satisfy (5.2), giving alternative feasible values for x_i. In the example, a shortest path from node 3 to node 0 is $3 \rightarrow 2 \rightarrow 1 \rightarrow 0$ with cost -20, giving the feasible value $x_3 = 20$. In fact these two scenarios represent the extreme realizable values for x_3, meaning that $20 \leq x_3 \leq 30$. In a similar way it can be shown [8] that the shortest path distances from node 0 and (negated) to node 0 provide the tightest possible bounds on each variable x_i.

7. Ratio Comparisons in Decision Analysis

In trying to gather information about outcome probabilities (or attribute priorities) held by a decision maker, the experimenter can gain preference information by eliciting ratio comparisons between pairs of outcomes. Often this preference information is imprecise [3, 17]. For example, the decision maker might indicate that outcome i is between two and three times more important than outcome j. If the unknown outcome probabilities (or normalized priorities) are denoted $y_1, y_2, \ldots, y_n$, one then has bounds on the relative importance of outcomes in the form

$$l_{ji} \leq y_i/y_j \leq u_{ji},$$

giving rise to inequalities of the form

$$y_i \leq a_{ji} y_j, \tag{5.3}$$

where the a_{ji} are observed positive values. For example, the following information might be elicited from an individual about his or her perceived probabilities of four outcomes:

$$y_1 \leq \frac{1}{2} y_3, \ y_1 \leq 2y_4, \ y_2 \leq y_1, \ y_2 \leq \frac{1}{3} y_3, \ y_3 \leq 4y_2, \ y_3 \leq 5y_4, \ y_4 \leq \frac{1}{2} y_2. \tag{5.4}$$

An important issue in this context is whether or not the individual's expressed information is consistent: i.e., whether there are indeed positive probabilities satisfying the conditions (5.3).

This problem is in fact a thinly disguised version of the problem considered in Section 5.6. To see this, associate a network G with the set of constraints (5.3) in the following way. There is a node of $G = (N, A)$ for each probability y_i and an arc (j, i) with weight a_{ji} corresponding to each constraint of the form $y_i \leq a_{ji} y_j$. Figure 5.6 shows the network associated with our sample problem (5.4).

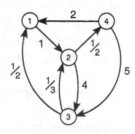

Figure 5.6. Network for assessing probabilities

By taking logarithms of the relation (5.3) and defining $x_i = \log y_i$, we obtain the equivalent formulation

$$x_i \leq x_j + c_{ji}, \tag{5.5}$$

where the cost $c_{ji} = \log a_{ji}$ is associated with arc $(j, i) \in A$. The relations (5.5) are simply the optimality conditions (5.2) for shortest path distances from a fixed root node r in G, which we previously encountered in Section 5.6. Therefore, the given system of constraints is consistent if and only if G contains no negative cycles with respect to the transformed costs c. However a negative cycle in G with respect to the (logarithmic) c-costs corresponds to a cycle whose product of a-weights is less than one. As a result, the original system is consistent if and only if its network G contains no cycle of (multiplicative) weight less than one. Examination of the network in Figure 5.6 reveals the cycle $2 \to 4 \to 3 \to 2$ whose product of arc weights is $(\frac{1}{2})(5)(\frac{1}{3}) = \frac{5}{6} < 1$ and so the system (5.4) is inconsistent. This cycle isolates the following string of inequalities

$$y_2 \le \frac{1}{3}y_3, \quad y_3 \le 5y_4, \quad y_4 \le \frac{1}{2}y_2,$$

which produce the contradiction

$$y_2 \le \frac{1}{3}y_3 \le \frac{1}{3}(5y_4) \le \frac{5}{3}(\frac{1}{2}y_2) = \frac{5}{6}y_2 < y_2.$$

The formulation in terms of the network G also provides information in the case that the system is consistent. To illustrate this possibility, replace the fourth constraint in (5.4) by $y_2 \le \frac{1}{2}y_3$. The revised network G, shown in Figure 5.7, now contains no cycles with multiplicative weight less than one so the new system is consistent. The shortest (multiplicative) path distance d_{ji} from node j to node i now provides the tightest bound on the ratio of y_i to y_j: namely, $y_i \le d_{ji}y_j$. In our example, the path of smallest multiplicative weight from node 1 to node 3 is $1 \to 2 \to 4 \to 3$ whose product of arc weights is $(1)(\frac{1}{2})(5) = \frac{5}{2}$, giving $y_3 \le \frac{5}{2}y_1$. Moreover, reasoning as in Section 5.6, one set of feasible probabilities $y_1, y_2, \ldots, y_n$ can be obtained by computing all shortest distances $d_{j1}, d_{j2}, \ldots, d_{jn}$ and then normalizing to probabilities, by dividing by $\sum_i d_{ji}$. In fact the value obtained in this manner for y_j is the *smallest* feasible probability for the jth outcome consistent with the given data. Similarly one can compute a set of feasible probabilities $y_1, y_2, \ldots, y_n$ by computing the reciprocals of all shortest distances $d_{1i}, d_{2i}, \ldots, d_{ni}$ and normalizing these to probabilities, by dividing by $\sum_j 1/d_{ji}$. The value obtained in this manner for y_i is the *largest* feasible probability for the ith outcome consistent with the given data.

We illustrate these results for the consistent system defined by the network in Figure 5.7. The shortest multiplicative path weights from node 1 are then $(d_{11}, d_{12}, d_{13}, d_{14}) = (1, 1, \frac{5}{2}, \frac{1}{2})$. When normalized, these produce the feasible probabilities $(\frac{1}{5}, \frac{1}{5}, \frac{1}{2}, \frac{1}{10})$. In addition, the values d_{1i} give the (tightest) bounds

$$y_1 \le 1 \cdot y_1, \quad y_2 \le 1 \cdot y_1, \quad y_3 \le \frac{5}{2} \cdot y_1, \quad y_4 \le \frac{1}{2} \cdot y_1.$$

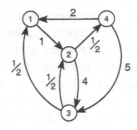

Figure 5.7.   Revised network for assessing probabilities

If we let $y_1 = \alpha$ and add the above inequalities, we obtain

$$1 = y_1 + y_2 + y_3 + y_4 \leq (1 + 1 + \frac{5}{2} + \frac{1}{2})\alpha = 5\alpha,$$

so the smallest value of y_1 is $\alpha = \frac{1}{5}$, as expected.

Likewise, the shortest multiplicative path weights into node 1 are $(d_{11}, d_{21}, d_{31}, d_{41}) = (1, 1, \frac{1}{2}, 2)$. The reciprocal distances are $(1, 1, 2, \frac{1}{2})$, which when normalized yield the feasible probabilities $(\frac{2}{9}, \frac{2}{9}, \frac{4}{9}, \frac{1}{9})$. The value $y_1 = \frac{2}{9}$ obtained here provides an upper bound on any feasible probability for the first outcome. Using the lower bound obtained earlier, the first probability is then constrained by the given relations to lie in the interval $[\frac{1}{5}, \frac{2}{9}]$.

8.   DNA Sequencing

Rapid progress is being made in determining the DNA sequence of the human genome, which will have great potential in screening for disease as well as developing treatments for genetic disorders. All DNA is composed of the four chemicals (bases) adenine (A), thymine (T), guanine (G), and cytosine (C). Thus a section of DNA can be considered a word in the 4-symbol alphabet (A, T, G, C). Human DNA contains approximately 3 billion pairs of these bases, sequenced in a very special manner.

One particular method for discovering the exact sequence of bases in a given DNA segment is called hybridization. It proceeds by first determining all length k sequences present in the given DNA string. For example, analyzing the string AGTACCG would produce the following (overlapping) length 3 substrings: AGT, GTA, TAC, ACC, and CCG. Of course these are not produced in any particular order. Thus the real problem is that of reconstructing the original string S from a knowledge of its set of length k substrings.

Suppose then that we are given a set $\mathcal{L}$ of length k substrings. To aid in reconstructing a potential string S from $\mathcal{L}$, we define a directed graph G in the following way. The nodes of G are all length $k-1$ sequences from the alphabet A, T, G, C. For each substring $\sigma \in \mathcal{L}$, with $\sigma = a_1 a_2 \ldots a_{k-1} a_k$, create an arc from $a_1 a_2 \ldots a_{k-1}$ to $a_2 \ldots a_{k-1} a_k$. This arc is labelled with the symbol a_k, indicating that we have transformed the length $k - 1$ sequence $a_1 a_2 \ldots a_{k-1}$ into the length $k - 1$ sequence $a_2 \ldots a_{k-1} a_k$ by appending a_k to the right of the first sequence (thus obtaining σ) and then deleting the leading symbol a_1. In the resulting directed graph G, any nodes that have no incident arcs can be removed.

As an illustration suppose that $\mathcal{L} = \{$ACT, AGC, ATG, CAG, CAT, CTG, GAC, GCA, GCT, TGA, TGC$\}$. The substring ACT from $\mathcal{L}$ produces an arc joining AC to CT with label T, and so forth, resulting in the labelled graph of Figure 5.8. All nodes of this graph have equal indegree and outdegree except nodes CA and CT. The former has outdegree one greater than its indegree, while the latter has outdegree one less than its indegree. This condition ensures the existence of an *Euler path* in G: namely, a (nonsimple) path in G that uses each arc of G exactly once. In fact, such a path must start at node CA (outdegree $=$ indegree $+ 1$) and end at node CT (indegree $=$ outdegree $+ 1$). Here, an Euler path P is given by the node sequence CA $\rightarrow$ AT $\rightarrow$ TG $\rightarrow$ GC $\rightarrow$ CA $\rightarrow$ AG $\rightarrow$ GC $\rightarrow$ CT $\rightarrow$ TG $\rightarrow$ GA $\rightarrow$ AC $\rightarrow$ CT. This path corresponds to the string CATGCAGCTGACT, obtained by successively appending the arc labels of P to the initial node CA of this Euler path. Thus we have reconstructed a possible string S that has precisely the given set $\mathcal{L}$ of length 3 substrings.

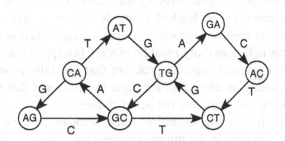

Figure 5.8. DNA sequencing network

In general, any string S will produce an Euler path P in the graph G generated from the set $\mathcal{L}$ of all its length k substrings. (It is possible that the starting and ending nodes of P might coincide.) Moreover, an Euler path in G will identify a possible generating string S. It is important to note that there can be many Euler paths in G, so that the "true" string S is not uniquely reconstructed by this procedure. Such is the case in our example, since G also contains the Euler path CA $\rightarrow$ AG $\rightarrow$ GC $\rightarrow$ CT $\rightarrow$ TG $\rightarrow$ GC $\rightarrow$ CA $\rightarrow$ AT $\rightarrow$ TG $\rightarrow$ GA $\rightarrow$ AC $\rightarrow$ CT, which yields the string CAGCTGCATGACT. However, there is often additional information about the proximity of certain substrings (of length k), and this leads to a weighted version of the Euler path problem. The interested reader is referred to [11] for details of this important practical extension.

Before closing, it should be mentioned that a string S might contain several identical substrings of length k. In this case, the constructed G will contain multiple arcs joining certain nodes. Using current experimental techniques, it is not possible to tell how many times a length k substring has been produced when analyzing a DNA segment. Therefore we have made the simplifying assumption that S does not contain any identical length k substrings: i.e., $\mathcal{L}$ is a set rather than a multiset. However, it is easy to check the validity of this assumption, since typically one knows the actual length of S and this can be compared with the length of the string generated by an Euler path of G.

9. Computer Memory Management

Computer memory typically contains a relatively small amount of fast memory as well as a relatively large amount of slower auxiliary memory. All memory is segmented into a number of units called *pages*. We assume here that the fast memory consists of k slots (or *frames*) each of which can contain a single page. During the execution of computer programs, both instructions and data need to be available in fast memory. If the information required is already available on a page in fast memory, then no additional work is needed. However, if a particular *request* involves locations in slower memory, then the associated page must first be fetched from slow memory and then placed in fast memory. If fast memory already contains k pages, then one of the pages currently residing in fast memory must be evicted (swapped out) to make room for the currently requested page. A problem that naturally arises in this context is that of devising an algorithm for page management that minimizes the number of evictions incurred over the set of requests.

In practice, such a paging algorithm must work on-line: i.e., without full knowledge of all requests in the future. However in order to gain insights into the real world problem, we study the off-line version, in which the entire set of requests is known in advance. Let this sequence of requests be denoted by $[r_1, r_2, r_3, \ldots, r_n]$, where $r_i = j$ means that the ith request involves information present on page j. As an illustration, suppose there are $k = 3$ frames, with requests given by $[1, 5, 1, 2, 5, 3, 1, 4, 2, 5, 3]$. Assuming that fast memory initially contains no pages whatsoever, the first requested page $r_1 = 1$ can be placed in the first frame, and the second requested page $r_2 = 5$ can be placed in the second frame. The third request $r_3 = 1$ is for a page already in fast memory, so the fourth requested page $r_4 = 2$ is next placed in frame 3. The fifth request $r_5 = 5$ is also for a page resident in fast memory. When the sixth request $r_6 = 3$ is encountered, then an eviction must take place. If we simply evict the page of smallest request index (the "oldest" resident page), then page $r_3 = 1$ is evicted, being replaced by page $r_6 = 3$. One possible history of processing this request sequence can be seen in Table 5.6. Thus frame 1 is successively filled with requests r_1, r_3, r_6, r_9 with evictions occurring between $r_3 = 1$ and $r_6 = 3$ and also between $r_6 = 3$ and $r_9 = 2$. Overall this history reveals six evictions, denoted by $(r_3, r_6), (r_6, r_9), (r_5, r_8), (r_8, r_{11}), (r_4, r_7), (r_7, r_{10})$. Our interest is in developing a page management scheme that achieves the minimum number of page evictions.

Table 5.6.   History of requests allocated to frames

frame	requests
1	$\emptyset, r_1, r_3, r_6, r_9$
2	$\emptyset, r_2, r_5, r_8, r_{11}$
3	$\emptyset, r_4, r_7, r_{10}$

We provide a network formulation of a paging problem with k frames and n requests $[r_1, r_2, r_3, \ldots, r_n]$. Define a bipartite network $G = (N, A)$ with node set $N = N_1 \cup N_2$, where $N_1 = \{s_1, s_2, \ldots, s_k, a_1, a_2, \ldots, a_n\}$ and $N_2 = \{t_1, t_2, \ldots, t_k, b_1, b_2, \ldots, b_n\}$. Each request r_i will be represented using nodes a_i and b_i. If (on some frame) request r_i can be followed next by request r_j, then we create an arc (a_i, b_j). Thus the arc set A contains all arcs of the form (a_i, b_j) for $1 \le i < j \le n$. The cost of arc (a_i, b_j) is 0 if $r_i = r_j$ and is 1 otherwise; the latter case represents an eviction of the form (r_i, r_j). The "start" nodes s_u are used to initially fill the empty frames, with arc $(s_u, b_j) \in A$ indicating that empty frame u is initialized with request r_j. Similarly, the "terminal" nodes t_v are used to indicate the last entry in a given frame: arc $(a_i, t_v) \in A$ signifies that r_i is the last request on frame v. The arcs of type (s_u, b_j) and (a_i, t_v) are assigned cost 0.

In this way any feasible history represents an assignment (or *perfect matching*) of all the nodes of N_1 with nodes of N_2. For example, the history in Table 5.6 is represented by the selection of "initial frame" arcs $(s_1, b_1), (s_2, b_2), (s_3, b_4)$, "final frame" arcs $(a_9, t_1), (a_{11}, t_2), (a_{10}, t_3)$, and "next request" arcs $(a_1, b_3), (a_3, b_6), (a_6, b_9), (a_2, b_5), (a_5, b_8), (a_8, b_{11}), (a_4, b_7), (a_7, b_{10})$. Since $(a_3, b_6), (a_6, b_9), (a_5, b_8), (a_8, b_{11}), (a_4, b_7), (a_7, b_{10})$ all have cost 1, this assignment has cost 6, representing the total number of evictions. Two different optimal assignments, each having cost 3, are displayed in Tables 5.7 and 5.8.

The very special structure of this assignment problem can in fact be exploited to produce a greedy algorithm for page management. The so-called *MIN algorithm* [5] decides on which page to evict by selecting that page hav-

Table 5.7.　An optimal assignment of requests

frame	requests
1	$\emptyset, r_1, r_3, r_7, r_8, r_9$
2	$\emptyset, r_2, r_5, r_{10}$
3	$\emptyset, r_4, r_6, r_{11}$

Table 5.8.　Another optimal assignment of requests

frame	requests
1	$\emptyset, r_1, r_3, r_7, r_8$
2	$\emptyset, r_2, r_5, r_6, r_{11}$
3	$\emptyset, r_4, r_9, r_{10}$

ing its next request furthest in the future. To return to our original example, after the fifth request, the three frames contain pages $1, 5, 2$ respectively. When request $r_6 = 3$ is processed, the first future occurrence of page 1 is request 7, while page 5 is next requested at r_{10} and page 2 is next requested at r_9. Thus page 5 is evicted, replaced by page $r_6 = 3$. The full history of applying this greedy approach is shown in Table 5.8, which achieves the minimum number of evictions. In fact, optimality of the MIN algorithm has been shown by several authors; Cohen and Burkhard [7] use duality theory of linear programming to prove optimality.

References

[1] R. K. Ahuja, T. L. Magnanti, and J. B. Orlin, *Network Flows: Theory, Algorithms, and Applications*, Prentice-Hall, Englewood Cliffs, NJ, 1993.

[2] R. K. Ahuja, T. L. Magnanti, J. B. Orlin, and M. R. Reddy, Applications of network optimization, in *Handbooks in Operations Research and Management Science*, Vol. 7, M. O. Ball et al. (eds.), Elsevier, Amsterdam, 1995, pp. 1–83.

[3] A. Arbel, Approximate articulation of preference and priority derivation, *European Journal of Operational Research* 43 (1989), 317–326.

[4] M. L. Balinksi, On a selection problem, *Management Science* 17 (1970), 230–231.

[5] L. A. Belady, A study of replacement algorithms for virtual storage computers, *IBM Systems Journal* 5 (1966), 78–101.

[6] R. E. Bellman and S. E. Dreyfus, *Applied Dynamic Programming*, Princeton University Press, Princeton, NJ, 1962.

[7] A. Cohen and W. A. Burkhard, A proof of the optimality of the MIN paging algorithm using linear programming duality, *Operations Research Letters* 18 (1995), 7–13.

[8] R. Dechter, I. Meiri, and J. Pearl, Temporal constraint networks, *Artificial Intelligence* 49 (1991), 61–95.

[9] L. R. Ford, Jr. and D. R. Fulkerson, *Flows in Networks*, Princeton University Press, Princeton, NJ, 1962.

[10] J. F. Gilsinn, P. B. Saunders, and M. H. Pearl, Path finding algorithms and data structures for point-to-point trip management, NBSIR 75-676, National Bureau of Standards, Washington, DC, 1975.

[11] D. Gusfield, R. Karp, L. Wang, and P. Stelling, Graph traversals, genes and matroids: an efficient case of the travelling salesman problem, *Discrete Applied Mathematics* 88 (1998), 167–180.

[12] D. S. Hochbaum, A new-old algorithm for minimum-cut and maximum-flow in closure graphs, *Networks* 37 (2001), 171–193.

[13] A. J. Hoffman and T. J. Rivlin, When is a team "mathematically" eliminated?, in *Princeton Symposium on Mathematical Programming*, 1967, H. W. Kuhn (ed.), Princeton University Press, Princeton, NJ, 1970, pp. 391–401.

[14] J. C. Picard, Maximal closure of a graph and applications to combinatorial problems, *Management Science* 22 (1976), 1268–1272.

[15] J. M. W. Rhys, A selection problem of shared fixed costs and network flows, *Management Science* 17 (1970), 200–207.

[16] L. W. Robinson, Baseball playoff eliminations: an application of linear programming, *Operations Research Letters* 10 (1991), 67–74.

[17] A. A. Salo and R. P. Hämäläinen, Preference programming through approximate ratio comparisons, *European Journal of Operational Research* 82 (1995), 458–475.

[18] B. L. Schwartz, Possible winners in partially completed tournaments, *SIAM Review* 8 (1966), 302–308.

[19] R. E. Tarjan, *Data Structures and Network Flows*, SIAM, Philadelphia, 1983.

Chapter 6

MODELING AND OPTIMIZATION OF VEHICLE ROUTING AND ARC ROUTING PROBLEMS

Jean-Francois Cordeau and Gilbert Laporte

Canada Research Chair in Distribution Management and GERAD
HEC Montréal
3000 chemin de la Côte-Sainte-Catherine
Montreal, Canada H3T 2A7

{ cordeau,gilbert } @crt.umontreal.ca

Abstract   This chapter describes some of the most important models and algorithms for the classical vehicle routing problem and for several families of arc routing problems. Exact methods (mostly based on branch-and-cut) and heuristics (mostly based on tabu search) are described, and computational results are presented.

Keywords: vehicle routing problem, arc routing problems, models, algorithms, branch-and-cut, metaheuristics, tabu search, variable neighbourhood search.

1.   Introduction

Transportation and logistics hold a central place in modern economies. Several studies have established that transportation costs account for a proportion of 11% to 13% of the total production cost of goods (Owoc and Sargious, 1992). It is estimated that in Canada and the United Kingdom, transportation expenses represent 15% of the total national expenses (Button, 1993). These figures do not account for expenditures in the public sector for operations like mail delivery, street maintenance, etc. Such operations also generate sizeable costs, certainly several billion dollars each year. The potential for rationalization and savings is therefore considerable and operational research can play a major role in generating economies. Applications of operational research to the field of transportation are widespread and well documented: see, e.g., Barnhart, Desrosiers and Solomon (1998a) and McGill and van Ryzin (1999) for air transportation; Barnhart, Desrosiers and Solomon (1998b) for rail transportation; Psaraftis (1999) for maritime transportation; Barnhart and Crainic (1998) and Crainic and Sebastian (2002) for freight transportation.

This chapter is about models and algorithms developed in the field of routing problems. Such problems are defined on a graph $G = (V, E \cup A)$ where V is a vertex set, E is a set of (undirected) edges and A is a set of (directed) arcs. Routing problems consist of designing optimal routes including given subsets of V, E or A subject to side constraints. It is common to distinguish between *Vehicle Routing Problems* (VRPs) in which the requirements are expressed in terms of vertex visits, and *Arc Routing Problems* (ARPs) which are defined in terms of the edges or arcs of the graph. The VRP is thus concerned with visiting isolated locations as in commercial pickup and delivery operations, door-to-door transportation of handicapped people, site inspections, etc. (see, e.g., Toth and Vigo, 2002a). Applications of ARPs arise naturally in mail delivery, garbage collection, street and highway maintenance, school bus routing, etc. (see, e.g., Eiselt, Gendreau and Laporte, 1995b; Assad and Golden 1995; Dror, 2000). *Generalized Routing Problems* combine the features of VRPs and ARPs (see, e.g., Blais and Laporte, 2003).

This chapter describes some of the most important models and algorithms for routing problems. In Section 6.2, we concentrate on the *classical* VRP with capacity and duration constraints. In Section 6.3, we cover the *Chinese Postman Problem* (CPP) which serves as a basis for the study of constrained ARPs. Two such cases are introduced in Section 6.4: the *Rural Postman Problem* (RPP), and the *Capacitated Arc Routing Problem* (CARP). Conclusions follow in Section 6.5.

2.   The Vehicle Routing Problem

The classical VRP, in its undirected (and most common) version is defined on a graph $G = (V, E)$, where $V = \{v_0, v_1, \ldots, v_n\}$ and $E = \{(v_i, v_j) : v_i, v_j \in V, i < j\}$. Vertex v_0 represents a *depot* while the remaining vertices correspond to *customers*. A fleet of m identical vehicles of capacity Q is based at the depot. The number of vehicles is either known in advance or expressed as a decision variable. A service time t_i and a demand q_i are associated with customer v_i and a travel time c_{ij} is associated with edge (v_i, v_j). Travel costs are normally proportional to distance or travel times so that these terms are often used interchangeably. The classical VRP consists of determining a set of m vehicle routes of least total cost such that 1) each route starts and ends at the depot, 2) each customer belongs to exactly one vehicle route, 3) the total demand of each route does not exceed Q, 4) the total duration of each route, including service and travel times, does not exceed a preset limit D. This definition implicitly assumes that all demands are either collected or delivered, but not both. Common variants include problems with heterogeneous vehicle fleets (Gendreau et al., 1999), the VRP with backhauls (Toth and Vigo, 2002c), the site dependent VRP (Cordeau and Laporte, 2001), the periodic and multi-depot VRPs (Cordeau, Gendreau and Laporte, 1997), the VRP with split deliveries (Dror, Laporte and Trudeau, 1994), and the VRP with time windows (Cordeau et al., 2002), stochastic VRPs (Laporte and Louveaux, 1998) and dynamic pickup and delivery problems (Mitrović-Minić, Krishnamurti and Laporte, 2004).

The VRP lies at the heart of distribution management. It is encountered in several industrial settings, and the potential for savings is considerable. In a recent article, Golden, Assad and Wasil (2002) described applications arising in areas as diverse as solid waste collection, food and beverage distribution, as well as operations arising in the dairy industry, and newspaper delivery. When such problems are tackled by means of operational research techniques, savings in the range of 10% to 15% on collection or delivery operations are not uncommon.

The VRP is also an NP-hard combinational optimization problem since it generalizes the Traveling Salesman Problem which is itself NP-hard (Garey and Johnson, 1979). To this day, there does not exist any exact algorithm capable of *consistently* solving instances involving more than 50 vertices (Toth and Vigo, 1998a; Naddef and Rinaldi, 2002), although a number of larger instances have been solved (Naddef and Rinaldi, 2002). Another difficulty with exact approaches is their high performance variability on instances of similar sizes. Heuristics are therefore the preferred solution technique in practice. We now review the most important exact and approximate algorithms for the VRP.

2.1 Exact Algorithms for the Vehicle Routing Problem

Several families of exact algorithms have been proposed for the classical VRP. These methods include direct tree search procedures (Christofides, 1976a; Christofides, Mingozzi and Toth, 1981a; Hadjiconstantinou, Christofides and Mingozzi, 1995; Toth and Vigo, 2002b), dynamic programming (Christofides, Mingozzi and Toth, 1981b), set partitioning algorithms (Agarwal, Mathur and Salkin, 1989; Bramel and Simchi-Levi, 2002), vehicle flow algorithms (Laporte, Nobert and Desrochers, 1985; Cornuéjols and Harche, 1993; Fisher, 1995; Ralphs, 1995; Naddef and Rinaldi, 2002), and commodity flow algorithms (Gavish and Graves, 1982; Baldacci, Mingozzi and Hadjiconstantinou, 1999). Here we concentrate on vehicle flow formulations and algorithms which have proved the most successful.

2.1.1 Vehicle flow formulation.   Vehicle flow formulations for the VRP are rooted in the work of Laporte, Nobert and Desrochers (1985). They work with integer variables x_{ij} equal to the number of vehicle trips between vertices v_i and v_j. Thus if $i = 0$, then x_{ij} can take the values 0, 1 or 2 (the latter case corresponds to a return trip between the depot and customer v_j); if $i > 0$, then x_{ij} is binary. Let f be the fixed cost of a vehicle, and define $\delta(S) = \{(v_i, v_j) \in E : v_i \in S, v_j \in V \setminus S \text{ or } v_i \in V \setminus S, v_j \in S\}$ for $S \subset V$, and $x(\delta(S)) = \sum_{(v_i, v_j) \in \delta(S)} x_{ij}$.

The integer programming formulation is then:

$$\text{(VRP)} \qquad \text{minimize} \quad \sum_{i<j} c_{ij} x_{ij} + fm \qquad (6.1)$$

subject to

$$x(\delta(\{v_0\})) = 2m \qquad\qquad (6.2)$$

$$x(\delta(\{v_i\})) = 2 \qquad\qquad (i = 1, \ldots, n) \quad (6.3)$$

$$x(\delta(S)) = 2V(S) \qquad (S \subseteq V \setminus \{v_0\}) \quad (6.4)$$

$$0 \le x_{0j} \le 2 \qquad\qquad (j = 1, \ldots, n) \quad (6.5)$$

$$0 \le x_{ij} \le 1 \qquad (i < j; i, j = 1, \ldots, n) \quad (6.6)$$

$$x_{ij} \text{ integer} \qquad (i < j; i, j = 0, \ldots, n). \quad (6.7)$$

In this formulation, $V(S)$ is a lower bound on the number of vehicles required to visit all vertices of S in a feasible solution. If the problem only contains capacity constraints, then $V(S)$ is set equal to $\left\lceil \sum_{v_k \in S} q_k / Q \right\rceil$. If the problem contains duration constraints as well, then a strengthening of $V(S)$ can be obtained by properly exploiting the bounds provided by the optimal subproblem solutions in an enumerative algorithm (Laporte, Nobert and Desrochers, 1985). Constraints (6.2) and (6.3) are degree constraints, whereas constraints (6.4) are connectivity constraints. They ensure that no solution will contain a subtour of customers disconnected from the depot and that all vehicle routes will be feasible.

2.1.2   Branch-and-cut algorithm.   It is customary to solve the VRP by means of a branch-and-cut algorithm, similar to what is done for the *Traveling Salesman Problem* (TSP) (Jünger, Reinelt and Rinaldi, 1995). The subproblem solved at the root node of the search tree is defined by (6.1), (6.2), (6.3), (6.5) and (6.6). Violations of (6.4) are detected by means of a separation procedure. In addition, other families of valid inequalities are used in the solution process. Several such inequalities are described by Naddef and Rinaldi (2002) for the capacity constrained VRP. Some of these inequalities are rather intricate, and not all of them have been successfully implemented within a branch-and-cut algorithm. We will therefore only provide a summary of the main known results.

In what follows, it is assumed that m is a known constant. A first valid inequality is obtained by strengthening the right-hand side of (6.4) to $2r(S)$, where $r(S)$ is the solution of a bin packing problem on S, with item weights equal to q_i and bin size equal to Q. While the bin packing is NP-hard, it is rather manageable for small to medium size instances (Martello and Toth, 1990). A further strengthening is obtained by first defining $\mathcal{P}$ as the set of all feasible partitions of $V \setminus \{v_0\}$ into m customer sets. Then for any non-empty subset $S \subseteq V$ and $P = \{S_1, \ldots, S_m\} \in \mathcal{P}$, define $\beta(P, S) = |\{k : S_k \cap S \neq \emptyset\}|$ as the number of vehicles visiting S in the solution corresponding to P, and $R(S) = \min_{P \in \mathcal{P}} \{\beta(P, S)\}$. Then the right-hand side of (6.4) can be set equal to $2R(S)$. Cornuéjols and Harche (1993) show that if $q_1 = 5$, $q_2 = q_3 = q_4 = 3$, $q_5 = q_6 = q_7 = 4$, $q_8 = 2$, $m = 4$, $Q = 7$ and $S = \{v_1, v_2, v_3, v_4\}$, then $r(S) = 3$ and $R(S) = 4$, while $\lceil (q_1 + q_2 + q_3 + q_4)/7 \rceil = 2$. In other words, both strengthenings are non-trivial.

In addition, Naddef and Rinaldi (1993) introduce *generalized capacity constraints*. Let $\mathcal{S} = \{S_1, \ldots, S_t\}$ be a collection of t (> 1) disjoint subsets of V and let $\bar{r}(H, \mathcal{S})$ be the cost of a solution to a bin packing problem with an item of size q_i for each $v_i \in H \setminus \bigcup_{k=1}^{t} S_k$ and $H \subseteq V$, and one item of size $q(S_k) = \sum_{i \in S_k} q_i$ for $k = 1, \ldots, t$. Then the constraints

$$\sum_{k=1}^{t} x(\delta(S_k)) \geq 2R(\mathcal{S}) \tag{6.8}$$

and

$$\sum_{k=1}^{t} x(\delta(S_k)) \geq 2t + 2(\bar{r}(V, \mathcal{S}) - m) \tag{6.9}$$

are valid inequalities for the VRP.

In addition, if H is a subset of V including $S_1, \ldots, S_t$, then the *framed capacity constraint*

$$x(\delta(H)) + \sum_{k=1}^{t} x(\delta(S_k)) \geq 2t + 2\bar{r}(H, \mathcal{S}) \tag{6.10}$$

is a valid inequality for the VRP (Augerat, 1995).

Naddef and Rinaldi (2002) also show under which conditions valid inequalities for the TSP (Jünger, Reinelt and Rinaldi, 1995) are valid for the VRP, the simplest of which being *VRP comb inequalities*. Let H and $T_1, \ldots, T_t$ be subsets of V satisfying $T_k \setminus H \neq \emptyset$ and $T_k \cap H \neq \emptyset$ for $k = 1, \ldots, t$, $T_k \cap T_h \neq \emptyset$

for $k, h = 1, \ldots, t$, $k \neq h$, $t \geq 3$ and odd. Then the comb inequality

$$x\big(\delta(H)\big) \geq (t+1) - \sum_{k=1}^{t} \big(x\big(\delta(T_k)\big) - 2\big) \qquad (6.11)$$

is valid for the VRP. Also, if no T_k contains v_0 and $r(T_k \setminus H) + r(T_k \cap H) > r(T_k)$ for all k, then

$$x\big(\delta(H)\big) + \sum_{k=1}^{t} x\big(\delta(T_k)\big) \geq t + 1 + 2 \sum_{k=1}^{t} r(T_k) \qquad (6.12)$$

is a valid inequality for the VRP (Laporte and Nobert, 1984). More involved comb inequalities are provided by Cornuéjols and Harche (1993) and by Naddef and Rinaldi (2002).

A number of algorithms are available for the separation of constraints (6.4) and (6.9) within the framework of a branch-and-cut algorithm (Ralphs, 1995; Augerat et al., 1999) and several branching strategies are possible (Clochard and Naddef, 1993). More elaborate multistar inequalities have recently been proposed by Letchford, Eglese and Lysgaard (2002) and have been embedded within a branch-and-cut algorithm (Lysgaard, Letchford and Eglese, 2003).

2.1.3   Computational results.   Computational results presented by Naddef and Rinaldi (2002) show that on instances of the capacity constrained VRP (with $22 \leq n \leq 135$), the optimality gap at the root node of the search tree lies between 0% and 8.20%. Six instances containing up to 45 vertices and two 135-vertex instances were solved optimally. As Naddef and Rinaldi (2002) rightly observe, polyhedral research on the VRP is still in its early stages, particularly in comparison with the TSP. While the TSP and the VRP share a number of features, it appears that the VRP is considerably more intricate.

2.2   Classical Heuristics for the Vehicle Routing Problem

Because the VRP is so hard to solve exactly and algorithmic behaviour is highly unpredictable, a great deal of effort has been invested on the design of heuristics. We distinguish between *classical heuristics* usually consisting of a construction phase followed by a relatively simple postoptimization phase, and *metaheuristics* based on new optimization concepts developed over the past fifteen to twenty years. Since the performance of heuristics can only be assessed experimentally it is common to make comparisons on a set of fourteen benchmark instances proposed by Christofides, Mingozzi and Toth (1979) (CMT) which range from 50 to 199 cities. The best known solution values for these instances have been obtained by Taillard (1993) and Rochat and Taillard

(1995). We first describe some of the most representative classical heuristics. An extensive survey is provided by Laporte and Semet (2002).

2.2.1   The Clarke and Wright heuristic.   One of the first and probably the most famous VRP heuristic is the Clarke and Wright (1964) *savings* method. Initially, n back and forth routes are constructed between the depot and each customer. At a general iteration, two routes $(v_0, \ldots, v_i, v_0)$ and $(v_0, v_j, \ldots, v_0)$ are merged into $(v_0, \ldots, v_i, v_j, \ldots, v_0)$ as long as this is feasible and the saving $s_{ij} = c_{i0} + c_{0j} - c_{ij}$ resulting from the merge is positive. The process is repeated until no more merges are possible. There are two standard ways to implement this algorithm. In the *parallel* version, the merge yielding the largest saving is implemented. In the *sequential* version, one route is expanded at a time. It is common to apply a 3-opt local search postoptimization phase (Lin, 1965) to each route resulting from this process. In practice, the best results are obtained with the parallel implementation followed by 3-opt. Laporte and Semet (2002) have tested this heuristic on the fourteen CMT instances. They have obtained an average deviation of 6.71% on the best known solution values. Average computation times were insignificant (0.13 seconds on a Sun Ultrasparc 10 Workstation, 42 Mflops).

Several enhancements have been proposed to the Clarke and Wright heuristic, namely aimed at speeding up the initial sorting of savings (Nelson et al., 1985; Paessens, 1988). Also, some authors (e.g., Yellow, 1970; Golden, Magnanti and Nguyen, 1977) have suggested the use of modified savings of the form $s_{ij} = c_{i0} + c_{0j} - \lambda c_{ij}$, with $\lambda > 0$ to prevent the formation of circumferential routes which tend to be created by the original heuristic. Another modification of the savings algorithm consists of optimizing route merges at each iteration by means of a matching algorithm (Desrochers and Verhoog, 1989; Altinkemer and Gavish, 1991; Wark and Holt, 1994). The best implementation, in terms of accuracy, is due to Wark and Holt (1994). The best of five runs on each instance yields an average accuracy of 0.63% on the CMT instances, but with an average total time (for the five runs) of 13,371 seconds on a Sun 4/630 MP.

The Clarke and Wright algorithm remains popular to these days because of its simplicity and low execution time. However, its accuracy leaves alot to be desired and it is not very flexible in the sense that its performance is low when additional constraints are introduced into the problem (see, e.g. Solomon, 1987).

2.2.2   The sweep heuristic.   Another well-known VRP heuristic is the *sweep* method commonly attributed to Gillett and Miller (1974), although previous implementations are known (Wren, 1971; Wren and Holliday, 1972). The method is best suited to planar instances with a centrally located depot. In this algorithm, vehicle routes are gradually constructed by rotating a ray centered at the depot until it becomes clear that adding an extra customer to the current route would result in an infeasibility. The current route is then closed and the process is iteratively reapplied until all customers have been routed. The routes can then be reoptimized by applying a 3-opt scheme to each route or by performing customer moves or swaps between consecutive routes. The accuracy of this algorithm is similar to that of Clarke and Wright. Thus, Renaud, Boctor and Laporte (1996) have obtained an average deviation of 7.09% on the CMT instances with an average computation time of 105.6 seconds on a Sun Sparcstation 2 (4.2 Mflops). A natural extension is to produce a large set of routes (using a sweep mechanism) and combine them optimally by means of a set partitioning algorithm (Foster and Ryan, 1976; Ryan, Hjorring and Glover, 1993; Renaud, Boctor and Laporte, 1996). The latter implementation appears to be the best. By suitably generating vehicle routes, Renaud, Laporte and Boctor have obtained an average accuracy of 2.38% within an average computing time of 208.8 seconds on a Sun Sparcstation 2 (4.2 Mflops).

Overall, the sweep algorithm seems dominated by the Clarke and Wright savings heuristic. It is slightly less simple to implement (because of the post-optimization phase) and its execution time is higher. Finally, it is not well suited to urban contexts with a grid street network.

2.2.3   The Fisher and Jaikumar heuristic.   The Fisher and Jaikumar (1981) heuristic is best described as a cluster-first, route-second procedure. In a first phase, clusters of customers feasible to the capacity constraint are created. This is done by suitably selecting seeds and optimizing the allocation of customers to seeds by means of a *Generalized Assignment Problem* (GAP). A route is then created on each cluster by means of any TSP algorithm. This heuristic is natural in the sense that it mimics the sequential process of human schedulers. It can also benefit from any algorithmic advance for the GAP or the TSP. Solutions obtained by Fisher and Jaikumar are difficult to assess. Contrary to recent practice, Fisher and Jaikumar have reported integer solution values without specifying the rounding convention used for the c_{ij}'s in the algorithm. Also, it is not clear how duration constraints can be handled by this algorithm and the performance of the method seems highly related to the initial choice of seeds (see, Cordeau et al., 2002). Bramel and Simchi-Levi (1995) describe a variant of this algorithm in which the location of seeds is optimized. They obtain an average deviation of 3.29% on the seven capacity constrained instances of the CMT test bed.

2.3   Metaheuristics for the vehicle routing problem

Several metaheuristics have been proposed for the VRP. Gendreau, Laporte and Potvin (2002) have identified six main classes: simulated annealing (SA), deterministic annealing (DA), tabu search (TS), genetic search (GS), neural networks (NN) and ant systems (AS).

2.3.1   Local search heuristics.   The first three families of metaheuristics perform a search of the solution space while allowing intermediate deteriorating and even infeasible solutions. They operate by moving at each iteration t from the current solution x_t to a solution x_{t+1} in the neighbourhood $N(x_t)$ of x_t, defined as the set of all solutions reachable from x_t by performing a given type of transformation (such as moving a customer from its current route to another route, swapping two customers between two routes, etc.).

The main differences between SA, DA and TS are as follows. Let $f(x)$ be the cost of solution x. At iteration t of SA, a solution x is drawn randomly in $N(x_t)$. If $f(x) \leq f(x_t)$, then $x_{t+1} := x$. Otherwise, $x_{t+1} := x$ with probability p_t. It is common to define p_t as $\exp\left(-\left[f(x) - f(x_t)\right]/\theta_t\right)$, where θ_t denotes the *temperature* at iteration t and is usually a decreasing step function of t. The process stops when a termination criterion has been met. One of the best known SA implementations in the context of the VRP is due to Osman (1993).

Deterministic annealing is similar to SA, but here a deterministic rule is used to accept a move. In *threshold accepting* (Dueck and Scheuer, 1990) $x_{t+1} := x$ if $f(x) < f(x_t) + \theta_1$, where θ_1 is a user controlled parameter, and $x_{t+1} := x_t$ otherwise. In *record-to-record* travel (Dueck, 1993), $x_{x+1} := x$ if $f(x) < \theta_2 f(x^*)$, where θ_2 is a user-controlled parameter (usually larger than 1), and x^* is the best known solution value; $x_{t+1} := x_t$ otherwise. An application of DA to the VRP can be found in Golden et al. (1998).

In TS, x_{t+1} is normally the best solution in $N(x_t)$, with an important exception: to avoid cycling, any solution possessing a given attribute of x_t is declared *tabu*, or forbidden for a certain number of iterations. However, the tabu status of a solution can be revoked if it is clear that moving to it cannot cause cycling. Such a rule is called an *aspiration criterion* (see, e.g., Glover and Laguna, 1997). A common aspiration criterion is to move to any solution that corresponds to a new best known solution either in absolute terms, or with respect to a subset of solutions possessing a certain attribute (see, e.g., Cordeau, Gendreau and Laporte, 1997).

TS implementations also contain *diversification* and *intensification* features.

The purpose of diversification is to widen the set of solutions considered during the search. A common technique is to penalize moves frequently performed during the search process. Intensification consists of accentuating the search in promising regions of the solution space. Tabu search is without any doubt the most popular, and also the most successful metaheuristic for the VRP. For a more detailed analysis, see Cordeau and Laporte (2002).

2.3.2   Genetic search.   Genetic search maintains a population of good solutions and performs at each iteration k combinations choosing in each case two parents to produce two offspring, by using a *crossover operator*. A random mutation of each offspring (with a small probability) is then performed. Finally, the $2k$ worst elements of the population are removed and replaced with the $2k$ offspring. Van Breedam (1996) and Schmitt (1994,1995) report implementations to the VRP with capacity constraints only and to the VRP with capacity and duration constraints, respectively. Prins (2004) has also obtained excellent results through the application of a genetic algorithm.

2.3.3   Neural networks.   Neural networks can be viewed as a process that transforms inputs into outputs through a weighting mechanism. The output is then compared to a desired solution and weights are adjusted accordingly in order to obtain a better fit. This process is often employed as a learning mechanism that mimics the human brain. A natural application in the field of routing is where the desired solutions are produced by an expert such as in real-time pickup and delivery operations (Shen et al., 1995). Neural networks were not initially conceived as a combinatorial optimization tool. It was Hopfield and Tank (1985) who pioneered the use of NN in the field of combinatorial optimization. In the area of vehicle routing, the first efforts have concentrated on the TSP, through the use of elastic nets (Durbin and Willshaw, 1987) and self-organizing maps (Kohonen, 1988). These are deformable templates that adjust themselves to a set of vertices to form a Hamiltonian cycle. This idea has been extended to the VRP, taking multiple vehicles and side constraints into account (Ghaziri, 1991, 1993). Neural networks have produced good results on the VRP but are not competitive with the best known approaches.

2.3.4   Ant systems.   Ant systems are inspired from an analogy with ant colonies who lay pheromone on their path as they forage for food. Colorni, Dorigo and Maniezzo (1991) have suggested an optimization technique based on this idea. Artificial ants search the solution space and remember the trails followed to reach good quality solutions. The pheromone value associated with an edge (v_i, v_j) is increased whenever that edge belongs to good solutions. Several implementations proposed in the context of the VRP (Kawamura et al., 1998; Bullnheimer, Hartl and Strauss, 1998, 1999) have produced encouraging results, but again are not competitive with the best available metaheuristics.

These six families of VRP metaheuristics are surveyed in Gendreau, Laporte and Potvin (2002). In what follows, we will focus on some of the best algorithms, all in the field of tabu search.

2.3.5   Taillard's heuristic.   One of the best known TS heuristics for the VRP is due to Taillard (1993). It uses a simple neighbourhood structure consisting of swapping vertices belonging to two different routes. The reverse of a move is declared tabu for θ iterations, where θ is randomly selected in [5, 10]. The algorithm makes use of a continuous diversification scheme: the objective function value associated with a potential move contains a penalty term proportional to the past frequency of that move. Periodical route reoptimizations are performed by means of an exact TSP algorithm (Volgenant and Jonker, 1983). To accelerate the search, Taillard uses a parallelization strategy in which each processor operates on a different portion of the problem. In planar instances, this decomposition is obtained by partitioning the region into sectors centered at the depot, and then into concentric rings. In non-planar instances, the regions are defined by means of a shortest spanning arborescence routed at the depot. Twelve of the best known solutions for the fourteen CMT instances where obtained by means of this algorithm. Computation times associated with these solutions are not reported and are believed to be rather high.

2.3.6   Taburoute.   Another popular TS heuristic is Taburoute, developed by Gendreau, Hertz and Laporte (1994). Neighbour solutions are obtained by moving a vertex from its current route to another route containing one of its close neighbours. Insertions are performed by means of GENI (Gendreau, Hertz and Laporte, 1992), a generalized insertion procedure for the TSP that performs a local reoptimization of the route in which a vertex is inserted. This algorithm uses the same tabu mechanism and continuous diversification scheme as in the Taillard algorithm. A novel feature of Taburoute is the use of a penalized objective. If $f(x)$ denotes the value of solution x, then the penalized objective is defined as

$$f'(x) = f(x) + \alpha C(x) + \beta D(x), \qquad (6.13)$$

where $C(x)$ and $D(x)$ are the total over-capacity and excess duration of all routes, respectively, and α, β are two penalty factors. Initially set equal to 1, these parameters are periodically divided by 2 if all previous λ solutions were feasible, or multiplied by 2 if they were all infeasible. This way of proceeding produces a mix of feasible and infeasible solutions which acts as a diversification strategy. Three intensification mechanisms are used. The first consists of conducting a limited search on a small number of starting solutions. The full search then proceeds starting from the most promising solution. The second mechanism periodically reoptimizes vehicle routes using the US post-optimization heuristic of Gendreau, Hertz and Laporte (1992). Finally, the search is accentuated around the best known solution in the last part of the algorithm.

2.3.7   Adaptive memory.   The *Adaptive Memory Procedure* (AMP) of Rochat and Taillard (1995) is a highly portable concept that can be applied to a wide variety of combinatorial problems. Similar to what is done in GS, an AMP maintains a pool of high quality solutions which are periodically recombined to obtain new, hopefully better solutions, and the worst solutions are discarded from the pool. In the VRP context, a new solution is initially built by selecting several non-overlapping routes from the solutions of the pool. This will leave a number of unrouted vertices which will gradually be incorporated in one of the starting routes, or in new ones. This mechanism can be used in conjunction with any post-optimization heuristic such as TS. An AMP can be viewed as a diversification scheme during the course of the search process, or as an intensification technique in the final phase of the algorithm. The AMP of Rochat and Taillard (1995) produced the best solutions for all CMT instances.

2.3.8   **Ejection chains.**   Ejection chains are a mechanism for generating neighbourhoods involving more that two routes in a TS algorithm. Put simply, an ejection chain operates on ℓ different routes. In a chain-like fashion, it successively "ejects" a vertex from route k and inserts it in its best position in route $k+1$ ($k = 1, \ldots, \ell-1$). A vertex is also ejected from route ℓ and inserted in another route. The neighbourhood of a solution is made up of several solutions reachable by ejection chains. Two fine ejection chain based heuristics for the VRP are those of Rego and Roucairol (1996) and Rego (1998). The first of these two implementations uses parallel computing.

2.3.9   **Granularity.**   *Granularity* (Toth and Vigo, 1998b) is yet another powerful concept. It consists of initially removing unpromising elements from a problem in order to perform a more efficient search on the remaining part. Thus, very costly edges are unlikely to be part of a good VRP solution and can often be disregarded. In their implementation, Toth and Vigo only retain the edge set $E(\nu) = \{(v_i, v_j) \in E : c_{ij} \leq \nu\} \cup I$, where I is a set of "important" edges defined as all those incident to the depot. The value of ν is set equal to $\beta \bar{c}$, where β is a *sparsification parameter*, and $\bar{c}$ is the average cost associated with the edges of a good solution (quickly identified by means of the Clarke and Wright algorithm, for example). If $\beta \in [1.0, 2.0]$, then approximately 80% to 90% of all edges can be eliminated. Toth and Vigo have applied this idea in conjunction with Taburoute. They have obtained slightly better solutions than Gendreau, Hertz and Laporte (1994) in about one quarter of the computing time.

2.3.10   **The unified tabu search algorithm.**   Finally, another good TS implementation is the Unified Tabu Search Algorithm (UTSA) of Cordeau, Gendreau and Laporte (1997). First designed to solve periodic and multi-depot VRPs, it was later extended to the VRP with site dependencies (Cordeau and Laporte, 2001), to the periodic and multi-depot VRP with Time Windows

(Cordeau, Laporte and Mercier, 2001) and to the classical VRP (Cordeau et al., 2002). UTSA incorporates some of the features of Taburoute, including the same neighbourhood structures, GENI insertions, and intermediate infeasible solutions controlled through the use of a penalized objective function. However, it uses no intensification strategy and works with fixed tabu durations. In UTSA, the α and β penalty coefficients are either divided or multiplied by $1 + \delta$ (with $\delta > 0$) according to whether the previous solution was feasible (with respect to capacity or duration) or not. The tabu mechanism operates on an attribute set $\beta(x) = \{(i, k) : v_i$ is visited by vehicle k in solution $x\}$, and neighbourhood solutions are defined by removing an attribute (i, k) from x and replacing it with another attribute (i, k') for some $k' \neq k$. The attribute (i, k) is declared tabu for a number of iterations, but its tabu status may be revoked if reassigning v_i to route k would yield a new best solution. Two interesting aspects of UTSA are its simplicity and its flexibility. All its parameters except one took the same value in all applications on which the algorithm was tested. The flexibility of UTSA lies in its capability to perform well on several VRP variants. On the classical VRP, it produces excellent solutions within relatively short computing times.

2.3.11   Computational results.   We provide in Table 6.1 a comparison of the TS implementations just described on the fourteen CMT instances. These results show that the best heuristics are quite accurate and can typically reach solution values within 1% of the best known. Computation times are not reported in the table but can be quite high. Research efforts should now focus on producing faster algorithms possessing a leaner structure and fewer parameters, even if this causes a slight deterioration in accuracy.

3.   The Chinese Postman Problem

Arc routing problems have received less attention than vertex routing problems, probably because most of their applications lie in the public sector where the need of running cost-effective operations is not always felt as acutely as in the private sector. However, over the past ten years or so, budget cuts and privatization have led several public authorities to manage their resources more tightly. The history of arc routing is truly fascinating (Eiselt and Laporte, 2000) and can be traced to the early eighteenth century.

3.1   Euler Cycles and Circuits

The theory of arc routing is rooted in the work of Leonhard Euler (1936) who was called upon to solve the Königsberg bridges problem. The question put to him by the inhabitants of Königsberg (now Kaliningrad, Russia) was whether there existed a closed walk using each of the seven bridges of

Instance	Size	Type[1]	Taillard (1993)	Taburoute (1994)	Rochat-Taillard (1995)	Rego-Roucairol (1996)	Rego (1998)	Toth-Vigo (1998b)	UTSA (2002)
1	50	C	**524.61**	**524.61**	**524.61**	**524.61**	524.81	**524.61**	**524.61**
2	75	C	**835.26**	835.77	**835.26**	835.32	847.00	838.60	835.45
3	100	C	**826.14**	819.45	**826.14**	827.53	832.04	826.56	826.44
4	150	C	**1028.42**	1036.16	**1028.42**	1044.35	1047.21	1033.21	1038.44
5	199	C	1298.79	1322.65	**1291.45**	1334.55	1351.18	1318.25	1305.87
6	50	C,D	**555.43**	**555.43**	**555.43**	**555.43**	559.25	**555.43**	**555.43**
7	75	C,D	**909.68**	931.23	**909.68**	**909.68**	922.21	920.72	**909.68**
8	100	C,D	**865.94**	**865.94**	**865.94**	866.75	876.97	869.48	866.68
9	150	C,D	**1162.55**	1177.76	**1162.55**	1164.12	1191.30	1173.12	1171.81
10	199	C,D	1397.94	1418.51	**1395.85**	1420.84	1460.83	1435.74	1415.40
11	120	C,D	**1042.11**	1073.47	**1042.11**	**1042.11**	1052.04	1042.87	1074.13
12	100	C	**819.56**	**819.56**	**819.56**	819.56	821.63	819.56	**819.56**
13	120	C,D	**1541.14**	1547.93	**1541.14**	1550.17	1558.06	1545.51	1568.91
14	100	C,D	**866.37**	**866.37**	**866.37**	**866.37**	867.79	**866.37**	866.53
Average deviation from best			0.05%	0.88%	0.00%	0.55%	1.50%	0.64%	0.69%

[1]   C: Capacity restrictions;   D: Duration restrictions.

Table 6.1.   Solution values produced by several TS heuristics on the fourteen CMT instances. Best known solutions are shown in boldface.

the Pregel river exactly once. (Figure 6.1a). Using a graph representation (Figure 6.1b), Euler proved that there was none and put forward a necessary condition for the existence of a closed walk through all edges of an undirected graph: the graph must be connected and all its vertices must have an even degree, which is clearly not the case of the graph depicted in Figure 6.1b. Graphs that satisfy the necessary conditions are called *unicursal* or *Eulerian*. Hierholzer showed that the conditions stated by Euler are also sufficient in the sense that a closed walked can always be determined in a Eulerian graph.

3.1.1   Closed walks in undirected graphs: The end-pairing algorithm.
Hierholzer's polynomial time *end-pairing algorithm* has been rediscovered under several names and variants (Fleischner, 1990, 1991).

Step 1. Starting at an arbitrary vertex, gradually construct a cycle by considering an untraversed edge contiguous to the partial cycle at each iteration. Repeat until the starting vertex has been reached.

Step 2. If all edges have been traversed, stop.

Step 3. Otherwise, construct a second cycle in a similar fashion, starting with an edge incident to the first cycle at vertex v. Coalesce the two

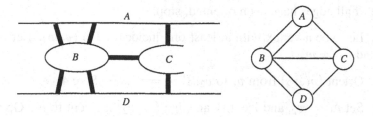

a) The seven bridges of   b) Undirected graph
   Königsberg   representation

Figure 6.1. The Königsberg bridges problem

cycles $(\ldots, v_i, v, v_j, \ldots)$ and $(\ldots, v_k, v, v_\ell, \ldots)$ into a single cycle $(\ldots, v_i, v, v_\ell, \ldots, v_k, v, v_j, \ldots)$. Go to Step 2.

3.1.2   Necessary and sufficient conditions for unicursality.   Necessary and sufficient conditions for the existence of an Eulerian circuit in a directed graph $G = (V, A)$ or mixed graph $G = (V, E \cup A)$ were later stated (Ford and Fulkerson, 1962). If G is directed, it must be strongly connected and the number of incoming arcs of each vertex must be equal to the number of outgoing arcs. If G is mixed, it must be strongly connected, every vertex must be incident to an even number of arcs and edges and for every $S \subset V$, the absolute difference in the number of arcs for S to $V \setminus S$ and the number of arcs from $V \setminus S$ to S cannot be less than the number of edges between S and $V \setminus S$.

3.1.3   Closed walks in directed and mixed graphs.   Determining a closed walk in a Eulerian directed graph can be achieved by adapting the end-pairing algorithm or by using a specialized technique based on the computation of a spanning arborescence (see van Aardenne-Ehsenfest and de Bruijn (1951) for the original article, and Eiselt, Gendreau and Laporte (1995a) for a recent description). If G is a Eulerian mixed graph, then all its edges must first be transformed into arcs by means of a network flow algorithm (Ford and Fulkerson, 1962).

Step 1.   Replace each edge of G with a pair of opposite arcs, thus obtaining a directed graph $G' = (V, A')$. Assign to each arc of $A' \cap A$ a lower bound of 1 and to each arc of $A' \setminus A$ a lower bound of 0. Assign to each arc of A' an upper bound of 1.

Step 2.   Determine a feasible flow in G'. Let x_{ij} be the flow on arc (v_i, v_j).

Step 3.   If $(v_i, v_j) \in A' \setminus A$ and $x_{ij} = 0$, orient edge (v_i, v_j) from v_j to v_i.

Step 4. If all edges have been oriented, stop.

Step 5. Let v be a vertex with at least one incident edge (v, w). Let $v_1 := v$ and $v_2 := w$.

Step 6. Orient (v_1, v_2) from v_1 to v_2. If $v_2 = v$, go to Step 4.

Step 7. Set $v_1 := v_2$, and identify an edge (v_1, v_2) incident to v_1. Go to Step 6.

As noted by Ford and Fulkerson, this algorithm can be applied to a mixed graph not known to be Eulerian. If it is Eulerian, a feasible flow will be produced in Step 2; otherwise, the algorithm will fail.

3.2   The Undirected Chinese Postman Problem

The *Chinese Postman Problem* (CPP) was first defined by Guan (1962) in the context of undirected graphs. It consists of determining a shortest closed traversal of a graph. If the graph is not Eulerian, this amounts to determining a least cost *augmentation* of a graph, i.e., a replication of some of its edges or arcs to make it Eulerian. If $G = (V, E)$ is undirected, the odd-degree vertices arc first identified (there is always an even number of such vertices) and the cost $\bar{c}_{ij}$ of a shortest chain is computed for each pair of odd degree vertices v_i and v_j. Formally, the augmentation problem can be cast as an integer linear program as follows (Edmonds and Johnson, 1973). Let x_{ij} ($i < j$) be an integer variable equal to the number of copies of edge (v_i, v_j) that must be added to G in order to make it Eulerian. A non-empty subset S of V is called *odd* if it contains an odd number of odd degree vertices. Let $\delta(S) = \{(v_i, v_j) \in E : v_i \in S, v_j \in V \setminus S \text{ or } v_i \in V \setminus S, v_j \in S\}$. The problem is then:

$$\text{(UCPP)} \qquad \text{minimize} \sum_{(v_i, v_j) \in E} \bar{c}_{ij} x_{ij} \tag{6.14}$$

subject to

$$\sum_{(v_i, v_j) \in \delta(S)} x_{ij} \geq 1 \qquad (S \subset V, S \text{ is odd}) \tag{6.15}$$

$$x_{ij} \geq 0 \qquad \qquad \left((v_i, v_j) \in E\right) \tag{6.16}$$

$$x_{ij} \text{ integer} \qquad \left((v_i, v_j) \in E\right). \tag{6.17}$$

Edmonds and Johnson show that the polyhedron of solutions to (6.15) and (6.16) is equal to the convex hull of feasible solutions to (6.15)–(6.17). They also show how (6.14)–(6.16) can be solved as a *Matching Problem* on the odd degree vertices of V, with matching costs $\bar{c}_{ij}$.

3.3   The Directed Chinese Postman Problem

A similar approach exists for the solution of the augmentation problem on a directed graph (Edmonds and Johnson, 1973; Orloff, 1974; Beltrami and Bodin, 1974). Let I denote the set of vertices v_i for which the number of incoming arcs exceeds the number of outgoing arcs by s_i, and let J be the set of vertices v_j for which the number of outgoing arcs exceeds the number of incoming arcs by d_j. In other words, s_i can be seen as a supply, and d_j as a demand. Define $\bar{c}_{ij}$ as the length of a shortest path from v_i to v_j.

Again, x_{ij} is the number of extra copies of arc (v_i, v_j) that must be added to the graph in order to make it Eulerian. The augmentation problem can be cast as a *Transportation Problem*:

$$(\text{DCPP}) \qquad \text{minimize} \sum_{v_i \in I} \sum_{v_j \in J} \bar{c}_{ij} x_{ij} \qquad\qquad (6.18)$$

subject to

$$\sum_{v_j \in J} x_{ij} = s_i \qquad (v_i \in I) \qquad\qquad (6.19)$$

$$\sum_{v_i \in I} x_{ij} = d_j \qquad (v_j \in J) \qquad\qquad (6.20)$$

$$x_{ij} \geq 0 \qquad (v_i \in I, v_j \in J). \qquad (6.21)$$

3.4   The Mixed Chinese Postman Problem

The Chinese Postman Problem on a mixed graph $G = (V, E \cup A)$ is NP-hard. It was formulated by Grötschel and Win (1992) and by Nobert and Picard (1996) in two different ways. Here we present the Nobert and Picard formulation. For every proper subset S of V, define

$$A^+(S) = \big\{ (v_i, v_j) \in A : v_i \in S, v_j \in V \setminus S \big\},$$
$$A^-(S) = \big\{ (v_i, v_j) \in A : v_i \in V \setminus S, v_j \in S \big\},$$
$$\delta(S) = \big\{ (v_i, v_j) \in E : v_i \in S, v_j \in V \setminus S \text{ or } v_i \in V \setminus S, v_j \in S \big\},$$
$$w(S) = |A^+(S)| - |A^-(S)| - |\delta(S)|.$$

In addition, let p_k be a binary constant equal to 1 of and only if the degree of vertex v_k is odd, and let z_k be an integer variable. Also define integer variables x_{ij} and y_{ij} equal to the number of copies of arcs (v_i, v_j) and of edges (v_i, v_j), respectively, that must be added to the graph to make it Eulerian. The problem is then:

$$(\text{MCPP}) \qquad \text{minimize} \sum_{(v_i, v_j) \in A} c_{ij} x_{ij} + \sum_{(v_i, v_j) \in E} c_{ij} y_{ij} \qquad (6.22)$$

subject to

$$\sum_{(v_i,v_j)\in A} x_{ij} + \sum_{(v_i,v_j)\in E} y_{ij} = 2z_k + p_k \qquad (v_k \in S) \qquad (6.23)$$

$$\sum_{(v_i,v_j)\in A^-(S)} x_{ij} - \sum_{(v_i,v_j)\in A^+(S)} x_{ij} + \sum_{(v_i,v_j)\in\delta(S)} y_{ij} \geq u(S) \quad (S \subset V, S \neq \emptyset)$$

$$(6.24)$$

$x_{ij}, y_{ij}, z_k \geq 0$ and integer for all edges (v_i, v_j), and all vertices v_k. (6.25)

In this formulation, constraints (6.23) and (6.24) are the necessary and sufficient conditions of Ford and Fulkerson (1962) for the unicursality of a strongly connected graph. Nobert and Picard solve MCPP by means of a branch-and-cut technique in which constraints (6.24) are initially relaxed and introduced as they are found to be violated. They also make use of the redundant constraints

$$\sum_{(v_i,v_j)\in A^+(S)} x_{ij} + \sum_{(v_i,v_j)\in A^-(S)} x_{ij} + \sum_{(v_i,v_j)\in\delta(S)} y_{ij} \geq 1 \quad (S \subset V, S \text{ odd})$$

$$(6.26)$$

to strengthen the linear relaxation of MCPP, and add a number of Gomory cuts to help gain integrality. The algorithm was successfully applied to several randomly generated mixed CPPs with $16 \leq |V| \leq 225$, $2 \leq |A| \leq 5569$, and $15 \leq |E| \leq 4455$. Out of 440 instances, 313 were solved without any branching. The number of constraints generated in the course of the algorithm was of the order of $|V|$.

4.   Constrained Arc Routing Problems

Most arc routing applications encountered in practice involve several side constraints, giving rise to a large number of variants. The two more common, are the *Rural Postman Problem* (RPP) and the *Capacitated Arc Routing Problem* (CARP). The RPP has been studied both heuristically and exactly, while only heuristics are known to exist for the CARP. Both problems can be defined in their undirected, directed or mixed versions. For the sake of brievity, we will concentrate on the undirected case.

4.1   Heuristics for the Rural Postman problem

In several arc routing contexts, it is not necessary to traverse all edges of a graph $G = (V, E)$ but only a subset $R \subseteq E$ of *required edges*. The p subgraphs induced by the edges of R are called *connected components*. The RPP consists of determining a minimum cost closed walk through a subset of E that includes all required edges. This problem was introduced by Orloff (1974). It is NP-hard (Lenstra and Rinnooy Kan, 1976), unless $p = 1$, in which case it reduces to the undirected CPP.

4.1.1   Frederickson's heuristic.   One of the best known heuristics for the symmetric RPP is a simple construction procedure proposed by Frederickson (1979). It works along the lines of the Christofides (1976b) heuristic for the symmetric TSP.

Step 1. Construct a shortest spanning tree on a graph H having one vertex for each connected component of G, and edges linking these vertices. The cost of edge (v_i, v_j) is the length of a shortest chain between components i and j in G. Denote by T the edge set of the tree.

Step 2. Solve a minimum cost matching problem (using shortest chain costs) on the odd-degree vertices of the graph induced by $R \cup T$. Denote by M the set of edges induced by the matching.

Step 3. The graph induced by $R \cup T \cup M$ is Eulerian. Compute a Eulerian cycle on it by means of the end-pairing algorithm.

The worst-case performance ratio of this heuristic is $3/2$, provided the edge costs in G satisfy the triangle inequality. Figure 6.2 illustrates a case where Frederickson's heuristic fails to produce an optimal solution.

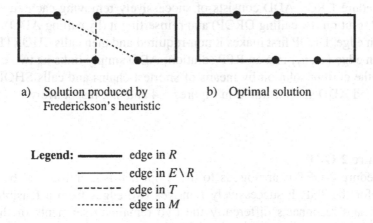

a) Solution produced by   b) Optimal solution
    Frederickson's heuristic

Legend: ———— edge in R
        ———— edge in $E \setminus R$
        ------- edge in T
        ·········· edge in M

Figure 6.2. Example for the Frederickson's heuristic does not yield an optimal solution.

4.1.2   Post-optimization heuristics.   A number of post-optimization heuristics for the RPP were developed by Hertz, Laporte and Nanchen-Hugo (1999). They show that standard edge exchange operations, so common for the TSP, can become quite intricate in an ARP context. The three procedures are SHORTEN, DROP-ADD and 2-OPT.

Procedure SHORTEN

Procedure SHORTEN can be described as follows. An RPP solution S can be written as a circular vector $S = (v_{i_1}, \ldots, v_{i_r}, \ldots, v_{i_t} = v_{i_1})$. Using in turn each vertex v_{i_r} as a starting point for the circular vector, SHORTEN first attempts to create a route $S' = (v_{j_1}, \ldots, v_{j_s}, \ldots, v_{j_1})$ through the same edge sequence as S but with a different starting vertex, in which the chain $(v_{j_1}, \ldots, v_{j_s})$ contains no required edge. It then replaces this chain by a shortest chain between v_{j_1} and v_{j_s}. Two operations are used to transform S into S': POSTPONE and REVERSE. Operation POSTPONE can be applied whenever an edge (v_i, v_j) appears in S first as serviced, and later as traversed without being serviced. POSTPONE makes the edge first traversed and then serviced. Operation REVERSE simply reverses a chain starting with serviced edge (v_i, v_j) and ending with a non-serviced edge (v_k, v_i). Procedure SHORTEN is illustrated on the graph depicted in Figure 6.3a, where vertex 5 is used as a starting point. The RPP solution can be written as the circular vector (5,3,2,1,3,2,4,3,5) shown in Figure 6.3b. Applying POSTPONE makes the first edge (5,3) traversed and the last edge (3,5) serviced (Figure 6.3c). Applying REVERSE to the chain (2,1,3,2) yields the solution shown in Figure 6.3d, at which point the chain (5,3,2,3,1) is shortened into (5,3,1) (Figure 6.3e). The postoptimized RPP solution is illustrated in Figure 6.3f.

Procedure DROP-ADD

Procedure DROP-ADD consists of successively removing each edge from an RPP solution (by calling DROP) and reinserting it (by calling ADD). To remove an edge, DROP first makes it non-required and then calls SHORTEN. To insert an edge (v_i, v_j) into an RPP solution, ADD simply links its two extremities to the current solution by means of shortest chains and calls SHORTEN. DROP and ADD are illustrated in Figures 6.4 and 6.5, respectively.

Procedure 2-OPT

Procedure 2-OPT is analogous to the 2-opt heuristic introduced by Croes (1958) for the TSP. It successively removes two edges from a feasible RPP solution and reconnects differently the two remaining segments of the tour by means of shortest chains (Figure 6.6). SHORTEN is then called. If any required edges are missing from the solution (this may occur if one of the removed edges was required), these are reintroduced by means of ADD.

Computational tests carried out by Hertz, Laporte and Nanchen-Hugo (1999) on several classes of graphs indicate that Frederickson's heuristic is fast, but frequently produces suboptimal solutions (up to 10% worse than the best known solutions). Procedure 2-OPT reduces this gap considerably, often to nothing. Procedure DROP-ADD is much faster than 2-OPT (sometimes by two orders

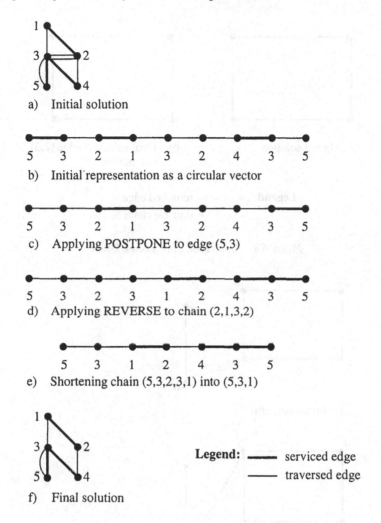

a) Initial solution

b) Initial representation as a circular vector

c) Applying POSTPONE to edge (5,3)

d) Applying REVERSE to chain (2,1,3,2)

e) Shortening chain (5,3,2,3,1) into (5,3,1)

f) Final solution

Legend: ━━━ serviced edge
        ─── traversed edge

Figure 6.3. Illustration of procedure SHORTEN

of magnitude) but is is not as effective. For best results, it is recommended to apply Frederickson's heuristic followed by 2-OPT.

4.2 A Branch-and-Cut Algorithm for the Undirected Rural Postman Problem

One of the first exact algorithms for the undirected RPP is a branch-and-cut procedure in which Lagrangean relaxation is used to compute lower bounds (Christofides et al., 1981). More recent approaches are the branch-and-cut algorithms of Corberán and Sanchis (1994) in which the separation problems are solved visually, and the Letchford (1996) enhancement of this algorithm

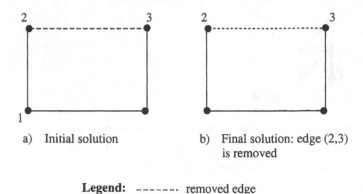

a)   Initial solution   b)   Final solution: edge (2,3)
                                    is removed

Legend: – – – – – – removed edge
          - - - - - - - - shortest chain

Figure 6.4.   Illustration of procedure DROP

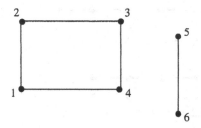

a)   Initial solution

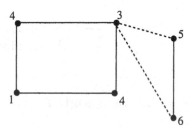

b)   Final solution before SHORTEN: edge (5,6) is added

Legend: - - - - - - - - shortest chain

Figure 6.5.   Illustration of procedure ADD

which adds new valid inequalities to the original formulation. However, this
author only reports results for the first node of the search tree. The Ghiani and

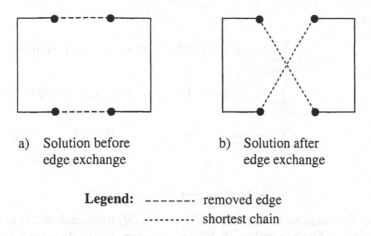

a) Solution before   b) Solution after
   edge exchange   edge exchange

Legend: - - - - - - - removed edge
            - - - - - - - shortest chain

Figure 6.6. Illustration of procedure 2-OPT

Laporte (2000) algorithm is a fully automated branch-and-cut method. It will now be described.

4.2.1   Graph reduction.   We first introduce a simple graph reduction technique due to Christofides et al. (1981). Let $C_k(k = 1, \ldots, p)$ be the kth connected component of G. Let V_R be the set of vertices v_i such that an edge (v_i, v_j) exists in R, and $V_k \subseteq V_R(i = 1, \ldots, p)$ the vertex set of C_k. Denote by c_e the cost of edge $e \in R$. A vertex $v_i \in V_R$ is R-odd (R-even) if and only if an odd (even) number of edges of R are incident to v_i. The reduction is then:

Step 1. Add to $G_R = (V_R, R)$ an edge between every pair of vertices of V_R having a cost equal to that of the corresponding shortest chain on G.

Step 2. Delete one of two parallel edges if they have the same cost, and all edges $(v_i, v_j) \notin R$ such that $c_{ij} = c_{ik} + c_{kj}$ for some v_k.

4.2.2   The Coberán and Sanchis formulation.   We now recall the Corberán and Sanchis (1994) formulation on which the Ghiani-Laporte formulation is based. Given $S \subset V$, let $\delta(S)$ be the set of edges of E with one extremity in S and one in $V \backslash S$. If $S = \{v\}$, then we write $\delta(v)$ instead of $\delta(S)$. Let $x_e = x_{ij}$ represent the number of additional (deadheaded) copies of edge $e = (v_i, v_j)$ that must be added to G to make it Eulerian. The integer programming formulation is then:

(URPP1)   minimize $\sum\limits_{e \in E} c_e x_e$ (6.27)

subject to

$$\sum_{e \in \delta(v)} x_e = 0 \;(\text{mod } 2) \quad (\text{if } v \in V_R \text{ is } R\text{-even}) \qquad (6.28)$$

$$\sum_{e \in \delta(v)} x_e = 1 \;(\text{mod } 2) \quad (\text{if } v \in V_R \text{ is } R\text{-odd}) \qquad (6.29)$$

$$\sum_{e \in \delta(S)} x_e \geq 2 \quad (S = \bigcup_{k \in P} V_k, P \subset \{1, \ldots, p\}, P \neq \emptyset)$$

$$(6.30)$$

$$x_e \geq 0 \text{ and integer} \quad (e \in E). \qquad (6.31)$$

In this formulation, constraints (6.28) and (6.29) force each vertex to have an even degree, while constraints (6.30) ensure connectivity. In what follows, we recall some dominance relations that will enable a reformulation of the problem containing only binary variables and without the non-linear constraints (6.28) and (6.29).

Dominance relation 1.   (Christofides et al., 1981).
Every optimal RPP solution satisfies the relations

$$x_e \leq 1 \qquad\qquad (\text{if } e \in R) \qquad\qquad (6.32)$$
$$x_e \leq 2 \qquad\qquad (\text{if } e \in E \backslash R) \qquad\qquad (6.33)$$

Dominance relation 2.   (Corberán and Sanchis, 1994).
Every optimal solution satisfies

$$x_e \leq 1 \qquad (\text{if } e = (v_i, v_j), \; v_i, \; v_j \text{ belong to the same component } C_k).$$
$$(6.34)$$

Dominance relation 3.   (Ghiani and Laporte, 2000).
    Let $x(e^{(1)}), x(e^{(2)}), \ldots, x(e^{(\ell)})$ be the variables associated with the edges $e^{(1)}, e^{(2)}, \ldots, e^{(\ell)}$ having exactly one end vertex in a given component C_k and exactly one end vertex in another given component C_h. Then, in an optimal solution, only the variables $x(e^{(r)})$ having a cost $c(e^{(r)}) = \min\{c(e^{(1)}), c(e^{(2)}), \ldots, c(e^{(\ell)})\}$ can be equal to 2.
    Now define a $0/1/2$ edge as an edge e for which x_e can be at most equal to 2 in (URPP1), and a $0/1$ edge as an edge e for which x_e can be at most equal to 1. Denote by E_{012} and E_{01} the corresponding edge sets.

Dominance relation 4.   (Ghiani and Laporte, 2000).
Let $G_C^* = (V_C, E_C)$ be an auxiliary graph having a vertex v_i' for each component C_i and, for each pair of components C_i and C_j, an edge (v_i', v_j') corresponding to a least cost edge between C_i and C_j. Then, in any optimal

(URPP1) solution, the only $0/1/2$ edges belong to a *Minimum Spanning Tree* on G_C^* (denoted by MST_C).

4.3   The Ghiani and Laporte formulation

Using Dominance relation 4, formulation (URPP1) can now be rewritten by replacing each $0/1/2$ edge e belonging to a given MST_C by two parallel $0/1$ edges e' and e''. Denote by $E'(E'')$ the sets of edges $e'(e'')$, and let $\overline{E} = E_{01} \cup E' \cup E''$. In formulation (URPP1), constraints (6.31) are simply replaced by

$$x_e = 0 \text{ or } 1 \qquad (e \in \overline{E}). \qquad (6.35)$$

Ghiani and Laporte (2000) also replace the modulo relations (6.28) and (6.29) by the following constraints called cocircuit inequalities by Barahona and Grötschel (1986):

$$\sum_{e \in \delta(v) \setminus F} x_e \geq \sum_{e \in F} x_e - |F| + 1 \quad (v \in V, F \subseteq \delta(v), |F| \text{ is odd if } v \text{ is } R\text{-even},$$

$$|F| \text{ is even if } v \text{ is } R\text{-odd}). \quad (6.36)$$

Thus the new undirected RPP formulation, called (URPP2) and defined by (6.27), (6.36), (6.30) and (6.35), is linear in the $0/1$ x_e variables. Constraints (6.36) can be generalized to any non-empty subset S of V:

$$\sum_{e \in \delta(S) \setminus F} x_e \geq \sum_{e \in F} x_e - |F| + 1 \quad (F \subseteq \delta(S), |F| \text{ is odd if } S \text{ is } R\text{-even},$$

$$|F| \text{ is even if } S \text{ is } R\text{-odd}), \quad (6.37)$$

which are valid inequalities for (URPP2). If S is R-odd and $F = \emptyset$, constraints (6.37) reduce to the known R-odd inequalities (Corberán and Sanchis, 1994):

$$\sum_{e \in \delta(S)} x_e \geq 1. \qquad (6.38)$$

If S is R-even and $F = \{e_b\}$, they reduce to the R-even inequalities introduced by Ghiani and Laporte (2000):

$$\sum_{e \in \delta(S) \setminus \{e_b\}} x_e \geq x_{e_b}. \qquad (6.39)$$

Ghiani and Laporte (2000) have shown that constraints (6.37) are facet inducing for (URPP2). They have also developed a branch-and-cut algorithm in which connectivity constraints (6.30) and generalized cocircuit inequalities (6.37) are dynamically generated. In practice, it is usually sufficient to generate constraints of type (6.36), (6.38) and (6.39) to identify a feasible RPP solution.

4.3.1   Computational results.   The branch-and-cut algorithm developed by Ghiani and Laporte (2000) was tested on several types of randomly generated instances and on two real-life instances. On planar graphs instances involving up to 350 vertices were solved optimally within modest computing times. The 350-vertex instances required an average of 22.4 branching nodes. The average ratio of the lower bound at the root of the search tree over the optimum was equal to 0.997 for these instances.

4.4   Heuristics for the undirected capacitated arc routing problem

In the undirected *Capacitated Arc Routing Problem* (CARP) introduced by Golden and Wong (1981), a non-negative demand q_{ij} is associated with each edge (v_i, v_j) and there is a fleet of m identical vehicles of capacity Q based at the depot. The aim is to determine a least cost traversal of all edges of the graph, using the m vehicles, in such a way that each vehicle starts and ends at the depot and the total demand associated with each vehicle route does not exceed Q. The RPP can be viewed as a particular case of the CARP in which $q_{ij} = 0$ or 1 for all edges, and only one vehicle of capacity $Q = |E|$ is used.

Early construction and improvement heuristics for the undirected CARP are path-scanning (Golden, de Armon and Baker, 1983), construct-strike (Christofides, 1973), modified construct-strike and modified path-scanning (Pearn, 1989). Recently Greistorfer (1994), Hertz, Laporte and Mittaz (2000), and Hertz and Mittaz (2001) have proposed more efficient algorithms based on tabu search and variable neighbourhood search.

4.4.1   Basic procedures.   We first present four procedures developed by Hertz, Laporte and Mittaz: PASTE, CUT, SWITCH, and POSTOPT. We then show how these can be combined into constructive and improvement heuristics for the undirected CARP, and integrated with two highly efficient local search heuristics: a tabu search algorithm called CARPET and a variable neighbourhood search algorithm.

Procedure PASTE

Given a CARP solution made up of m vehicle routes, PASTE merges all routes into a single RPP tour, possibly infeasible for the CARP. Procedure PASTE is illustrated in Figure 6.7.

Procedure CUT

This procedure is the reverse of PASTE. Given an RPP solution, CUT transforms it into a feasible CARP solution. Starting at the depot v_0, it determines a vertex v incident to a serviced edge on the RPP tour such that the total weight of the first chain (from v_0 to v) does not exceed Q and the weight of the second chain (from v to v_0) does not exceed $Q(\lceil d/Q \rceil - 1)$, where d is the total weight

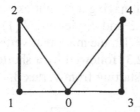

a) Initial solution consisting of two routes

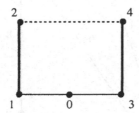

b) Final solution consisting of one route

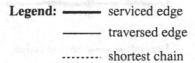

Legend: ———— serviced edge

———— traversed edge

-------- shortest chain

Figure 6.7. Illustration of procedure PASTE

of the RPP solution. When several choices are possible for v, the selection is made so as to minimize $L(v)$, the sum of the shortest chain lengths between v and v_0, and between v_0 and v', the first vertex of the first required edge after v on the RPP tour. Vertex v is then linked to v_0 by means of a shortest chain and some of the edges of this shortest chain may be serviced as long as the vehicle capacity is not exceeded. Vertex v_0 is then linked to v' by means of a shortest chain, and the same procedure is recursively applied to the remaining part of the tour until a feasible CARP solution has been reached. Procedure CUT is illustrated in the example depicted in Figure 6.8, where the depot is located at vertex 0. The numbers in square boxes are edge weights (for required edges only). The remaining numbers on the dashed lines or on the edges are shortest chain lengths. In this example, $Q = 11$ and $d = 24$. The procedure first computes $Q(\lceil d/Q \rceil - 1) = 22$, which means the first route must include edge $(0,1)$ (then the remaining weight is 20, which does not exceed 22), and can include up to three required edges without having a weight exceeding Q. Three choices are therefore possible for v: 1, 3 and 4, yielding $L(1) = 3 + 6 = 9$, $L(3) = 4 + 4 = 8$ and $L(4) = 6 + 6 = 12$, meaning that vertex 3 is selected. The first route includes the chain (0,1,2,3) followed by a shortest chain to the depot. The procedure is then reapplied starting from vertex 3.

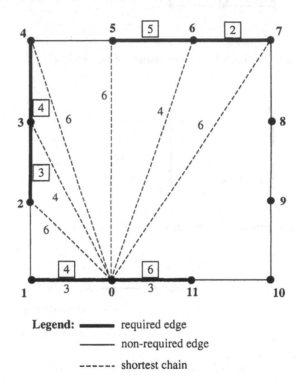

Figure 6.8. Illustration of procedure CUT

Procedure SWITCH

Given a route in which a vertex v appears several times, procedure SWITCH reverses all minimal subroutes starting and ending at v as long as these do not include the depot. The role of this procedure is to create a better mix of solutions within a search procedure. For example, if PASTE and CUT were repeatedly applied to a given CARP solution, then this solution may repeat itself indefinitely. The chances of this happening are reduced if SWITCH is periodically applied.

Procedure POSTOPT

Procedure POSTOPT can be applied to postoptimize any CARP solution. It successively calls PASTE, SWITCH, CUT and it then applies SHORTEN to each route until a termination criterion has been met, such as a given number of iterations.

4.4.2 Constructive heuristic. A feasible CARP solution can be obtained as follows. First generate an RPP solution by means of Frederickson's algorithm, and then a CARP solution by means of CUT. Improvements to the feasible solution can be obtained by applying POSTOPT.

4.4.3 Tabu search heuristic. A more elaborate way to postoptimize a CARP solution is to use tabu search. Hertz, Laporte and Mittaz (2000) have developed such a heuristic, called CARPET. It works with two objective functions: $F(x)$, the solution cost of solution x, and a penalized objective $F'(x) = F(x) + \beta E(x)$, where β is a positive parameter and $E(x)$ is the total excess weight of all routes in a possibly infeasible solution x. As in the tabu search heuristic TABUROUTE developed for the *Vehicle Routing Problem* (Gendreau, Hertz and Laporte, 1994), parameter β is self-adjusting. It is initially equal to 1. Every λ iterations, it is either halved if all previous λ solution were feasible, or doubled if they were all infeasible. The value of λ is equal to 5 in CARPET. Starting from an initial solution (see Section 6.4.4.2), CARPET performs a search over neighbour solutions, by moving at each iteration from the current solution to its best non-tabu neighbour, even if this causes a deterioration in the objective function. To define the neighbourhood $N(x)$ of solution x, consider a route S of x, an edge (v_j, v_k) serviced in S, and another route S' containing only the depot or an edge of R with an extremity distant from v_j or v_k by at most α, where α is a parameter. A neighbour of x is then obtained by dropping (v_j, v_k) from S adding it to S', using procedures DROP and ADD described in Section 6.4.1. Whenever an edge is removed from route S at a given iteration, it cannot be reintroduced into this route for θ iterations, where θ is randomly selected in [5,10]. The algorithm terminates after a set number of iterations without improvement in F or F', or after a preset time, or whenever a solution having the same value as that of a known lower bound is encountered.

4.4.4 Variable neighbourhood search heuristic. Variable neighbourhood search (VNS) is a technique put forward by Mladenović and Hansen (1997) which can be embedded within a local search algorithm such as tabu search. The basic idea is to define several neighbourhoods to perform the search, thus reducing the risk of becoming trapped in a local optimum. The VNS algorithm developed by Hertz and Mittaz (2001) for the CARP uses the same framework as CARPET, except that feasibility is preserved at all times (only $F(x)$ is used). A lower bound $\underline{m} = \left\lceil \sum_{(v_i,v_j) \in R} q_{ij}/Q \right\rceil$ on the number of routes in a feasible solution is first computed, and given a solution x and a value of k in $[2, \underline{m}]$, a neighbourhood $N_k(x)$ is defined by applying a procedure similar to POSTOPT to k of the routes. This procedure calls PASTE, SHORTEN, SWITCH and CUT. Procedure SHORTEN is then applied to each route. Since there are usually several ways to select k routes from a CARP solution and also several ways to choose the vertex v in order to reverse subroutes, only a limited number of choices are considered. Also, another neighbourhood $N_1(x)$ is obtained as in CARPET, but by forcing the solution to remain feasible at all times. For $k = 1, \ldots, \underline{m}$, the neighbourhoods $N_k(x)$ grow from fine to coarse in the sense that a small value of k will tend to yield a neighbour solution close to x, while a large value of k will produce a widely different solution. The idea is to use the various neighbourhoods in succession. Hertz and Mittaz tested three strategies: 1) from $\underline{m}$ to 1, 2) from 1 to $\underline{m}$; 3) in random order. The first strategy turned out to be the best.

4.4.5   Computational results.   CARPET was run on a wide variety of test problems. On a first set of small scale instances ($7 \leq |V| \leq 27; 11 \leq |E| \leq 55; R = E$) used by de Armon (1981), CARPET outperforms all previous heuristics mentioned at the beginning of Section 6.4, even if run for only one second on a Silicon Indigo 1 machine (195 MHz, IP 28 processor). With this running time, the average deviation to a lower bound is 3.4%. This can be compared with previous heuristics which produced average deviations ranging from 3.9% to 13.9%. When no time limit is imposed, CARPET achieves solution values exceeding that of the lower bound by an average of 0.17%, in times ranging from 0 to 330.6 seconds. CARPET was also applied to 25 instances proposed by Benavent (1997) with $24 \leq |V| \leq 50$ and $34 \leq |E| \leq 97$ where it produced solutions having an average deviation of 1.1% from a lower bound. Finally, CARPET was tested on 270 larger and denser instances containing 20, 40 or 60 vertices, with edge densities in [0.1, 0.3], [0.4, 0.6] or [0.7, 0.9] and $|R|/|E|$ in [0.1, 0.3], [0.4, 0.6] and [0.8, 1.0]. Average deviations from a lower bound were then equal to 5.5% when CARPET was run for up to 60 seconds and to 1.6% when it was run for up to 600 seconds. VNS produced results similar to those of CARPET on the de Armon instances (the deviation from optimality was identical and computation times were about 20% shorter), and it yielded much better results on the larger randomly generated instances. When no time limits were imposed, the average deviation from optimality and computing times on the 270 instances were 0.71% and 349.81 seconds for CARPET, and 0.54% and 42.50 seconds for VNS. In other words, on larger instances VNS is better than CARPET in terms of solution quality, and also faster.

5. Conclusions

Transportation is without any doubt one of the fields that most benefit from the application of operations research models and methods. In this chapter we have outlined some of the most significant results obtained for two fundamental problem classes: the vehicle routing problem and arc routing problems. Over the past years, significant progress has been achieved in the development of both exact and approximate algorithms. In the first case, branch-and-cut techniques similar to those that have been successfully applied to the *Traveling Salesman Problem* (Jünger, Reinelt and Rinaldi, 1995) have been developed for the *Vehicle Routing Problem* and for several difficult families of arc routing problems. In the area of heuristics, several powerful tools such as tabu search and variable neighbourhood search are now available. The solutions obtained by these techniques are generally of very high quality as measured by comparisons with lower bounds or with best known solutions.

Acknowledgments

This work was partly supported by the Canadian Natural Sciences and Engineering Research Council under grants 227837–00 and OGP0039682, and by the Quebec government FCAR research program under grant 2002–GR–73080. This support is gratefully acknowledged. Thanks are due to a referee for valuable comments.

References

[Agarwal, Mathur and Salkin] Y. Agarwal, K. Mathur and H.M. Salkin. A Set-Partitioning-Based Exact Algorithm for the Vehicle Routing Problem. *Networks*, 19:731–749, 1989.

[Altinkemer and Gavish] K. Altinkemer and B. Gavish. Parallel Savings Based Heuristic for the Delivery Problem. *Operations Research*, 39:456–469, 1991.

[Augerat] P. Augerat. Approche polyédrale du problème de tournées de véhicules. Institut National Polytechnique de Grenoble, France, 1995.

[Augerat et al.] P. Augerat, J.M. Belengner, E. Benavent, A. Corberán and D. Naddef. Separating Capacity Inequalities in the CVRP Using Tabu Search. *European Journal of Operational Research*, 106:546–557, 1999.

[Assad and Golden] A.A. Assad and B.L. Golden. Arc Routing Methods and applications. In: M.O. Ball, T.L. Magnanti, C.L. Monma and G.L. Nemhauser (eds), *Network Routing*, Handbooks in Operations Research and Management Science, North-Holland, Amsterdam, 8:375–483, 1995.

[Baldacci, Mingozzi and Hadjiconstantinou] R. Baldacci, A. Mingozzi and E.A Hadjiconstantinou. An Exact Algorithm for the Capacitated Vehicle Routing Problem Based on a Two-Commodity Network Flow Formulation. *Operations Research*, 2004. Forthcoming.

[Barahona and Grötschel] F. Barahona and M. Grötschel. On the Cycle Polytope of a Binary Matroid. *Journal of Combinatorial Theory*, 40:40–62, 1986.

[Barnhart and Crainic] C. Barnhart and T.G. Crainic (eds). Focused Issue on Freight Transportation. *Transportation Science*, 32:87–203, 1998.

[Barnhart, Desrosiers and Solomon] C. Barnhart, J. Desrosiers and M.M. Solomon (eds). Focused Issue on Airline Optimization. *Transportation Science*, 32:205–301, 1998a.

[Barnhart, Desrosiers and Solomon] C. Barnhart, J. Desrosiers and M.M. Solomon (eds). Focused Issue on Rail Optimization. *Transportation Science*, 32:303–404, 1998b.

[Beltrami and Bodin] E.L. Beltrami and L.D. Bodin. Networks and Vehicle Routing for Municipal Waste Pollution. *Networks*, 4:65–94, 1974.

[Benavent] E. Benavent. ftp://indurain.estadi.uv.es/pub/CARP, 1997.

[Blais and Laporte] M. Blais and G. Laporte. Exact Solution of the Generalized Routing Problem through Graph Transformations. *Journal of the Operational Research Society*, 54:906–910, 2003.

[Bramel and Simchi-Levi] J. Bramel and D. Simchi-Levi. A Location Based Heuristic for General Routing Problems. *Operations Research*, 43:649–660, 1995.

[Bramel and Simchi-Levi] J. Bramel and D. Simchi-Levi. Set-Covering-Based Algorithms for the Capacitated VRP. In: P. Toth and D. Vigo (eds), *The Vehicle Routing Problem*, SIAM Monographs on Discrete Mathematics and Applications, Philadelphia, 85–108, 2002.

[Bullnheimer, Hartl and Strauss] B. Bullnheimer, R.F. Hartl and C. Strauss. An Improved Ant System for the Vehicle Routing Problem. *Annals of Operations Research*, 89:319–328, 1999.

[Bullnheimer, Hartl and Strauss] B. Bullnheimer, R.F. Hartl and C. Strauss. Applying the Ant System to the Vehicle Routing Problem. In: S. Voss, S. Martello, I.H. Osman and C. Roucairol (eds), *Meta-Heuristics: Advances and Trends in Local Search Paradigms for Optimization*, Kluwer, Boston, 109–120, 1998.

[Button] K.J. Button. *Transport Economics*. Edward Elgar, Aldershot, United Kingdom, 1993.

[Christofides] N. Christofides. The Optimum Traversal of a Graph. *Omega*, 1:719–732, 1973.

[Christofides] N. Christofides. The Vehicle Routing Problem. *RAIRO (recherche opérationnelle)*, 10:55–70, 1976a.

[Christofides] N. Christofides. Worst-Case Analysis of a New Heuristic for the Traveling Salesman Problem. Report No 388, Graduate School of Industrial Administration, Carnegie Mellon University, Pittsburgh, 1976b.

[Christofides et al.] N. Christofides, V. Campos, A. Corberán and E. Mota. An Algorithm for the Rural Postman Problem. Report IC.O.R. 81.5, Imperial College, London, 1981.

[Christofides, Mingozzi and Toth] N. Christofides, A. Mingozzi and P. Toth. The Vehicle Routing Problem. In: N. Christofides, A. Mingozzi and P. Toth (eds), *Combinatorial Optimization*, Wiley, Chichester, 315–338, 1979.

[Christofides, Mingozzi and Toth] N. Christofides, A. Mingozzi and P. Toth. Exact Algorithms for the Vehicle Routing Problem, Based on Spanning Tree Shortest Path Relaxation. *Mathematical Programming*, 20:255–282, 1981a.

[Christofides, Mingozzi and Toth] N. Christofides, A. Mingozzi and P. Toth. State Space Relaxation Procedures for the Computation of Bounds to Routing Problems. *Networks*, 11:145–164, 1981b.

[Clarke and Wright] G. Clarke and J.W. Wright. Scheduling of Vehicles from a Central Depot to a Number of Delivery Points. *Operations Research*, 12:568–581, 1964.

[Clochard and Naddef] J.-M. Clochard and D. Naddef. Use of Path Inequalities for TSP. In: G. Rinaldi and L.A. Wolsey (eds), *Proceedings of the Third Workshop on Integer Programming and Combinatorial Optimization*. CORE, Université de Louvain-la-Neuve, Belgium, 1993.

[Colorni, Dorigo and Maniezzo] A. Colorni, M. Dorigo and V. Maniezzo. Distributed Optimization by Ant Colonies. In: F. Varela and P. Bourgine (eds), *Proceedings of the European Conference on Artificial Life*, Elsevier, Amsterdam, 1991.

[Corberán and Sanchis] A. Corberán and J.M. Sanchis. A Polyhedral Approach to the Rural Postman Problem. *European Journal of Operational Research*, 79:95–114, 1994.

[Cordeau et al.] J.-F. Cordeau, G. Desaulniers, J. Desrosiers, M. Solomon and F. Soumis. VRP with Time Windows. In: P. Toth and D. Vigo (eds), *The Vehicle Routing Problem*, SIAM Monographs on Discrete Mathematics and Applications, Philadelphia, 157–194, 2002.

[Cordeau, Gendreau and Laporte] J.-F. Cordeau, M. Gendreau and G. Laporte. A Tabu Search Heuristic for the Periodic and Multi-Depot Vehicle Routing Problems. *Networks*, 30:105–119, 1997.

[Cordeau and Laporte] J.-F. Cordeau and G. Laporte. A Tabu Search Heuristic for the Site Dependent Vehicle Routing Problem with Time Windows. *INFOR*, 39:292–298, 2001.

[Cordeau and Laporte] J.-F. Cordeau and G. Laporte. Tabu Search Heuristics for the Vehicle Routing Problem. *Cahiers du GERAD G-2002-15*, École des Hautes Études Commerciales, Montreal, 2002.

[Cordeau, Laporte and Mercier] J.-F Cordeau, G. Laporte and A. Mercier. A Unified Tabu Search Heuristic for Vehicle Routing Problems with Time Windows. *Journal of the Operational Research Society*, 52:928–936, 2001.

[Cordeau et al.] J.-F. Cordeau, M. Gendreau, G. Laporte, J.-Y. Potvin and F. Semet. A Guide to Vehicle Routing Heuristics. *Journal of the Operational Research Society*, 53:512–522, 2002.

[Cornuéjols and Harche] G. Cornuéjols and F. Harche. Polyhedral Study of the Capacitated Vehicle Routing Problem. *Mathematical Programming*, 60:21–52, 1993.

[Crainic and Sebastian] T.G. Crainic and H.-J. Sebastian (eds). Focused Issue on Freight Transportation. *Transportation Science*, 36:1–143, 2002.

[Croes] G.A. Croes. A Method for Solving Traveling - Salesman Problems. *Operations Research*, 6:791-812, 1981.

[de Armon] J.S. de Armon. A Comparison of Heuristics for the Capacitated Chinese Postman Problem. Master's Thesis, University of Maryland, College Park, MD, 1991.

[Desrochers and Verhoog] M. Desrochers and T.W. Verhoog. A Matching Based Savings Algorithm for the Vehicle Routing Problem. *Cahiers du GERAD G-89-04*, École des Hautes Études Commerciales, Montreal, 1989.

[Dror] M. Dror (ed.). *Arc Routing: Theory, Solutions and Applications*, Kluwer, Boston, 2000.

[Dror, Laporte and Trudeau] M. Dror, G. Laporte and P. Trudeau. Vehicle Routing with Split Deliveries. *Discrete Applied Mathematics*, 50:239–254, 1994.

[Dueck] G. Dueck. New Optimization Heuristics: The Great Deluge Algorithm and the Record-to-Record Travel. *Journal of Computational Physics*, 104:86–92, 1993.

[Dueck and Scheuer] G. Dueck and T. Scheuer. Threshold Accepting: A General Purpose Optimization Algorithm. *Journal of Computational Physics*, 90:161–175, 1990.

[Durbin and Willshaw] R. Durbin and D. Willshaw. An Analogue Approach to the Traveling Salesman Problem Using and Elastic Net Method. *Nature*, 326:689–691, 1987.

[Edmonds and Johnson] J. Edmonds and E.L. Johnson. Matching, Euler Tours and the Chinese Postman Problem. *Mathematical Programming*, 5:88–124, 1973.

[Eiselt, Gendreau and Laporte] H.A Eiselt, M. Gendreau and G Laporte. Arc Routing Problems, Part I: The Chinese Postman Problem. *Operations Research*, 43:231–242, 1995a.

[Eiselt, Gendreau and Laporte] H.A Eiselt, M. Gendreau and G Laporte. Arc Routing Problems, Part II: The Rural Postman Problem. *Operations Research*, 43:399–414, 1995b.

[Eiselt and Laporte] H.A Eiselt and G Laporte. A Historical Perspective on Arc Routing. In: M. Dror (ed.), *Arc Routing: Theory, Solutions and Applications*, Kluwer, Boston, 1–16, 2000.

[Euler] L. Euler. Solutio Problematis ad Geometriam Situ Pertinentis. *Commentarii Academiae Scientarum Petropolitanae*, 8:128–140, 1736.

[Fisher] M.L. Fisher. Optimal Solution of Vehicle Routing Problems Using Minimum k-Trees. *Operations Research*, 42:626–642, 1995.

[Fisher and Jaikumar] M.L. Fisher and R. Jaikumar. Generalized Assignment Heuristic for the Vehicle Routing Problem. *Networks*, 11:109–124, 1981.

[Fleischner] H. Fleischner. *Eulerian Graphs and Related Topics* (Part 1, Vol 1), *Annals of Discrete Mathematics*, North-Holland, Amsterdam, 1990.

[Fleischner] H. Fleischner. *Eulerian graphs and Related Topics* (Part 1, vol 2), *Annals of Discrete Mathematics*, North-Holland, Amsterdam, 1991.

[Frederickson] G.N. Frederickson. Approximation Algorithms for Some Routing Problems. *Journal of the Association for Computing Machinery*, 26:538–554, 1979.

[Ford and Fulkerson] L.R. Ford and D.R. Fulkerson. *Flows in Networks*. Princeton University Press, Princeton, NJ, 1962.

[Foster and Ryan] B.A. Foster and D.M. Ryan. An Integer Programming Approach to the Vehicle Scheduling Problem. *Operations Research*, 27:367–384, 1976.

[Garey and Johnson] M.R. Garey and D.S. Johnson. *Computers and Intractability. A Guide to the Theory of NP-Completeness*. Freeman, New York, 1979.

[Gavish and Graves] B. Gavish and S. Graves. Scheduling and Routing on Transportation and Distribution Systems: Formulations and New Relaxations. Working Paper, graduate School of Management, University of Rochester, NY., 1982.

[Gendreau, Hertz and Laporte] M. Gendreau, A. Hertz and G. Laporte. A Tabu Search Heuristic for the Vehicle Routing Problem. *Management Science*, 40:1276–1290, 1994.

[Gendreau, Hertz and Laporte] M. Gendreau, A. Hertz and G. Laporte. New Insertion and Postoptimization Procedures for the Traveling Salesman Problem. *Operations Research*, 40:1086–1094, 1992.

[Gendreau et al.] M. Gendreau, G. Laporte, C. Musaraganyi and É.D. Taillard. A Tabu Search Heuristic for the Heterogeneous Fleet Vehicle Routing Problem. *Computers & Operations Research*, 26:1153–1173, 1999.

[Gendreau, Laporte and Potvin] M. Gendreau, G. Laporte and Y.-Y. Potvin. Metaheuristics for the Capacitated VRP. In: P. Toth and D. Vigo (eds), *The Vehicle Routing Problem, SIAM Monographs on Discrete Mathematics and Applications*, Philadelphia, 129–154, 2002.

[Ghaziri] H. Ghaziri. Algorithmes connexionistes pour l'optimisation combinatoire. Thèse de doctorat, École Polytechnique Fédérale de Lausanne, Switzerland, 1993.

[Ghaziri] H. Ghaziri. Solving Routing Problems by a Self-Organizing Map. In: T. Kohonen, K. Makisara, O. Simula and J. Kangas (eds), *Artificial Neural Networks*, North-Holland, Amsterdam, 1991.

[Ghiani and Laporte] G. Ghiani and G. Laporte. A Branch-and-Cut Algorithm for the Undirected Rural Postman Problem. *Mathematical Programming*, Series A 87:467–481, 2000.

[Gillett and Miller] B.E. Gillett and L.R. Miller. A Heuristic Algorithm for the Vehicle Dispatch Problem. *Operations Research*, 22:340–349, 1974.

[Glover and Laguna] F. Glover and M. Laguna. *Tabu Search*, Kluwer, Boston, 1997.

[Golden, Assad and Wasil] B.L. Golden, A.A. Assad and E.A. Wasil. Routing Vehicles in the Real World: Applications in the Solid Waste, Beverage, Food, Dairy, and Newspaper Industries. In: P. Toth and D. Vigo (eds), *The Vehicle Routing Problem*, SIAM Monographs on Discrete Mathematics and Applications, Philadelphia, 245–286, 2002.

[Golden, de Armon and Baker] B.L. Golden, J.S. de Armon and F.K. Baker. Computational Experiments with Algorithms for a Class of Routing Problems. *Computers & Operations Research*, 10:47–59, 1983.

[Golden, Magnanti and Nguyen] B.L. Golden, T.L. Magnanti and H.G. Nguyen. Implementing Vehicle Routing Algorithms. *Networks*, 7:113–148, 1977.

[Golden et al.] B.L. Golden, E.A. Wasil, J.P. Kelly and I-M. Chao. Metaheuristics in Vehicle Routing. In: T.G. Crainic and G. Laporte (eds), *Fleet Management and Logistics*, Kluwer, Boston, 35–56, 1998.

[Golden and Wong] B.L. Golden and R.T. Wong. Capacitated Arc Routing Problems. *Networks*, 11:305–315, 1981.

[Greistorfer] P. Greistorfer. Computational Experiments with Heuristics for a Capacitated Arc Routing Problem. Working Paper 32, Department of Business, University of Graz, Austria, 1994.

[Grötschel and Win] M. Grötschel and Z. Win. A Cutting Plane Algorithm for the Windy Postman Problem. *Mathematical Programming*, 55:339-338, 1992.

[Guan] M. Guan. Graphic Programming Using Odd and Even Points. *Chinese Mathematics*, 1:273-277, 1962.

[Hadjiconstantinou, Christofides and Mingozzi] E.A. Hadjiconstantinou, N. Christofides and A. Mingozzi. A New Exact Algorithm for the Vehicle Routing Problem Based on q-Paths and k-Shortest Paths Relaxations. In: M. Gendreau and G. Laporte (eds), *Freight Transportation, Annals of Operations Research* Baltzer, Amsterdam, 61:21–43, 1995.

[Harre et al.] C.A. Harre, C. Barnhart, E.L. Johnson, R.E. Marston, G.L. Nemhauser and G. Sigismondi. The Fleet Assignment Problem: Solving a Large-Scale Integer Program. *Mathematical Programming*, 70:211–232.

[Hertz, Laporte and Mittaz] A. Hertz, G. Laporte and M. Mittaz. A Tabu Search Heuristic for the Capacitated Arc Routing Problem. *Operations Research*, 48:129–135, 2000.

[Hertz, Laporte and Nanchen-Hugo] A. Hertz, G. Laporte and P. Nanchen-Hugo. Improvement Procedures for the Undireded Rural Postman Problem. *INFORMS Journal on Computing*, 11:53–62, 1999.

[Hertz and Mittaz] A. Hertz and M. Mittaz. A Variable Neighbourhood Descent Algorithm for the Undirected Capacitated Arc Routing Problem. *Transportation Science*, 35:452–434, 2001.

[Hierholzer] C. Hierholzer. Uber die Möglichkeit, einen Linienzug ohne Wiederholung und ohne Unterbrechnung zu umfahren. *Mathematische Annalen*, VI:30–32, 1873.

[Hopfield and Tank] J.J. Hopfield and D.W. Tank. Neural Computation of Decisions in Optimization Problems. *Biological Cybernetics*, 52:141–152, 1985.

[Jünger, Reinelt and Rinaldi] M. Jünger, G. Reinelt and G. Rinaldi. The Traveling Salesman Problem. In: M.O. Ball, T.L. Magnanti, C.L. Monma and G.L. Nemhauser (eds), *Network Models*, Handbooks in Operations Research and Management Science, North-Holland, Amsterdam, 7:225–330, 1995.

[Kawamura et al.] H. Kawamura, M. Yamamoto, T. Mitamura, K. Suzuki and A. Ohuchi. Cooperative Search on Pheromone Communication for Vehicle Routing Problems. *IEEE Transactions on Fundamentals*, E81-A:1089–1096, 1998.

[Kohonen] T. Kohonen. *Self-Organizing and Associative Memory*, Springer, Berlin, 1988.

[Laporte and Louveaux] G. Laporte and F.V. Louveaux. Solving Stochastic Routing Problems. In: T.G. Crainic and G. Laporte (eds), *Fleet Management and Logistics*, Kluwer, Boston, 159–167, 1998.

[Laporte and Nobert] G. Laporte and Y. Nobert. Comb Inequalities for the Vehicle Routing Problem. *Methods of Operations Research*, 51:271–276, 1984.

[Laporte, Nobert and Desrochers] G. Laporte, Y. Nobert and M. Desrochers. Optimal Routing under Capacity and Distance Restrictions. *Operations Research*, 33:1050–1073, 1985.

[Laporte and Semet] G. Laporte and F. Semet. Classical Heuristics for the Capacitated VRP. In: P. Toth and D. Vigo (eds), *The Vehicle Routing Problem*, SIAM Monographs on Discrete Mathematics and Applications, Philadelphia, 109–128, 2002.

[Lenstra and Rinnooy Kan] J.K. Lenstra and A.H.G. Rinnooy Kan. On General Routing Problems. *Networks*, 6:273–280, 1976.

[Letchford] A.N. Letchford. Polyhedral Results for Some Constrained Arc-Routing Problems. Ph.D. Thesis, Department of Management Science, Lancaster University, United Kingdom, 1996.

[Letchford, Eglese and Lysgaard] A.N. Letchford, R.W. Eglese and J. Lysgaard. Multistars, Partial Multistars and the Capacitated Vehicle Routing Problem. *Mathematical Programming*, 94:21–40, 2002.

[Lin] S. Lin. Computer Solutions of the Traveling Salesman Problem. *Bell System Technical Journal*, 44:2245–2269, 1965.

[Lysgaard, Letchford and Eglese] J. Lysgaard, A.N. Letchford and R.W. Eglese. A New Branch-and-Cut Algorithm for Capacitated Vehicle Routing Problems. *Mahtematical Programming*, 2003. Forthcoming.

[Martello and Toth] S. Martello and P. Toth. *Knapsack Problems*, Wiley, Chichester, 1990.

[McGill and van Ryzin] J.I. McGill and G.J. van Ryzin (eds). Focused Issue on Yield Management in Transportation. *Transportation Science*, 33:135–256, 1999.

[Mitrović-Minić, Krishnamurti and Laporte] S. Mitrović-Minić, R. Krishnamurti and G. Laporte. Double-Horizon Based Heuristics for the Dynamic

Pickup and Delivery Problem with Time Windows. *Transportation Research B*, 2004. Forthcoming.

[Mladenović and Hansen] N. Mladenović and P. Hansen. Variable Neighborhood Search. *Computers & Operations Research*, 34:1097–1100, 1997.

[Naddef and Rinaldi] D. Naddef and G. Rinaldi. The Graphical Relaxation: A New Framework for the Symmetric Traveling Salesman Polytope. *Mathematical Programming*, 58:53–88, 1993.

[Naddef and Rinaldi] D. Naddef and G. Rinaldi. Branch-and-Cut Algorithms for the Capacitated VRP. In: P. Toth and D. Vigo (eds), *The Vehicle Routing Problem*, SIAM Monographs on Discrete Mathematics and Applications, Philadelphia, 53–84, 2002.

[Nobert and Picard] Y. Nobert and J.-C. Picard. An Optimal Algorithm for the Mixed Chinese Postman Problem. *Networks*, 27:95–108, 1996.

[Nelson et al.] M.D. Nelson, K.E. Nygard, J.H. Griffin and W.E. Shreve. Implementation Techniques for the Vehicle Routing Problem. *Computers & Operations Research*, 12:273–283, 1985.

[Orloff] C.S. Orloff. A Fundamental Problem in Vehicle Routing. *Networks*, 4:35–64, 1974.

[Osman] I.H. Osman. Metastrategy Simulated Annealing and Tabu Search Algorithms for the Vehicle Routing Problem. *Annals of Operations Research*, 41:421–451, 1993.

[Owoc and Sargious] M. Owoc and M.A. Sargious. The Role of Transportation in Free Trade Competition. In: N. Waters (ed.), *Canadian Transportation: Competing in a Global Context*, Banff, Alberta, Canada, 23–32, 1992.

[Paessens] H. Paessens. The Savings Algorithm for the Vehicle Routing Problem. *European Journal of Operational Research*, 34:336–344, 1988.

[Pearn] W.-L. Pearn. Approximate Solutions for the Capacitated Arc Routing Problem. *Computers & Operations Research*, 16:589–600, 1989.

[Prins] C. Prins. A Simple and Effective Evolutionary Algorithm for the Vehicle Routing Problem. *Computers & Operations Research*, 2004. Forthcoming.

[Psaraftis] H.N. Psaraftis (ed.). Focused Issue on Maritime Transportation. *Transportation Science*, 33:3–123, 1999.

[Ralphs] T.K. Ralphs. Parallel Branch and Cut for Vehicle Routing. Ph.D. Thesis, Cornell University, Ithaca, NY, 1995.

[Rego] C. Rego. A Subpath Ejection Method for the Vehicle Routing Problem. *Management Science*, 44:1447–1459, 1998.

[Rego and Roucairol] C. Rego and C. Roucairol. A Parallel Tabu Search Algorithm Using Ejection Chains for the Vehicle Routing Problem. In: I.H. Osman and J.P. Kelly (eds), *Meta-Heuristics: Theory and Applications*, Kluwer, Boston, 661-675, 1996.

[Renaud, Boctor and Laporte] J. Renaud, F.F. Boctor and G. Laporte. An Improved Petal Heuristic for the Vehicle Routing Problem. *Journal of the Operational Research Society*, 47:329–336, 1996.

[Rochat and Taillard] Y. Rochat and É.D. Taillard. Probabilistic Diversification and Intensification in Local Search for Vehicle Routing. *Journal of Heuristics* 1:147–167, 1995.

[Ryan, Hjorring and Glover] D.M. Ryan, C. Hjorring and F. Glover. Extensions of the Petal Method for Vehicle Routing. newblock *Journal of the Operational Research Society*, 47:329–336, 1993.

[Schmitt] L.J. Schmitt. An Empirical Computational Study of Genetic Algorithms to Solve Order Based Problems: An Emphasis on TSP and VRPTC. Ph.D. Dissertation, Fogelman College of Business and Economics, University of Memphis, TN, 1994.

[Schmitt] L.J. Schmitt. An Evaluation of a Genetic Algorithm Approach to the Vehicle Routing Problem. Working Paper, Department of Information Technology Management, Christian Brothers University, Memphis, TN, 1995.

[Shen et al.] Y. Shen, J.-Y. Potvin, J.-M. Rousseau and S. Roy. A computer Assistant for Vehicle Dispatching with Learning Capabilities. In: M. Gendreau and G. Laporte (eds), *Freight Transportation, Annals of Operations Research*, 61:189–211, 1995.

[Solomon] M.M. Solomon. Algorithms for the Vehicle Routing and Scheduling Problems with Time Window Contraints. *Operations Research*, 35:254–265, 1987.

[Taillard] É.D. Taillard. Parallel Iterative Search Methods for Vehicle Routing Problems. *Networks*, 23:661–6673, 1993.

[Toth and Vigo] P. Toth and D Vigo. Exact Solution of the Vehicle Routing Problem. In: T.G. Crainic and G. Laporte (eds), *Fleet Management and Logistics*, Kluwer, Boston, 1-31, 1998a.

[Toth and Vigo] P. Toth and D Vigo. The Granular Tabu Search and its Application to the Vehicle-Routing Problem. *INFORMS Journal on Computing*, 2003. Forhtcoming.

[Toth and Vigo] P. Toth and D Vigo (eds). *The Vehicle Routing Problem*, SIAM Monographs on Discrete Mathematics and Applications, Philadelphia, 2002a.

[Toth and Vigo] P. Toth and D Vigo. Branch-and-Bound Algorithms for the Capacitated VRP. In: P. Toth and D. Vigo (eds), *The Vehicle Routing Problem*, SIAM Monographs on Discrete Mathematics and Applications, Philadelphia, 29–51, 2002b.

[Toth and Vigo] P. Toth and D Vigo. VRP with Backhauls. In: P. Toth and D. Vigo (eds), *The Vehicle Routing Problem*, SIAM Monographs on Discrete Mathematics and Applications, Philadelphia, 195–224, 2002c.

[van Aardenne-Ehsenfest and de Bruijn] T. van Aardenne-Ehsenfest and N.G. de Bruijn. Circuits and Trees in Oriented Linear Graphs. *Simon Stevin*, 28:203–217, 1951.

[Van Breedam] A. Van Breedam. An Analysis of the Effect of Local Improvement Operators in Genetic Algorithms and Simulated Annealing for the Vehicle Routing Problem. RUCA Working Paper 96/14, University of Antwerp, Belgium, 1996.

[Volgenant and Jonker] A. Volgenant and R. Jonker. The Symmetric Traveling Salesman Problem. *European Journal of Operational Research*, 12:394–403, 1983.

[Wark and Holt] P. Wark and J. Holt. A repeated Matching Heuristic for the Vehicle Routing Problem. *Journal of the Operational Research Society*, 45:1156–1167, 1994.

[Wren] A. Wren. *Computers in Transport Planning and Operation*, Ian Allan, London, 1971.

[Wren and Holliday] A. Wren and A. Holliday. Computer Scheduling of Vehicles from One or More Depots to a Number of Delivery Points. *Operational Research Quarterly*, 23:333–344, 1972.

[Yellow] P. Yellow. A Computational Modification to the Savings Method of Vehicle Scheduling. *Operational Research Quarterly*, 21:281–283, 1970.

II

APPLICATIONS

Chapter 7

RADIO RESOURCE MANAGEMENT

in CDMA Networks

Katerina Papadaki[1] and Vasilis Friderikos[2]

[1] *Operational Research Department*
London School of Economics
k.p.papadaki@lse.ac.uk

[2] *Centre for Telecommunications Research*
King's College London
vasilis.friderikos@kcl.ac.uk

Abstract   Over the last few years wireless communications have experienced an unprece-
dented growth providing ubiquitous services irrespective of users' location and
mobility patterns. This chapter reviews different resource management strategies for wireless cellular networks that employ Code Division Multiple Access
(CDMA). The main aim of the formulated optimization problems is to provide
efficient utilization of the scarce radio resources, namely power and transmission
rate. Due to the fact that practical systems allow discrete rate transmissions, the
corresponding optimization problems are discrete in their domains. Additionally, these optimizations problems are also non-linear in their nature. We focus
on power minimization and throughput maximization formulations both for single time period and multiple period models. We provide a new method for solving resource management problems in CDMA networks, formulated as dynamic
programming problems. A linear approximate dynamic programming algorithm
is implemented that estimates the optimal resource allocation policies in real
time. The algorithm is compared with baseline heuristics and it is shown that it
provides a significant improvement in utilizing the radio network resources.

Keywords: wireless cellular systems, radio resource management, power and rate control,
approximate dynamic programming

1. Introduction

Over the last decade the cumulative impact of the success of wireless networks and the Internet sparked a significant research interest on the convergence aspects of these two technologies. Due to the increased available bandwidth, mobile users are becoming increasingly demanding and sophisticated in their expectations from mobile wireless networks. This stimulates the development of advanced resource management tasks for optimal usage of the scarce resources of wireless networks.

In this chapter we will describe two resource allocation problems that arise in Code Division Multiple Access (CDMA) networks. These problems arise in wireless telecommunication systems where serving mobile users is constrained by system capacity. Mobile users need two services, transmitting information to them and from them. The two problems are generally studied separately: Both are within the setting of a *mobile wireless network* that we describe in detail in the following.

1.1 Mobile wireless network

The main concept behind a mobile wireless system is that a geographical area is divided into a number of slightly overlapping circular "cells." Each cell contains a base station, which is responsible in transmitting (receiving) information traffic to (from) multiple mobile users within the cell.

There are two types of transmission that occur within each cell: the *downlink communication*, where the information arrives at the base station for each mobile user and queues to be transmitted to the respective user; and the *uplink communication*, where each mobile user transmits information to the base station.

Currently, we are witnessing a paradigm shift in mobile wireless networks from commodity voice delivery to mobile multimedia packet based transmission. Packet based transmission, which means that information (in bits) is packaged in discrete bundles of information packets, occurs in Internet like applications such as Voice over Internet Protocol (VoIP), Video Streaming, WWW and e-mail to mention just a few. For example, mobile users could be downloading files from the internet (downlink communication) or uploading files to the internet (uplink communication).

There are different access systems for performing the transmission. In this chapter we are concerned with Code Division Multiple Access (CDMA) systems where multiple transmissions can occur simultaneously in each time slot. In Time Division Multiple Access (TDMA) systems the transmissions are scheduled over time since only one transmission can occur per time slot. In Frequency Division Multiple Access (FDMA) systems only transmission within different frequency bands is allowed in each time slot.

Constraints

Under light traffic conditions the issue of packet scheduling from or to different mobile users is trivial as far as there are enough capacity resources to satisfy the requests by the users. The problem arises when there is heavy traffic of information and there are scarce resources for the transmission process. The main resource needed for a device to perform a transmission is power. The power needed to perform a transmission in a time slot depends on the transmission rate (bites per second) chosen for this transmission as well as the transmission rates of other transmissions that occur simultaneously. This is due to the interference that occurs between transmissions that take place in the same time slot. The interference caused by simultaneous transmissions is a drawback for CDMA systems, but nonetheless it is balanced by the higher throughput of information. Given the interference of simultaneous transmissions, managing the power resource becomes challenging since it is a non-separable function of all the transmission rates that occur concurrently.

Further, the power levels needed for a transmission depend on the *channel gain* between the user and the base station; the channel gain is a metric that quantifies the quality of the connection between the user and the base station: *Higher channel gain means better quality of the connection.* It is determined by a stochastic process whose mean is a function of the distance between the user and the base station. As a mobile user moves away from the base station his channel gain decreases in such a manner that the power needed for transmission (without interference from other users) increases as a quadratic function of the distance.

There is another constraint concerning the transmission process. This occurs due to the frequency bandwidth available for the transmission. This limitation results in an upper bound on the aggregate transmission rate from the transmitting device.

In order to efficiently serve the mobile users, *Quality of service* (QoS) requirements are defined and introduced into the problem. Quality of service of mobile users can be measured by the total transmission rate achieved by users, the total delay of information packets waiting in the queue, the average queue length for each user etc. These factors can either be introduced into the objective of the problem as quantities to be maximized/minimized, or they can appear as constraints to guarantee satisfactory levels of quality of service.

1.2 The single cell problems

We describe the two problems that arise in a single cell where transmission rates need to be allocated to users in order to efficiently utilize the resources needed for the transmission process.

In the *downlink problem* information packets arrive over time at the base station for each mobile user and are queued to be transmitted over a discrete time horizon, taking into account power and aggregate rate constraints. In the *uplink problem* information packets are queued at each mobile device to be transmitted to the base station over a discrete time horizon.

In both problems the objective is to perform the transmission efficiently by either maximizing system throughput or minimizing power used while keeping quality of service at satisfactory levels.

These two problems are generally studied separately. Their main difference stems from the origin of resources for the transmission. In the downlink, it is the base station that performs the transmission and thus there is one capacity constraint for each resource, whereas in the uplink it is the mobile devices that perform the transmission and thus there is a capacity constraint for each mobile user for each resource.

1.3   Literature review of CDMA downlink and uplink problems

Because in CDMA systems all mobile users share the same frequency, the interference produced during concurrent transmission is the most significant factor that determines overall system performance and per user call quality. Therefore the most crucial element in radio resource management is power control. As mobile users move within the cell their channel changes and therefore dynamic power control is required to limit the transmitted power while at the same time maintain link quality irrespectively of channel impairments. In that respect, power control has been extensively examined over the last years, and various schemes have been proposed in order to maintain a specific communication link quality threshold for each mobile user [1], [2], [3], [4]. For practical CDMA systems where there is a physical constraint on the maximum transmission power, it has been shown that by minimizing power transmission (i.e. reducing the interference between transmitting users) the maximum number of users can be served [5], [6]. These power control schemes have mainly focused on single service voice oriented communication rather than on multiservice wireless communication, where variable bit-rate packet transmission is also supported (data services). For data services, such as Internet like applications, where there is not a specific data rate requirement, not only power control but also rate control can be exploited to efficiently utilize the radio resources [6], [7]. The optimization of data packet transmission in terms of minimum time span, maximum throughput and minimum power transmission, through joint power and rate control have been studied in [8], [12]. Oh and Wasserman [7] consider a joint power and rate allocation problem. They consider a single cell system and formulate an optimization problem for a single class of mobiles. They show that for the optimal solution, mobiles are selected for transmission according to their channel state and if a mobile is selected, it can transmit at the maximum transmission power.

A textbook dynamic programming (classical Bellman DP [13]) formulation for non-real time transmission on the downlink of a CDMA system has been developed in [12], where the link between throughput maximization and minimum time span has been discussed. A similar DP formulation appeared in [16], where the authors discussed optimal joint transmission mode selection and power control. DP for resource management on the downlink of a CDMA system has also been used in [14]. The authors focused only on throughput maximization without considering queue length evolution and showed that by optimally allocating bit-rates, using a textbook DP algorithm, tangible gains can be achieved. Furthermore, the authors in [15] used classical backward recursion DP to obtain a fair-queueing algorithm for CDMA networks, but with the assumption that all codes are available on the downlink channel.

The above related work has mainly focused on DP formulation rather than implementation aspects for real-time decisions. The computational challenges that arise for the actual solution of these optimization problems have not been meticulously discussed. Textbook DP algorithms (i.e., value iteration or policy iteration) typically require computational time and memory that grow exponentially in the number of users. The inherent computational complexity render such algorithms infeasible for problems of practical scale. In this chapter, we present the work of Papadaki and Friderikos [19] that tackle the intractability of large-scale dynamic programs through approximation methods. A linear approximate DP architecture is used for near-optimal data transmission strategies on the downlink of CDMA systems. The work in this paper is closely related to algorithms used in [18]; other types of approximate DP algorithms have been discussed and developed in [20] and [21].

2. Problem Definition

In this section we define the capacity constraints and explain the relationship between power and transmission rates. We define the parameters and variables and set the notation for the rest of the chapter for both the uplink and the downlink problems.

Uplink Consider a single cell in a CDMA cellular radio system where N active mobile users share the same frequency channel. Each mobile user can transmit to the base station with power p_i that is in the range $0 \leq p_i \leq \bar{p}_i$, where $\bar{p}_i$ is the power capacity for mobile user i. We denote the channel gain or link gain between user i and the base station by g_i: *the higher the value of g_i, the better the quality of the connection between the user and the base* *station.* We define the problem for a short time interval, a *time slot*, such that g_i is stationary.

Given that the mobile users are using power given by the power vector $\mathbf{p} = (p_1, p_2, ..., p_N)$ and their channel gain is given by the channel vector $\mathbf{g} = (g_1, g_2, ..., g_N)$, we define the **S**ignal-to-**I**nterference-**R**atio (SIR), denoted by γ_i, for mobile i, as:

$$\gamma_i(\mathbf{p}) \equiv \frac{g_i p_i}{\sum_{j \neq i} g_j p_j + I + \nu}, \quad 1 \leq i \leq N \tag{7.1}$$

where $I > 0$ is the inter-cell interference noise (i.e. interference that occurs from neighboring cells) and $\nu > 0$ is the power of background and thermal noise at the base station. Both I and ν are considered to be fixed for the current analysis. The realized data rate of a mobile user depends on many factors, however the SIR is an important measure that is directly related to the transmission rate of a mobile. We let $r_i \in \mathcal{R}$ denote the transmission rate for mobile user i, where $\mathcal{R}$ is the set of allowable rates and $\mathbf{r} = (r_1, ..., r_N)$ is the corresponding vector of transmission rates. Then assuming a linear relationship between the SIR (γ_i) and the transmission rate (r_i) we have:

$$\frac{r_i}{\gamma_i} = \frac{W}{\Gamma} \tag{7.2}$$

where W denotes the spreading bandwidth, and $\Gamma = E_b/I_0$ is the bit-energy-to-interference-power-spectral-density ratio for each mobile that we assume to be the same for all mobile users. Γ gives the probability of a bit transmitted erroneously.

Equations (7.1) give γ_i in terms of the power vector $\mathbf{p}$. Given that the rate allocation decisions are based on the SIR's we would like to have the power of each user p_i in terms of the γ_i's. We can rewrite (7.1) as:

$$\gamma_i = \frac{g_i p_i}{\sum_{j=1}^{N} g_j p_j - g_i p_i + I + \nu}$$

which is equivalent to:

$$g_i p_i = \left(\sum_{j=1}^{N} g_j p_j + I + \nu \right) \frac{\gamma_i}{1 + \gamma_i} \tag{7.3}$$

Summing over i and rearranging terms yields:

$$\sum_{j=1}^{N} g_j p_j = \frac{\sum_{j=1}^{N} \frac{\gamma_j (I + \nu)}{1 + \gamma_j}}{1 - \sum_{j=1}^{N} \frac{\gamma_j}{1 + \gamma_j}} \tag{7.4}$$

Substituting (7.4) into (7.3) gives:

$$p_i = \frac{I + \nu}{1 - \sum_{j=1}^{N} \frac{\gamma_j}{1+\gamma_j} g_i (1 + \gamma_i)} \gamma_i \qquad (7.5)$$

The above expression gives the required power to perform the transmission for mobile user i when the transmission targets are set to $\gamma = (\gamma_1, ..., \gamma_N)$. It is easy to show [9] that there exist feasible powers **p** to perform the transmission, given by (7.5), when the following condition holds:

$$\sum_{j=1}^{N} \frac{\gamma_j}{1 + \gamma_j} < 1 \qquad (7.6)$$

Note from (7.5) that the power needed from mobile i to perform a transmission with SIR target γ_i does not solely depend on γ_i but on the SIR targets of all the users i.e. the entire SIR vector γ. This is due to the interference caused in a CDMA system when simultaneous transmissions take place. As can be seen from (7.5), the channel gain g_i of user i plays an important role in the power needed to perform a transmission. A user with a good channel (high g_i) needs less power than a user with a bad channel (low g_i). Further, when a single user transmits in a time slot then to achieve a target SIR γ_i his allocated power needs to be $p_i = \gamma_i \frac{I+\nu}{g_i}$ (see equation (7.1)). However, as can also be seen from (7.1) when other users transmit in the same time slot, user i needs to increase his power considerably to counteract the interference caused by other transmissions, in order to achieve the same γ_i target.

Because of the maximum power limit at each mobile,

$$p_i \leq \bar{p}_i \qquad (7.7)$$

we can rewrite the power capacity constraints given in (7.7) by using (7.5) and (7.6) as follows:

$$\sum_{j=1}^{N} \frac{\gamma_j}{1 + \gamma_j} \leq 1 - \frac{\frac{\gamma_i}{1+\gamma_i}}{\frac{g_i \bar{p}_i}{I+\nu}}, \quad 1 \leq i \leq N \qquad (7.8)$$

or, equivalently, we can express the N power constraints as:

$$\sum_{j=1}^{N} \frac{\gamma_j}{1 + \gamma_j} \leq 1 - \max_{1 \leq j \leq N} \left[\frac{\frac{\gamma_j}{1+\gamma_j}}{\frac{g_j \bar{p}_j}{I+\nu}} \right] \qquad (7.9)$$

We can express the power constraint (7.9) in terms of the transmission rates **r** by using (7.2):

$$\sum_{j=1}^{N} \frac{r_j}{\frac{W}{\Gamma} + r_j} \leq 1 - \max_{1 \leq j \leq N} \left[\frac{\frac{r_j}{\frac{W}{\Gamma}+r_j}}{\frac{g_j \bar{p}_j}{I+\nu}} \right] \qquad (7.10)$$

This defines the uplink uplink problem.

Downlink

In case of the downlink the base station transmits to N mobile users in a single cell. The power used by the base station for the transmission to the mobiles is given by the power vector $\mathbf{p} = (p_1, ..., p_N)$, where p_i is the power used for the transmission to user i. The SIR for the transmission to user i, γ_i, for the downlink, is defined as follows:

$$\gamma_i \equiv \frac{g_i p_i}{\theta \cdot g_i \sum_{j \neq i} p_j + I_i + \nu}, \quad 1 \leq i \leq N \qquad (7.11)$$

Here $I_i > 0$ is the inter-cell interference at the mobile user i, $\nu > 0$ is the lump-sum power of background and thermal noise. The orthogonality factor $0 \leq \theta \leq 1$, is taken to be a constant and can be considered as the proportion of power of transmitting users that contributes to the intracell interference. Typical values for the orthogonality factor are between $[0.1, 0.6]$, see [22] (p.163).

Similar to the uplink, we assume a linear relationship between the SIR and the transmitted rate: $\frac{r_i}{\gamma_i} = \frac{W}{\Gamma}$.

With a predefined assignment of γ_i targets, the power vector that supports every mobile can be found by solving the linear system of equations in (7.11):

$$p_i = \frac{\gamma_i}{1 + \theta \gamma_i} \left(\frac{\sum_{j=1}^{N} \frac{\theta \gamma_j}{1 + \theta \gamma_j} \frac{I_j + \nu}{g_j}}{1 - \sum_{j=1}^{N} \frac{\theta \gamma_j}{1 + \theta \gamma_j}} + \frac{I_i + \nu}{g_i} \right) \qquad (7.12)$$

Again, it can be shown that there exists a feasible power vector $\mathbf{p}$ if the following condition holds:

$$\sum_{j=1}^{N} \frac{\theta \gamma_j}{1 + \theta \gamma_j} < 1 \qquad (7.13)$$

Let us assume that the base station has a maximum power limit P. Then the constraint of the maximum transmission power on the downlink can be formally written as:

$$\sum_{i=1}^{N} p_i \leq P \qquad (7.14)$$

It is easily shown that:

$$\sum_{i}^{N} p_i = \frac{1}{1 - \sum_{i=1}^{N} \frac{\theta \gamma_i}{1 + \theta \gamma_i}} \sum_{i=1}^{N} \frac{\gamma_i}{1 + \theta \gamma_i} \cdot \frac{I_i + \nu}{g_i} \qquad (7.15)$$

Substituting the above in (7.14) and using (7.13) we derive the following power constraint at the base station:

$$\sum_{i=1}^{N} \frac{\gamma_i}{1 + \theta\gamma_i} \left(\frac{I_i + \nu}{g_i} + \theta P \right) \leq P \qquad (7.16)$$

The power constraint (7.16) can also be written with respect to the transmission rates. Substituting $\gamma_i = r_i \frac{\Gamma}{W}$ into (7.16) we get:

$$\sum_{i=1}^{N} \frac{r_i}{\frac{W}{\Gamma} + \theta r_i} \left(\frac{I_i + \nu}{g_i} + \theta P \right) \leq P \qquad (7.17)$$

This concludes the definition of the downlink problem.

3. Myopic Problem Formulations

The problem of radio resource management has various formulations. The main differences occur in the objective, which could be to maximize throughput (packets transmitted per unit time, or transmission rate) or to minimize power used. Kim and Jantti [12] formulate the problem with the objective to minimize the queue length and prove that this problem is equivalent to maximizing throughput. Further, formulations depend on the quality of service requirements considered, for example minimum transmission rate requirements, maximum packet delay requirements, or maximum queue length requirements. The above could appear in the objective function or in the form of constraints. Another important difference is whether the model is for a single time slot (*myopic*) or over a discrete time horizon (*dynamic*). In this section we present myopic formulations for the uplink and briefly describe their extensions to the downlink. A dynamic formulation of the downlink problem is described in section 7.4.

We start with a simple myopic formulation for the uplink communication where the objective is to minimize power subject to a total transmission rate requirement.

3.1 Minimum power transmission - uplink

The following formulation of the radio resource management problem aims to find SIR targets $(\gamma_1, ..., \gamma_N)$ for N mobile users in the uplink communication, that minimize the total power used. We call the following problem Minimize Power uplink (MPU):

$$\min_{\gamma_j} \frac{I + \nu}{1 - \sum_{j=1}^{N} \frac{\gamma_j}{1+\gamma_j}} \sum_{j=1}^{N} \frac{\gamma_j}{g_j (1 + \gamma_j)} \qquad (7.18)$$

$$\text{s.t} \quad \sum_{j=1}^{N} \gamma_j = \bar{\gamma} \tag{7.19}$$

$$\sum_{j=1}^{N} \frac{\gamma_j}{1 + \gamma_j} \leq 1 - \max_{1 \leq j \leq N} \left[\frac{\frac{\gamma_j}{1+\gamma_j}}{\frac{g_j \bar{p}_i}{I + \nu}} \right] \tag{7.20}$$

$$\gamma_j \in \mathcal{L} \tag{7.21}$$

where we let $\mathcal{L}$ be the set of allowable target SIRs. The objective function which is the sum of powers of all users follows from (7.5). Constraint (7.20) is the power constraint of all N mobile users. Given that we are minimizing power we need to guarantee a certain throughput otherwise the solution of problem will be zero (i.e. not to transmit for any users). Constraint (7.19) requires that the total SIR is set at some level $\bar{\gamma}$.

This formulation is used by Berggren and Kim [10] for a discrete set of SIR targets $\mathcal{L}$. They show the following result:

LEMMA 7.1 *Let us assume that $g_1 \geq g_2 \geq \ldots \geq g_N$ and that there exists a feasible vector $\gamma = (\gamma_1, \ldots, \gamma_N)$ that satisfies (7.19)-(7.21). For each feasible vector γ, the objective (7.18) is minimized by reassigning γ_i's such that $\gamma_1 \geq \gamma_2 \geq \ldots \geq \gamma_N$*

Lemma 7.1 becomes intuitively plausible when we observe the structure of the objective function (7.18). The objective is to minimize both the sum in the numerator and the denominator of the objective. In the sum of the numerator the terms $\frac{\gamma_j}{1+\gamma_j}$ have coefficients $\frac{I+\nu}{g_j}$ where as in the sum of the denominator the coefficients are all the same. Thus, apart from their channels g_j the users are indistinguishable in the objective. Given a feasible assignment of SIRs it is always better to reassign the users' SIRs according to the order of their channels. However, the question is what is the optimal SIR vector irrespective of the order of its elements. Berggren and Kim propose the following heuristic, based on the above result, that they call Greedy Rate Packing (GRP):

Greedy Rate Packing
LET: $\gamma_i^* = 0$, for all $i = 1, \ldots, N$.
FOR $i = 1$ TO N DO:

$$\gamma_i^* = \max \left\{ \gamma_i \in \mathcal{L} : \sum_{j=1}^{i} \frac{\gamma_j}{1 + \gamma_j} \leq 1 - \max_{1 \leq j \leq i} \left[\frac{\frac{\gamma_j}{1+\gamma_j}}{\frac{g_j \bar{p}_i}{I + \nu}} \right] \right\} \tag{7.22}$$

Starting with the user with the highest link gain, GRP iteratively allocates the maximum SIR that is feasible with respect to the current power constraints. The total SIR achieved from GRP is $\gamma^* = \sum_{i=1}^{N} \gamma_i^*$. The GRP heuristic ignores the constraint (7.19) of meeting the SIR target $\bar{\gamma}$. However, as long as γ^* is high enough the heuristic provides a good solution. Berggren in [9] shows that out of all feasible SIR assignments that will give total throughput equal to γ^*, the GRP assignment requires the minimum power.

Thus, they show that by replacing the SIR target $\bar{\gamma}$ by γ^* in (7.19):

$$\sum_{j=1}^{N} \gamma_j = \gamma^* \tag{7.23}$$

the GRP heuristic gives the optimal solution to the problem of minimizing (7.18) subject to (7.23), (7.20) and (7.21). If we let $MUP(\bar{\gamma})$ be the set of the SIR vectors that solve the problem (7.18)-(7.21), then the question becomes what is the SIR vector that solves the MUP problem and maximizes $\bar{\gamma}$. We can write this as follows:

$$\max_{\gamma} \sum_{j=1}^{N} \gamma_j \tag{7.24}$$

$$\text{s.t} \quad \gamma \in MUP \left(\sum_{j=1}^{N} \gamma_j \right) \tag{7.25}$$

Berggren and Kim do not provide a solution to the above problem but they do consider a special case where the power capacity of mobile i, $\bar{p}_i$ is large compared to the interference plus noise, $I + \nu$, and thus the power constraints (7.20) are reduced to the following simple form:

$$\sum_{j=1}^{N} \frac{\gamma_j}{1 + \gamma_j} \leq 1 \tag{7.26}$$

They also assume that the transmission allowable rates, and thus the corresponding SIRs, are geometrically related. Suppose the set of allowable rates $\mathcal{R} = \{0, r^{(1)}, \ldots, r^{(K)}\}$ where $r^{(i+1)} = \mu r^{(i)}$, $1 \leq i \leq K$, $\mu > 0$ and $r^{(1)}$. They show that for a rate allocation that maximizes throughput (and thus SIR) subject to (7.26), it is suboptimal to use the rates $r^{(1)}, \ldots, r^{(K)}$ more than $\mu - 1$ times each. Thus when $\mu = 2$, which is often the case in practice, it is suboptimal to use any of the rates more than once. Thus the GRP can be modified to the MGRP to allocate rates (SIRs) to users in a similar manner but to never use the same rate twice. Berggren and Kim using this result show that the MGRP gives maximal throughput subject to (7.26) and the assumption of geometric rates.

A similar heuristic is formulated for the downlink problem. The GRP heuristic is very power efficient but it does not take into account the packet delays in the queues. As we will see in the numerical results of section 7.4, where we compare it with a queue sensitive scheme, it consumes low power, but it cannot cope with service requirements when the system operates close to capacity.

3.2   Maximum Throughput - uplink

An interesting formulation for the Uplink is considered by Kumaran and Qian [11]. They consider the problem of maximizing throughput subject to power constraints. However, they assume that the rate r_i is concave with respect to γ_i, contrary to our assumption in (7.2) that they have a linear relationship. Their assumption is based on the Shannon formula for the AWGN Gaussian channel (see [23]) where:

$$r_i = \mu \log(1 + \gamma_i) \tag{7.27}$$

We use the substitution $a_i = \frac{\gamma_i}{1+\gamma_i}$ on the power constraints (7.8) to derive constraints:

$$\sum_{j=1}^{N} a_j + \frac{a_i}{g_i}\left(\frac{I+\nu}{\bar{p}_i}\right) \le 1, \quad 1 \le i \le N \tag{7.28}$$

Using the same substitution on r_i as defined in (7.27) we get $r_i = g(a_i)$, where g is as follows:

$$g(a_i) = \mu \log\left[\frac{1}{1-a_i}\right] \tag{7.29}$$

Thus the objective of maximizing throughput $\max_r \sum_{i=1}^{N} w_i r_i$ for some weights w_i, becomes one of maximizing a convex function $\max_a \sum_{i=1}^{N} w_i g(a_i)$. Thus the Maximum Throughput uplink (MTU) problem is defined as follows:

$$\max_a \sum_{i=1}^{N} w_i g(a_i) \tag{7.30}$$

$$\text{s.t} \quad \sum_{j=1}^{N} a_j + \frac{a_i}{g_i}\left(\frac{I+\nu}{\bar{p}_i}\right) \le 1, \quad 1 \le i \le N \tag{7.31}$$

$$a_i \ge 0, \quad 1 \le i \le N \tag{7.32}$$

Kumaran and Qian [11] consider solutions to the above problem for continuous values of a. This relaxation gives an insight to the structure of the solution. The above problem is one of maximizing a convex function with linear constraints. From convex maximization theory, due to joint convexity of the objective function in the variables a_i, the optimal solution lies at a corner point of the feasible region. Thus N of the above $2N$ constraints are effective. This means that for a group of users, $i \in J^0$, $a_i = 0$, and thus no transmission takes place, and for the rest, $i \in J^1$ we have:

$$\sum_{j=1}^{N} a_j + \frac{a_i}{g_i}\left(\frac{I+\nu}{\bar{p}_i}\right) = 1 \tag{7.33}$$

We can rewrite the above as:

$$\frac{(I + \nu)\frac{a_i}{g_i}}{1 - \sum_{j=1}^{N} a_j} = \bar{p}_i \tag{7.34}$$

Comparing the above with the original equation for power (7.5), we see that the left side of (7.34) is equal to p_i. Thus for the set $i \in J^1$ having the constraints (7.31) effective means transmitting at full power. Thus Kumaran and Qian prove the following result:

THEOREM 7.2 *The optimal transmission scheme has the following structure: if a user is transmitting at non-zero rate, then the user transmits at full power. Further, without power constraints only one user transmits in one time slot.*

Thus, given that we know that each transmitting user transmits at maximum power and thus at maximum rate, the question remains of finding the set of users J^1 that performs the transmission in each time interval. Kumaran and Qian, define $\Lambda = \sum_{j=1}^{N} a_j$, $0 \leq \Lambda \leq 1$ and solve the following problem that we denote MTU(Λ):

$$\max_a \sum_{i=1}^{N} w_i g(a_i) \tag{7.35}$$

$$\sum_{j=1}^{N} a_j = \Lambda, \quad 1 \leq i \leq N \tag{7.36}$$

$$0 \leq a_i \leq \frac{g_i \bar{p}_i}{I + \nu}(1 - \Lambda), \quad 1 \leq i \leq N \tag{7.37}$$

They use a greedy algorithm to solve the MTU(Λ) problem for a value of $\Lambda \in [0, 1]$, then they search for the optimal value of Λ in the interval $[0, 1]$.

A similar result has been reported in Oh and Wasserman [7]. Their objective is to maximize the total system throughput of mobiles with a constraint on the maximum transmission power of each mobile but without imposing any constraint on the transmission rate. They show that only a subset of users should transmit at each time slot at full power. This can be considered as a hybrid CDMA/TDMA scheduling scheme.

The work in [7] and [11] gives insights to the structure of the solution. As indicated in [11], the weights w_i used in the objective of the MTU problem can be the queue lengths of each user i. In this manner QoS of users is taken into account by giving the highest rates to the users with the longest queues. However, the problem is considered for continuous values of the transmission rates (and SIRs), which is not the case in practice. Further, it has the limitations of a myopic model where throughput and queue lengths are optimized for a single slot, not taking into the account the preceding time slots.

4.   The dynamic downlink problem

In the literature there are various formulations of the downlink problem depending on the choice of objectives, quality of service requirements, and time horizon under study. We provide a dynamic and stochastic formulation of the problem where the objective is to maximize throughput and minimize queue lengths within resource constraints. The work in this section is based on the paper by Papadaki and Friderikos [19].

4.1   System model

We now describe the model of the system under study in more detail. We consider a single cell consisting of a base station and N active mobile users where data packets for each user arrive stochastically at the base station and are queued to be transmitted.

General Description and Notation

Time is divided into equal length intervals called slots, which are indexed by $t \in \mathbb{N}$. Each slot is the time needed to send a packet with the minimum allowable rate. The arrival process of data packets for user $i \in \{1, .., N\}$ at the base station is denoted by $\{a_t(i)\}$ and it is a sequence of independent and identically distributed random variables. Let $\mathcal{D}$ be the set of allowable arrivals for each user; then $a_t \in \mathcal{D}^N$. We let $\lambda_i = \mathbb{E}[a_t(i)]$, which we assume to be stationary over time. At time slot t a transmission rate $r_t(i) \in \mathcal{R}$ (in Kbits per second - Kbps) is allocated to mobile user i, where $\mathcal{R}$ is the set of discrete allowable rates. We let the *queue state* for user i be the number of packets in the queue and denote it as $s_t(i) \in \mathcal{Q}$, where $\mathcal{Q}$ is a discrete set of all possible queue lengths. The queue state vector for all users is $s_t = (s_t(1), \ldots s_t(N)) \in s = \mathcal{Q}^N$.

We let $h(r_t(i))$ denote the maximum possible number of packets that can be transmitted in a slot for user i when the allocated rate is $r_t(i)$, and the user has unlimited packets in the queue; note that $h(r_t(i))$ is a linear function of $r_t(i)$ and thus can be written as $h(r_t(i)) = \beta r_t(i)$, where β is a coefficient that converts Kbits per second to number of packets. In the presence of the queue state parameter in h, $h(s_t(i), r_t(i))$ is taken to denote the maximum number of packets that can be transmitted in a slot for user i when the allocated rate is $r_t(i)$ and the queue length are $s_t(i)$. Therefore we can write $h(s_t(i), r_t(i)) = \min(h(r_t(i)), s_t(i))$. We also let $u(s_i)$ denote the minimum rate needed for user i to transmit all the packets when the queue state is $s_t(i)$, i.e., $u(s_t(i)) = \min\{r_t(i) \in \mathcal{R} : s_t(i) \leq h(r_t(i))\}$. We let $h(r_t)$, $h(s_t, r_t)$, and $u(s_t)$ represent the respective N-dimensional vectors.

In time slot t we assume that packet arrivals (a_t) are observed just before the queue length (s_t) is measured, and the decision to allocate transmission rates

r_t is taken just after s_t is measured. Thus the queue state process satisfies the following recursion,

$$s_{t+1} = s_t - h(s_t, r_t) + a_{t+1} \qquad (7.38)$$

We assume that in each slot the channel gain between user i and the base station is the *channel state* given by $g_t(i) \in \mathcal{G}$, where $\mathcal{G}$ is the set of allowable channels and $\mathcal{B} = \mathcal{G}^N$.

Stochastic Processes

There are two independent stochastic processes in the system: the packet arrival process and the channel process. We let $\Omega = \mathcal{D}^N \times \mathcal{G}^N$ to be the set of all outcomes. We assume that both processes are stationary for the small time horizon T that we investigate the system in. We denote the joint probability density function of the arrival process for all users with: $p^a(a_t)$. Using this probability distribution and equation (7.38) we can derive the transition probabilities for the queue state (which satisfies the Markov property):

$$p^s(s_{t+1}|s_t, r_t) = p^a(s_{t+1} - s_t + h(s_t, r_t)), \qquad (7.39)$$

where $p^s(s_{t+1}|s_t, r_t)$ is the probability that the queue state at time $t+1$ is s_{t+1} given that at time t the queue state was s_t and decision r_t was taken. We assume that the channel process is a Markov Process. We denote the conditional joint probability density function of the channel process for all users with: $p^b(g_{t+1}|g_t)$.

Feasible Region for the Decision Variables

At each time slot the transmission rates r_t are decided based on the queue state s_t and the channel state g_t. We define a decision rule r_t^π to be a function that takes as input the current states s_t and g_t and returns a feasible decision r_t: $r_t^\pi(s_t, g_t) = r_t$. We define a policy π to be a set of decision rules over all time periods: $\pi = (r_1^\pi, r_2^\pi, ...)$. Let Π be the set of all feasible policies.

We call the set of all feasible rate allocation vectors at a single slot the *action space*, and we denote it by: $\mathcal{A}(s_t, g_t)$. In order to characterize the action space at each time slot we consider possible constraints on the rate allocation vectors. There are two feasibility constraints in the transmission process: the power capacity constraint and the maximum aggregate rate constraint. Given that we are interested in optimal decision rules, we rule out suboptimal decisions by further imposing a set of queue length constraints.

power constraint

We introduce time in the power constraint (7.17) derived earlier:

$$\sum_{i=1}^{N} \frac{r_t(i)}{\frac{W}{T} + \theta r_t(i)} \left(\frac{I_i + \nu}{g_t(i)} + \theta P \right) \le P \qquad (7.40)$$

To simplify notation we let:

$$\phi_i\left(r_t(i), g_t(i)\right) \equiv \frac{r_t(i)}{\frac{W}{\Gamma} + \theta r_t(i)} \left(\frac{I_i + \nu}{g_t(i)} + \theta P\right) \tag{7.41}$$

Thus the power constraint for the downlink for each time slot t can be written:

$$\sum_{i=1}^{N} \phi_i\left(r_t(i), g_t(i)\right) \leq P \tag{7.42}$$

Aggregate Rate Constraint
Let R be the maximum limit on the aggregate rate transmitted from the base station. This gives us the following aggregate rate constraint for each time slot t:

$$\sum_{i=1}^{N} r_t(i) \leq R \tag{7.43}$$

Queue Length Constraints
To ensure that the network does not allocate rates to users that exceed the minimum rates needed to transmit all the packets in the queue, we impose the following N queue length constraints,

$$r_t(i) \leq u\left(s_t(i)\right) \text{ for } i = 1 \ldots N \tag{7.44}$$

Action Space
The action space at time slot t is defined as follows:

$$\mathcal{A}\left(s_t, g_t\right) = \{r_t \in \mathcal{R} \text{ such that } (7.42), (7.43), (7.44) \text{ are satisfied } \} \tag{7.45}$$

An example of the action space (feasible region) for a specific time slot for two users is shown in figure 7.1.

Objective
The objective is to allocate rates to users over time in order to maximize throughput while at the same time penalizing increased queue lengths. Thus we define the *one-step reward* function, c, to be a weighted sum of allocated rates (measured in packets sent) minus the weighted sum of the queue lengths in the next time period. Given the current queue state s_t, the current channel state g_t and the allocated rates r_t we define the one-step reward to be:

$$c_t(s_t, g_t, r_t) = \bar{g}_t^T \cdot h(r_t) - \alpha s_t^T \cdot (s_t - h(s_t, r_t)) \tag{7.46}$$

In the first term of the one-step reward function, the weights used for the allocated rates are the normalized channels for each user. With these weights there is an incentive to allocate higher rates to users with better channels.

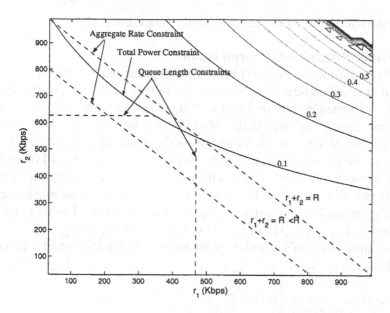

Figure 7.1. Feasible region for two users

The second term is a queue penalty term, where the packets held until the next time period $(s_t - h(s_t, r_t))$ are weighted with the current queue length (s_t). We impose a penalty for each packet that remains in the queue for the next time period, which is equal to the current queue length of the user. Therefore, the imposed penalty on the queue lengths can be thought of as a holding cost, where the holding cost per user depends on the current queue length. For example, consider a system with two users where the queue state is $s_t = (5, 20)$ for users $i = 1, 2$. Then the penalty for each packet not transmitted in the current slot is 5 for user 1 and 20 for user 2. If no packets are transmitted in time slot t, then the total queue penalty for user 1 will be $5^2 = 25$ and for user 2 it will be $20^2 = 400$. Thus the queue penalty is not linear with respect to the queue length but rather quadratic. With this penalty there is an incentive not only to keep queue lengths at low levels, but also (when capacity constraints do not permit this) to keep queue lengths at *even* levels. Further, in order to balance the two sums we use a normalizing coefficient α.

Our aim is to maximize the expected total reward over time. Thus the objective function is as follows:

$$\max_{\pi \in \Pi} \mathbb{E}\left[\sum_{t=1}^{T} c_t(s_t, g_t, r_t^{\pi}(s_t, g_t))\right] \qquad (7.47)$$

where the maximization is over all feasible policies $\pi \in \Pi$.

Dynamic Programming Formulation

The objective function in (7.47) becomes computationally intractable to solve even for a small number of users. In order to reduce the computational complexity, we formulate the problem as a dynamic program (we discuss computational complexity later in this section). We introduce the value function $V_t(s_t, g_t)$, which gives us the value (optimal reward) of being in state (s_t, g_t) at time t. The quantity $V_t(s_t, g_t)$ gives us the total optimal reward from time t until the end of the time horizon, which can be computed by finding the best feasible rate allocation decision at time t, $r_t \in \mathbb{A}(s_t, g_t)$, that maximizes the sum of the immediate reward, $c_t(s_t, g_t, r_t)$, and the expected value in the next time period, $\mathbb{E}_{a_{t+1}, g_{t+1}}\left[V_{t+1}\left(s_t - h(s_t, r_t) + a_{t+1}, g_{t+1}\right) \mid g_t\right]$.

This is called Bellman's equation or the optimality equations and it is written as follows:

$$V_t(s_t, g_t) = \max_{r_t \in \mathbb{A}(s_t, g_t)} \left\{ c_t(s_t, g_t, r_t) + \right.$$

$$\left. \mathbb{E}_{a_{t+1}, g_{t+1}}\left[V_{t+1}\left(s_t - h(s_t, r_t) + a_{t+1}, g_{t+1}\right) \mid g_t\right] \right\}$$

$$(7.48)$$

for $t = 0, 1, 2, ..., T$. Alternatively, if instead of the expectation we use the transition probabilities the optimality equations can be written,

$$V_t(s_t, g_t) = \max_{r_t \in \mathbb{A}(s_t, g_t)} \left\{ c_t(s_t, g_t, r_t) + \sum_{j \in \mathcal{S}} p^s\left(j \mid s_t, r_t\right) \sum_{l \in \mathcal{B}} p^b(l \mid g_t) V_{t+1}(j, l) \right\}$$

$$(7.49)$$

Further, we assume that $V_T(s_T, g_T) = c_T(s_T, g_T, 0)$. Note that the value function depends on s_t, g_t, which is the required information to make the rate allocation decision r_t at time slot t.

The optimality equations can be solved by a backward recursion (backward dynamic programming). Even though solving the optimality equations using backward recursion provides an improvement in computational complexity compared to the "brute force" enumeration of the objective function (7.47), the problem still becomes intractable for large multidimensional state spaces, $\mathcal{Q}^N \times \mathcal{G}^N$, such as the one under consideration. This is why we resort to approximate dynamic programming .

4.2   Non-Monotone Decision Rules

The solution of the optimization problem described in section 7.4.1 entails finding in each time period the optimal decision rule $r_t^*(s_t, g_t)$, which is a function of the queue and channel state. The question that naturally arises is whether the optimal decision rules are monotone with respect to the queue state. Monotonicity is an important structural property of the optimal policies, because it can drastically reduce the computational burden of decision making (see [17] and references therein). If we had monotone decision rules with respect to the queue state, then $s_t^+ \succeq s_t$, would imply that $r_t^*(s_t^+, g_t) \succeq r_t^*(s_t, g_t)$. Thus, knowing the optimal decision rules at s_t gives information for the optimal decision rule at s_t^+.

The optimization problem under consideration does not belong to the family of monotone decision problems and therefore computationally efficient threshold type policies can not be applied. In fact, it is easy to establish the following counterexample where monotonicity is reversed. Assume two mobile users with channel states $g_1 = 10^{-8}$ and $g_2 = 10^{-9}$ ($\bar{g}_1 = 0.909$, $\bar{g}_2 = 0.090$) and aggregate rate constraint $r_1 + r_2 \leq 64$, where $\mathcal{R} = \{0, 8, 16, 32, 64\}$. With the number of backlogged packets equal to $s_1 = 8$, $s_2 = 25$, the optimal rate allocation for two users in one time period (i.e., taking into account only the immediate reward function) would be $r_1^* = 32$Kbps and $r_2^* = 32$Kbps. If the state of the first user increased from $s_1 = 8$ to $s_1 = 14$ packets and all other parameters remained the same, then it can be shown that the optimal rate allocation would be $r_1^* = 64$Kbps and $r_2^* = 0$Kbps. Therefore the structure of the optimal decision rules is not monotone with respect to the queue state of the system.

4.3 Approximate Dynamic Programming

In this section we propose an approximate DP algorithm that solves the dynamic rate allocation problem in real-time. Further, we discuss the benefits of approximate DP methods.

The proposed algorithm is a functional approximation algorithm, in the sense that the value at each state is estimated as a closed form function rather than a set of discrete values. Functional approximate DP algorithms require only a small set of parameters to be estimated, thus reducing the complexity of textbook DP. We use a linear approximation of the value function which is the most computationally efficient approximation as it requires only N values (slopes) to be estimated. Furthermore, the proposed algorithm avoids evaluating the expectations of the optimality equations (7.48) by using Monte-Carlo sampling and adaptively estimating the value function.

In order to use Monte-Carlo sampling, we measure the queue state variable in each time slot before the realization of the arrivals and channel gains, and we call this the pre-stochastic state variable. We discuss this in section 7.4.3.1 but refer the reader to [18] for more details. In sections 7.4.3.1, 7.4.3.2 we introduce and discuss the algorithm and finally in section 7.4.3.3 we provide a brief comparative analysis of computational complexity of DP algorithms.

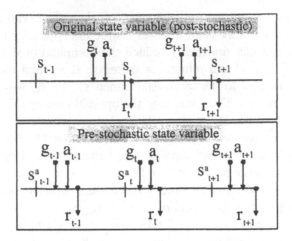

Figure 7.2. System events in the time domain for the original state variable and pre-decision state variable in time periods t and $t+1$

4.3.1   Linear Approximate Dynamic Programming (LADP) Algorithm.

In this section we introduce the proposed algorithm.

We want to consider an approximation using Monte Carlo sampling. For a sample realization of the stochastic processes $(a_1, g_1, a_2, g_2, ..., a_T, g_T)$ we propose the following approximation:

$$\tilde{V}_t(s_t, g_t) = \max_{r_t \in \mathcal{A}(s_t, g_t)} \left\{ c_t(s_t, g_t, r_t) + \tilde{V}_{t+1}(s_{t+1}, g_{t+1}) \right\} \qquad (7.50)$$

However, the state at time $t+1$, $(s_{t+1}, g_{t+1}) = (s_t - h(s_t, r_t) + a_{t+1}, g_{t+1})$, is a function of a_{t+1}, g_{t+1} and thus when calculating the suboptimal r_t in the above maximization we are using future information from the stochastic processes at time $t+1$. To correct this problem we revisit the problem definition.

We define the pre-arrival queue state variable, s_t^a, which measures the queue state just before the realization of the arrival process a_t as shown in figure 7.2. Using this new queue state variable we reformulate the dynamic program. We refer the reader to Papadaki and Powell [18] that have done similar work and describe this state variable change in detail.

The optimality equations, immediate reward function and transition function for the pre-decision state variable are as follows:

$$V_t(s_t, g_{t-1}) = \mathbb{E}_{a_t, g_t} \left[\max_{r_t \in \mathcal{A}(s_t + a_t, g_t)} \left\{ c_t(s_t, g_t, a_t, r_t) + V_{t+1}(s_{t+1}, g_t) \right\} \Big| g_{t-1} \right]$$

$$c_t(s_t, g_t, a_t, r_t) = \bar{g}_t^T \cdot h(r_t) - \alpha \left(s_t + a_t \right)^T \cdot \left(s_t + a_t - h(s_t + a_t, r_t) \right)$$

$$s_{t+1} = s_t + a_t - h(s_t + a_t, r_t)$$

$$(7.51)$$

where henceforth s_t will denote the pre-decision state variable.

Given that the model under investigation is stationary, in an infinite time horizon the value of being at a particular state s, $V_t(s)$, would be the same for all time periods t. Therefore, there is only one value function, $V(s)$, to be estimated. Thus, we drop the time index and use an infinite horizon DP model. If we assume that s, a, g, r are measured at the same time slot according to the order dictated by the pre-stochastic state variable (see figure 7.2) and let g' be the channel state at the previous time slot, then the equations (7.51) become:

$$V(s, g') = \mathbb{E}_{a,g} \left[\max_{r \in \mathcal{A}(s+a,g)} \{c(s, g, a, r) + V(s + a - h(s + a, r))\} \, | g' \right]$$

$$c(s, g, a, r) = \bar{g}^T \cdot h(r) - \alpha (s + a)^T \cdot (s + a - h(s + a, r))$$

$$(7.52)$$

The above are the optimality equations and one-step reward for the infinite horizon model. As we can see from equation (7.52) the value function $V(s, g')$ depends on g' solely due to the fact that the channel process at the current time period, g, depends on the channel of the last time period, g' (under the assumption of an underlying Markov process). If g is known then the value function does not depend on g'. Suppose (a, g) is a sample realization of the stochastic processes, where g is sampled given g'. We estimate the value function using a linear approximation $\hat{V}(s, g') = s^T \hat{v}$ (independent of g'), where $\hat{v}$ are the estimates of the slopes of the value function. Given the sample realization (a, g), we propose the following approximation to the recursion of equation (7.52):

$$\tilde{V}(s) = \max_{r \in \mathcal{A}(s+a,g)} \left\{ c(s, g, a, r) + (s + a - h(s + a, r))^T \hat{v} \right\} \quad (7.53)$$

Note that the information of g' is implicit in the value of g.

We can estimate $\hat{v}$ iteratively, where $\hat{v}^k$ are the slopes of the value function approximation at iteration k. If (a^k, g^k) is a sample realization of the stochastic processes at iteration k, where g^{k-1} was used to sample g^k, then let:

$$\tilde{V}^k(s) \equiv \max_{r \in \mathcal{A}(s+a^k,g^k)} \left\{ c(s, g^k, a^k, r) + \left(s + a^k - h(s + a^k, r) \right)^T \hat{v}^{k-1} \right\}$$

$$(7.54)$$

Given the value function slope estimates $\hat{v}^{k-1}$ from the previous iteration and the current state s^k, we can update the slope estimates in iteration k, $\hat{v}^k$, by taking discrete derivatives of $\tilde{V}^k(s)$ at the current state s^k. The estimates are then smoothed using the sequence $0 < \xi_k < 1$ that satisfies the following properties: $\sum_{k=0}^{\infty} \xi_k = \infty$ and $\sum_{k=0}^{\infty} (\xi_k)^2 < \infty$ to ensure convergence [20]. This is done in step 3 of the algorithm. To proceed to the next iteration we find r^k, which is the rate allocation vector that maximizes $\tilde{V}^k(s^k)$ (step 1). Given r^k we use the transition function: $s^{k+1} = s^k + a^k - h(s^k + a^k, r^k)$ to find the state in the next iteration (step 2). With $\hat{v}^k$ and s^{k+1} at hand we can proceed to the next iteration. The pseudo code of the proposed algorithm is as follows:

LADP Algorithm

STEP 0: Initialize an approximation for the slopes of the value function $\hat{v}^0$; initialize $g^0 \in \mathcal{B}$ and $s^1 \in s$ and let $k = 1$. Let e_i be the unit vector in the ith component.

STEP 1: Sample $\omega^k = (a^k, g^k)$, where g^{k-1} is used to sample g^k. Find the suboptimal decision rule at state s^k:

$$r^k \in \arg \max_{r \in \mathcal{A}(s^k + a^k, g^k)} \left\{ c(s^k, g^k, a^k, r^k) + \left(s^k + a^k - h(s^k + a^k, r)\right) \cdot \hat{v}^{k-1} \right\}$$

$$(7.55)$$

STEP 2: Find the next state visited,

$$s^{k+1} = s^k + a^k - h(s^k + a^k, r^k) \qquad (7.56)$$

STEP 3: Update the slopes of the value function at the current state (s^k) by using the function $\widetilde{V}$: Let,

$$\widetilde{V}^k(s) \equiv \max_{r \in \mathcal{A}(s+a^k, g^k)} \left\{ c(s, g^k, a^k, r) + \left(s + a^k - h(s + a^k, r)\right) \cdot \hat{v}^{k-1} \right\}$$

$$(7.57)$$

For all users $i = 1, \dots, N$ compute the slopes at state s^k:

$$\overline{v}^k(i) = \widetilde{V}^k(s^k + e_i) - \widetilde{V}^k(s^k) \qquad (7.58)$$

Smooth the above update using the slopes of the previous iteration:

$$\hat{v}^k = \xi_k \overline{v}^k + (1 - \xi_k)\hat{v}^{k-1} \qquad (7.59)$$

STEP 4: If $k < K$, let $k = k + 1$ and go to step 1, else stop.

It is worthwhile noting that the slopes $\hat{v}_i$, $i = 1 \dots N$ have a unique intuitive interpretation: $\hat{v}_i$ is an estimate of the value/reward (or lack of reward) of one packet in the queue of user i.

4.3.2   A Greedy Heuristic for the Embedded Integer Program (IP) in LADP .

We now propose a heuristic for the IP optimization problem that occurs in steps 1 and 3 of the LADP algorithm. For each iteration of the LADP algorithm, $N + 1$ such problems need to be solved. Thus, we provide a greedy-type heuristic to estimate the rate allocation vectors that solve this IP. The IP problem can be written as follows:

$$\max_r \sum_{i=1}^{N} \overline{g}_i h(r_i) - \alpha \cdot \sum_{i=1}^{N} (s_i + a_i - h(s_i + a_i, r_i)) (s_i + a_i) +$$

$$\sum_{i=1}^{N} v_i (s_i + a_i - h(s_i + a_i, r_i)) \tag{7.60}$$

$$\text{s.t} \quad r_i \leq u(s_i + a_i), \quad i = 1, ..., N$$

$$\sum_{i=1}^{N} r_i \leq R, \quad \sum_{i=1}^{N} \phi_i(r_i, g_i) \leq P$$

Substituting $h(r) = \beta r$ and $h(s + a, r) = \min(s + a, h(r))$ into the above objective function, the term corresponding to user i can be written as follows,

$$w_i(r_i) = \begin{cases} w_i^1(r_i) & \text{if } \beta r_i \leq s_i + a_i \\ w_i^2(r_i) & \text{if } \beta r_i \geq s_i + a_i \end{cases} \tag{7.61}$$

where $w_i^1(r_i) = [\beta \overline{g}_i + \beta(s_i + a_i)(\alpha - v_i)] r_i - \alpha(s_i + a_i)^2 + v_i(s_i + a_i)$ and $w_i^2(r_i) = \beta \overline{g}_i r_i$.

Figure 7.3 shows the function $w_i(r_i)$, which for continuous rates would be the depicted by line segments AC (line $w_i^1(r_i)$) and CD (line $w_i^2(r_i)$). Because rates are discrete, point C may not be at a feasible rate and the closest feasible rates r_i, r_i^+ would be at points B and D respectively. Thus, for discrete rates the function $w_i(r_i)$ would consists of line segments AB and BD. Therefore for each user i there are two different types of slopes: l_i^1 (slope of line 1) and l_i^2 (slope of line 2), as shown in figure 7.3. It can be easily shown that l_i^2 is always less than l_i^1. We let $L = \left(l_1^1, l_2^1, \dots l_N^1, l_1^2, l_2^2, \dots l_N^2\right)$ be the vector of all slopes. The heuristic sorts L and iteratively allocates rates to users starting with the one that has the highest slope. If the chosen slope is of type 1, then it allocates the highest possible rate that satisfies $r_i \leq (s_i + a_i)/\beta$; if the chosen slope is of type 2 then the highest possible rate is given. In both cases rates are allocated taking into account power, aggregate rate and queue length constraints. It is worth noting that since $l_i^1 \geq l_i^2$ for all i, each user is first considered for a rate less than or equal to $(s_i + a_i)/\beta$, meaning that not all backlogged packets will be transmitted, and is possibly considered a second time for full transmission i.e. a rate of $u(s_i + a_i)$.

4.3.3 Computational Complexity.

In this section we investigate the computational complexity of the previously discussed algorithms and evaluate their inherent complexity in a finite horizon setting.

The computational complexity of the classical DP depends upon (i) the size of the state space $\|\mathcal{S} \times \mathcal{B}\| = \|\mathcal{Q}^N\| \|\mathcal{G}^N\|$, (ii) the size of the outcome space

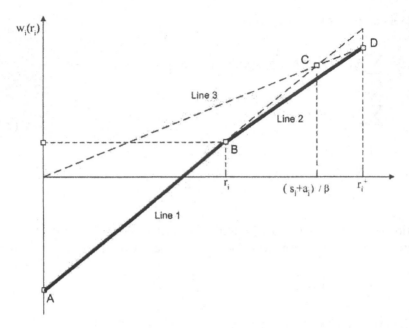

Figure 7.3. Geometrical interpretation of the heuristic used for the embedded IP optimization problem for user i. The next feasible rate to r_i is $r_i^+ = u(s_i + a_i)$

$\|\Omega\| = \|\mathcal{D}^N \times \mathcal{G}^N\|$, and (iii) the size of the action space which is at most $\|\mathcal{R}^N\|$. The size of the state, outcome and action space grows exponentially with the number of users N, which is the dimension of these spaces. These three "curses of dimensionality" render textbook DP algorithms infeasible for problems of practical scale. When solving the optimality equations (7.51) using backward dynamic programming (an exact DP algorithm): for each time $t = T - 1, \ldots 0$ we need to find $V_t(s_t, g_{t-1})$ for each $(s_t, g_{t-1}) \in \mathcal{S} \times \mathcal{B}$; to find each $V_t(s_t, g_{t-1})$ we need to evaluate the expectation over all the possible values of $(a_t, g_t) \in \Omega$; for each value of (a_t, g_t) we need to solve the maximization over all the values of $r_t \in \mathcal{A}(s_t + a_t, g_t)$. Thus the required number of calculations of textbook DP is $T \|\mathcal{S}\| \|\mathcal{B}\| \|\Omega\| \|\mathcal{A}\|$. Clearly, this complexity has been the basic handicap for the successful application of dynamic programming to large-scale problems since its inception. This is still an improvement compared to the complexity of the exhaustive search which is $[\|\mathcal{S}\| \|\mathcal{B}\| \|\mathcal{A}\| \|\Omega\|]^T$.

In the LADP algorithm we tackle these three curses of dimensionality as follows: First, we avoid evaluating the value function throughout the state space by using a functional approximation which only requires $N + 1$ evaluations of $\widetilde{V}^k(s)$ at each iteration k (step 3) to estimate the N slopes of the

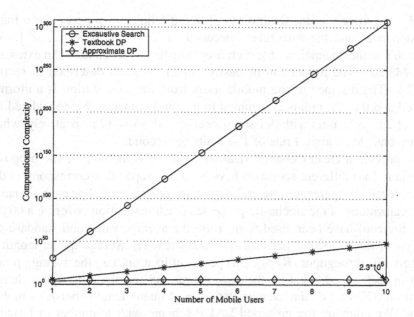

Figure 7.4. Computational complexity of the LADP, textbook DP, and exhaustive search in a scenario where the outcome space consist of an eight state Markov channel, the arrivals have been truncated to less than twelve packets per user

value function. Second, we avoid evaluating the expectation by using Monte Carlo sampling which requires K iterations of adaptively estimating the value function. Third, we avoid evaluating the action space throughout by using a greedy-type heuristic to solve the embedded integer programming maximization problem that arises in each evaluation of $\widetilde{V}^k(s)$ in step 3 of the algorithm. This heuristic takes $2N \log(2N)$ steps to sort and allocate rates.

Thus the computational complexity of the linear approximate DP (finite horizon) algorithm is of the order $N^2 \log(N)KT$, which is polynomial with respect to the number of users. As a comparison figure 7.4 shows the massive improvement in computational complexity of the the linear approximate DP compared to textbook DP and the exhaustive search.

4.4 Numerical Investigations

The aim of the conducted experiments which are discussed in this section is to quantify the performance gains of the proposed LADP scheme in a multirate CDMA system under a number of different network conditions in comparison with baseline heuristics.

4.4.1   Experimental Design.   The radio link can support seven transmission rates and the zero rate if necessary i.e. $\mathcal{R} = \{0, 2^3, \ldots, 2^9\}$. We assume that the channel, g_i, for each user is history independent. An extension to a Markov type model can be easily implemented as described in section 7.4.3.1. The distance of the mobile users from the base station is uniformly picked and the cell radius is assumed to be one kilometer. We considered the case of 7 mobile users with Poisson packet arrivals, $\lambda = 17$ respectively, which corresponds to an arrival rate of 136 Kbits per second.

To provide a clearer characterization on the performance of the proposed algorithm four different scenarios have been developed that correspond to different levels of aggregate rate and maximum power limit. For each scenario, 200 realizations of the stochastic processes (each realization covering a 60 slot time horizon) have been used to measure the average value and standard deviation of the following performance measures: (i) average queue length in packets, (ii) throughput (Kbps), (iii) power utilization, i.e., the average power used in each slot divided by the power limit and (iv) the cross queue length variation (CQV), i.e. the average variation of queue lengths between mobile users. We compare the proposed LADP scheme with a number of baseline greedy heuristics that are described in the following.

4.4.2   Baseline Heuristics.   We compare the proposed LADP scheme with a number of baseline greedy heuristics that are described in the following. **Greedy Rate Packing – GRP** The GRP algorithm assigns high rates to users that have large $g_i/(I_i + \nu)$ factor, i.e. high channel gains and low interference. Users are ordered based on the $g_i/(I_i + \nu)$ values and rates are assigned iteratively. We should note that the algorithm never reallocates rates of users previously assigned. This process is continued until the user with the worst channel conditions is allocated a feasible rate or the available rates are exhausted, whichever happens first (see section 7.3.1). Clearly, the GRP algorithm takes no more than $N \log(N)$ steps (in order to sort) for complete rate assignment. It has been shown [9] that the GRP algorithm can reduce the total power transmission, while maintaining the same throughput, by reassigning SIR-targets in a decreasing manner.

Greedy Channel – GC The GC algorithm, which is based on the Greedy Rate Packing (GRP), is a channel and interference aware scheme that also assigns high rates to users that have large $g_i/(I_i + \nu)$ factor. In contrast to the GRP scheme the GC can allocate the same rate to a different user at the same time slot. Users are ordered based on the $g_i/(I_i + \nu)$ values and rates are assigned iteratively. Therefore, for each user i the allocated rate is the maximum feasible rate. The GC algorithm takes no more than $N \log(N)$ steps.

Greedy Queue – GQ The GQ algorithm can be considered as a channel agnostic scheme where rate allocation depends only on the length of the backlog packets in the queue of each user. More specifically, users are ordered with respect to the instantaneous queue lengths in each time slot and rates are assigned iteratively with the highest feasible rate to the user that has the longest queue length. The complexity of the GQ algorithm is $N \log(N)$.

Constant Rate − CR The CR algorithm has the minimum implementation complexity as it allocates to all users the same rate, which is the maximum feasible rate at each time slot. Because of it's simplicity the CR scheme is considered to be used in early stages of 3G networks where the central focus is on operational rather than optimization aspects of the infrastructure.

4.4.3   Simulation Results.   We discuss the computational efficiency of the LADP algorithm and compare its performance with the greedy baseline heuristics described above with respect to several predefined metrics.

Figure 7.5 shows the computational times (in msec) of the proposed LADP algorithm as the number of supported mobile users increases from 5 to 40. The computational times were measured on a computer system equipped with an Intel Pentium-IV processor operating at 2.39 GHz with 512MB of RAM. The required computation time needed for rate allocation should be correlated with system constraints of real systems.

In practice data transmission occurs in fixed-size time frames, i.e., the rate is allowed to change between frames but remains constant within a single frame. In the 3GPP WCDMA system (see [22]) these time frames have minimum duration of 10 msec. Based on the above real-time constraints on the transmission time interval, the computational times depicted in figure 7.5 clearly show that the LADP algorithm can operate in real-time as the average computation time for even 40 mobile users is well below 2.5msec. Taking into account that dedicated hardware (i.e. custom processor chips or FPGAs) can yield several orders of magnitude faster performance over software simulation, this workload, in reality, would translate in tens to hundreds of microseconds execution time.

Table 7.1 shows the performance of the LADP algorithm, in terms of the previous defined metrics, in comparison with the alternative baseline heuristics. As can be seen from this table, in all different scenarios (aggregate rate and total power limit constraints) the LADP outperforms the alternative solutions except on the power utilization metric where the GRP algorithm has − as expected − the best performance. For example, concerning the average number of packets in the queue the LADP provides in all different scenarios a performance improvement between 3.6 to 35 times compared to the alternative heuristics. Also, as can be seen in the 7th and 8th column of the table the LADP algorithm provide the smallest average CQV and CQV standard deviation compared to the other techniques, which intuitively means that users with bad channel conditions are not penalized in terms of their average queue

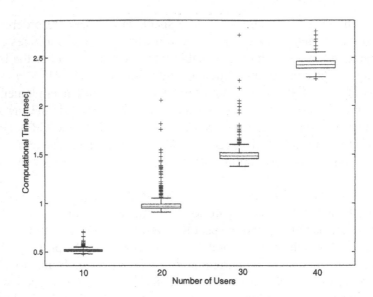

Figure 7.5. Computational times of the LADP algorithm in terms of CPU-time as a function of the number of mobile users

length. It is interesting to note that the GQ algorithm that seeks to minimize the maximum queue length in every transmission time interval achieves significant performance improvements over the other heuristics. Because the GQ scheme is channel agnostic it is especially efficient when the per users queue lengths and aggregate rate constraints are mostly effective in comparison to the constraint of the power limit. The average queue length of the LADP algorithm in comparison to GQ in all sets and scenarios has on average 3 times better performance with almost the same power utilization. Concerning the CR algorithm, by averaging across all different set of experiments and scenarios it can be seen that the LADP algorithm has on average 13 times performance improvement on the average queue length.

Summarizing, the benefits of the LADP algorithm are twofold: it achieves a reasonable estimate of the value function whilst it provides real-time operation. This is due to its simple nature since only N parameters need to be estimated.

5.   Concluding Remarks

The problem of radio resource management in CDMA networks for a single cell has been studied and various existing formulations have been reviewed. We considered myopic formulations of both the uplink and the downlink problems. For the downlink problem we provide a detailed description of a dynamic and stochastic model, where the problem is formulated as a dynamic

$N=7$ $\lambda=17$	Av. queue	sd Av. queue	Through--put	Through--put sd	Power Util.	Power Util. sd	CQV	CQV sd
	Rate 2000Kbps, P=6W							
LADP	3.05	0.51	947.58	11.12	0.611	0.036	3.00	0.50
GC	54.96	4.16	849.61	7.82	0.724	0.025	53.34	4.07
GRP	236.27	5.71	517.49	6.19	0.022	0.000	228.64	5.57
GQ	8.83	1.17	941.78	11.13	0.886	0.021	8.68	1.16
CR	35.83	5.70	888.07	15.74	0.255	0.001	34.72	5.56
	Rate 2000Kbps, P=9W							
LADP	2.46	0.41	948.18	11.23	0.562	0.037	2.42	0.40
GC	52.81	4.33	853.84	7.62	0.573	0.020	51.25	4.23
GRP	236.27	5.71	517.49	6.19	0.015	0.000	228.64	5.57
GQ	7.01	1.07	943.70	10.98	0.835	0.025	6.89	1.06
CR	35.83	5.70	888.07	15.74	0.170	0.000	34.72	5.56
	Rate 1000Kbps, P=15W							
LADP	1.32	0.24	949.26	11.41	0.667	0.025	1.30	0.24
GC	45.98	5.84	866.16	7.24	0.554	0.029	44.63	5.70
GRP	236.27	5.71	517.49	6.19	0.009	0.000	228.64	5.57
GQ	5.88	1.07	944.70	10.69	0.802	0.025	5.78	1.06
CR	35.83	5.70	888.07	15.74	0.102	0.000	34.72	5.56
	Rate 1000Kbps, P=15W							
LADP	15.33	5.12	928.54	6.15	0.113	0.009	14.94	4.96
GC	106.86	7.26	754.91	7.13	0.037	0.003	103.51	7.09
GRP	249.35	5.71	493.49	6.19	0.001	0.000	241.29	5.57
GQ	23.27	1.13	926.35	10.74	0.085	0.002	22.84	1.12
CR	35.83	5.70	888.07	15.74	0.102	0.000	34.72	5.56

Table 7.1. Analysis on the performance of different algorithms

program and a linear approximate dynamic programming algorithm is used to estimate the optimal allocation policies in real time. A similar approach can be used to study the dynamic and stochastic uplink problem. Simple formulations for these problems have been studied exhaustively in the literature. However, the issue of providing near optimal solutions to problems with differentiated quality of service requirements and stochastic mobility patterns still remain a challenge.

References

[1] J. Zander, Distributed cochannel interference control in cellular radio systems, *IEEE Transactions on Vehicular Technology, vol. 41, no. 3, 1992.*

[2] G. J. Foschini, Z. Miljanic, A simple distributed autonomous power contol algorithm and its convergence, *IEEE Transactions on Vehicular Technology, vol. 42, no. 4, 1993.*

[3] R. Yates, A framework for uplink power control in cellular systems, *IEEE J. Selected Areas in Commun., vol. 13, no. 7, 1995.*

[4] T. Alpcan, T. Bascedilar, R. Srikant, E. Altman, CDMA Uplink Power Control as a Noncooperative Game, *Wireless Networks, vol. 8, no. 6, 2002.*

[5] A. Sampath, P. Kumar, J. Holtzman, Power control and resource management for a multimedia CDMA wireless system, *In Proc. IEEE PIMRC'95, pp.21-25, 1995.*

[6] S. Ramakrisha and J. M. Holtzman, A Scheme for Throughput Maximization in a Dual-Class CDMA System, *IEEE J. Selected Areas in Commun., vol 16, pp. 830-844, June 1998.*

[7] S. J. Oh and K. M. Wasserman, , Optimality of greedy power control and variable spreading gain in multi-class CDMA mobile networks, *ACM Mobicom'99, pp. 102-112, 1999.*

[8] F. Berggren ,S-L. Kim, R. Jantti, and J. Zander , Joint power control and intra-cell scheduling of DS-CDMA nonreal time data, *IEEE J. Selected Areas in Commun., vol.19, no.10, pp. 1860-1870, Oct 2001.*

[9] F. Berggren, Power Control, transmission rate control and schedulling in cellular radio systems, *Licentiate Thesis, Royal Institute of Technology, 2001.*

[10] F. Berggren ,S-L. Kim, Energy-efficient control of rate and power in DS-CDMA systems, *IEEE Transactions on Wireless Communications, Vol. 3, Issue. 3, pp. 725- 733, May 2004.*

[11] K. Kumaran, L. Qian, Uplink Scheduling in CDMA Packet-Data Systems, *In Proc. of the IEEE INFOCOM, 2003.*

[12] R. Jantti, S.L Kim, Transmission Rate Scheduling for the Non-Real Time Data in a Cellular CDMA System, *IEEE Communication Letters, vol.5, no. 5, pp.200-2002, May 2001*.

[13] Martin Puterman, Markov Decision Processes: Discrete Stochastic Dynamic Programming, *John Wiley & Sons, New York, NY, 1994*.

[14] S. Kahn, M. K. Gurcan, and O. O. Oyefuga, Downlink Throughput Optimization for Wideband CDMA Systems, *IEEE Communications Letters, vol. 7, no. 5, pp. 251-253, May 2003*.

[15] A. Stamoulis, N. Sidiropoulos, G. B. Giannakis, Time-Varying Fair Queueing Scheduling for Multicode CDMA Based on Dynamic Programming, *IEEE Trans. on Wireless Communications, vol.3, no. 2, pp.512-523, March 2004*.

[16] S. Kandukuri, N. Bambos, Multimodal Dynamic Multiple Access (MDMA)in Wireless Packet Networks, *IEEE INFOCOM '01, Anchorage, Alaska, April 22-26, 2001*.

[17] K. Papadaki, W. B. Powell, A monotone adaptive dynamic programming algorithm for a stochastic batch service problem, *European Journal of Operational Research 142(1), 108-127, 2002*.

[18] K. Papadaki, W. B. Powell, An Adaptive Dynamic Programming Algorithm for a Stochastic Multiproduct Batch Dispatch Problem, *Naval Research Logistics, vol. 50, no. 7, pp. 742-769, 2003*.

[19] K. Papadaki, V. Friderikos, Resource Management in CDMA Networks based on Approximate Dynamic Programming, *IEEE 14th IEEE Workshop on Local and Metropolitan Area Networks, 2005*.

[20] D. Bertsekas, J. Tsitsiklis, Neuro Dynamic Programming, *Athena Scientific, Belmont, MA, 1996*.

[21] J. Tsitsiklis, B. Van Roy, An analysis of temporal-difference learning with function approximation, *IEEE Transactions on Automatic Control, vol. 42, pp. 674-690, 1997*.

[22] H. Holma, A. Toskala, WCDMA for UMTS, *Wiley, New York, 2000*.

[23] Proakis J.G., Digital Communications (4th Edition), *McGraw-Hill, 2001*.

Chapter 8

STRATEGIC AND TACTICAL PLANNING MODELS FOR SUPPLY CHAIN: AN APPLICATION OF STOCHASTIC MIXED INTEGER PROGRAMMING

Gautam Mitra[1], Chandra Poojari[1] and Suvrajeet Sen[2]

[1] *CARISMA*
School of Information Systems, Computing and Mathematics
Brunel University, London
United Kingdom
{ gautam.mitra, chandra.poojari } @brunel.ac.uk

[2] *Department of Systems & Industrial Engineering*
University of Arizona, Tuscon, AZ 85721
United States
sen@sie.arizona.edu

Abstract   Stochastic mixed integer programming (SMIP) models arise in a variety of applications. In particular they are being increasingly applied in the modelling and analysis of financial planning and supply chain management problems. SMIP models explicitly consider discrete decisions and model uncertainty and thus provide hedged decisions that perform well under several scenarios. Such SMIP models are relevant for industrial practitioners and academic researchers alike. From an industrial perspective, the need for well-hedged solutions cannot be overemphasized. On the other hand the NP-Hard nature of the underlying model (due to the discrete variables) together with the curse of dimensionality (due to the underlying random variables) make these models important research topics for academics. In this chapter we discus the contemporary developments of SMIP models and algorithms. We introduce a generic classification scheme that can be used to classify such models. Next we discuss the use of such models in supply chain planning and management. We present a case study of a strategic supply chain model modelled as a two-stage SMIP model. We propose a heuristic based on Lagrangean relaxation for processing the underlying model. Our heuristic can be generalized to an optimum-seeking branch-and-price algorithm, but we confine ourselves to a simpler approach of ranking the first-stage decisions which use the columns generated by the Lagrangean relaxation. In addition to providing integer feasible solutions, this approach has the advantage of mimicking the use of scenario analysis, while seeking good "here-and-now" solutions. The approach thus provides industrial practitioners an approach that they are able to incorporate within their decision-making methodology. Our computational investigations with this model are presented.

1.   Introduction and Background

Optimization models in general, and linear and integer programming models in particular have made considerable contribution to real world planning and scheduling problems. Unfortunately, their success has also demonstrated their limitations in certain situations. Whereas optimum decision making as captured by deterministic optimisation models are well understood and extensively adopted (Williams 1999, Schrage 2000), there are many problem contexts where the assumptions are difficult to justify. The world of LP, IP models are highly deterministic, and this assumption has raised questions of the following type: "But in real life, since you never know all ... values with perfect certainty, you may ask distrustfully, "Are ... deterministic models and techniques really practical?" (Wagner 1969). There are of course many applications in which data is reliable, and deterministic models are adequate. However, certain planning applications involve decisions whose impact can only be evaluated in the future. Such situations arise in supply chain applications in which a value has to be assigned to contracts for future services. Since, the impact of such contracts can only be evaluated in the future, a deterministic model will be unable to assess the true impact of uncertainty. For such situations, a decision-making paradigm that accommodates uncertainty must be adopted. Stochastic Programming (SP) models provide the appropriate framework for optimization models under uncertainty.

With some exceptions, the vast majority of the SP literature is dedicated to continuous optimization models and algorithms (Birge and Louveaux 1997). As a result, most of the applications in the published literature are also concerned with continuous SP models. However, discrete choices are prevalent in many supply chain applications; for instance, modelling batch sizes, fixed (setup) costs, network design decisions, and many others lead to discrete choices that are modelled using integer variables. Under uncertainty, such decision problems naturally lead to stochastic mixed-integer programming (SMIP) models.

A distinguishing feature of two and multi-stage SP models is that the randomness of the 'state of the world' are represented by a discrete scenario tree (see Figure 8.1) in which nodes represent the states, this is further supplemented by a linear dynamical system of decision vectors associated with these states.

A general multi-stage recourse problem with T stages can be written as Dempster (1980), Birge (1988).

$$\min_{x_1 \in \Re^{n_1}} \{ f_1 x_1 + E_{\omega_1} [\min_{x_2} \{ f_2(x_2) + E_{\omega_2|\omega_1}$$

$$[\min_{x_3} \{ f_3(x_3) + \ldots E_{\omega_T|\omega_1 \ldots \omega_{T-1}} [\min_{x_T} f_T(x_T)] \ldots \} \}] \}$$

$$A_1 x_1 = b_1$$
$$G_t x_{t-1} + A_t x_t = b_t \text{ a.s}$$
$$l_t \le x_t \le u_t \text{ a.s}, t = 2, \ldots T, x_t \ge 0; \tag{8.1}$$
$$\forall t = 1, 2, \ldots T \quad x_t := x_t(x_1, \ldots, x_{t-1}, |\omega_1, \ldots, \omega_t)$$

In (8.1) the separable objective is defined by the period functionals f_t, for $t = 1, \ldots, T$; $A_1 \in \Re^{m_1 \times n_1}$ and $b_1 \in \Re^{m_1}$ define deterministic constraints on the first stage decision x_1, while, for $t = 2, \ldots, T$, $A_t : \Omega \to \Re^{m_t \times n_t}$, $G_t : \Omega \to \Re^{m_{t-1} \times n_{t-1}}$, and $b_t : \Omega \to \Re^{m_t}$ define stochastic constraint regions for the recourse decisions $x_2 \ldots x_T$; and $E_{\omega_t|\omega_1 \ldots \omega_{t-1}}$ denotes conditional expectation of the state ω_t of the data process ω at time $t$with respect to the history $\{\omega_1, \omega_2, \ldots, \omega_t\}$ of the process up to time t, where the data process ω may be regarded as a random vector defined in a canonical probability space (Ω, F, P), with $\omega_t := (\xi_t, A_t, b_t)$in which ξ_t ω is a the random parameter in the objective functional given by $f_t(\xi_t, x_t)$. This model embodies the main features of a decision problem under uncertainty.

1.1 Recent Developments

Over the past decade, growing number of applications have been reported which are based on SMIP. Perhaps the most widely discussed application of SMIP is the stochastic unit-commitment model (Takriti, Birge and Long 1996, Escudero et al 1996, Carpentier et al 1996, and Nowak and Roemisch 2000). These models lead to multi-stage SMIP problems that provide well-hedged generator scheduling policies in the face of uncertain weather conditions. The special structure of these problems allows effective algorithms based on Lagrangian relaxation. Another class of applications in which weather-related uncertainties lead to a multi-stage SMIP arises in air-traffic flow management problems (Ball et al 2003, Alonso-Ayuso, Escudero, and Orteno 2000). Again the integer variables arise from scheduling flight progression through the airspace, together with arrival and departure scheduling at airports. Because of uncertainties associated with weather patterns, these scheduling models lead to SMIP problems. Uncertainty also arises due to the temporal separation between business decision epochs and the discovery of data regarding operational parameters (capacities, demands, prices etc.). For instance in telecommunications systems, addition of capacity must be carried out several months (or years) before demand data is available (Riis ,Skriver and Lodahi 2002). These types of problems involve 0-1 variables representing which nodes and links of a network should be added to achieve a specified level of service. A slightly different type of combinatorial optimization problem arises in financial planning (e.g. Drijver, Klein Haneveld and van der Vlerk 2003) where probabilistic constraints with discrete random variables which require the use of 0-1 variables to choose the subset of scenarios that provide sufficient coverage to meet the probabilistic constraint. Such considerations are also necessary in certain inventory control models (e.g. Lulli and Sen 2002) where probabilistic constraints are included in conjunction with a recourse model.

In this chapter, our focus is on stochastic integer programming models for supply chain planning problems. Important features of the model are contracting considerations, and setup costs, these in turn lead to the introduction of 0-1 variables. Such models have been presented in MirHassani et al 2000, and Alonso-Ayuso et al 2002. These models are discussed in detail subsequently.

1.2   The problem setting: stochastic integer programs with recourse

Following the classical two-stage stochastic linear programming model (Birge and Louveaux 1997), we begin with a statement of a two-stage Stochastic Mixed-Integer Program (SMIP). Let $\tilde{\omega}$ denote a random variable defined on a probability space $(\Omega, \Im, \mathrm{P})$. The model we consider may be stated as follows.

$$
\begin{aligned}
Z = \min \quad & cx + E\left[Q\left(x, \tilde{\omega}\right)\right] \\
\text{subject to} \quad & Ax \geq b \\
& x \geqslant 0, \ x := \left(x_j, j \in J_1\right), J_1 = C_1 \cup D_1 \cup B_1
\end{aligned}
\tag{8.2}
$$

where J_1 is the index set of first-stage variables consisting of three index subsets: C_1 continuous, D_1 integer and B_1 binary decision variables respectively. The recourse function $Q\left(x, \omega\right)$ is itself defined as the objective of a constrained optimisation problem:

$$
\begin{aligned}
Q\left(x, \omega\right) = \quad & \min \ f(\omega)y \\
\text{subject to} \quad & W(\omega)y \geqslant d(\omega) - T(\omega)x \\
& y \geqslant 0, \ y := (y_j, j \in J_2), J_2 = C_2 \cup D_2 \cup B_2
\end{aligned}
\tag{8.3}
$$

where J_2 is the index set of second-stage variables consisting of the subset of continuous, integer and binary decision variables, the corresponding index subsets are denoted as C_2, D_2 and B_2 respectively. The matrix A and the vector b are appropriately dimensioned, and are known with certainty. In contrast, the technology matrix $W(\omega)$, the right-hand side $d(\omega)$, the tenders matrix $T(\omega)$, and the objective function coefficients $f(\omega)$ of this problem are allowed to be random. Assuming a risk-neutral decision maker, the objective function in (8.1) optimizes the cost of first-stage decisions, and the expectation of second-stage decisions. Here the notation $E\left[Q\left(x, \tilde{\omega}\right)\right]$ denotes expectation with respect to $\tilde{\omega}$. Since most applications are modelled using discrete random variables, it is easiest to interpret the expectation operator as a summation over a finite set of outcomes (scenarios).

In general then, we will designate the set of continuous decision variables in stage t by the notation C_t, and the set of discrete and binary decision variables by D_t and B_t respectively. From the above notation it is clear that we obtain a classical multi-stage stochastic linear program when the index sets $V_{B_t} = V_{D_t} = \{\varphi\}$ $\forall t \in T$. By the same token, a multi-stage SMIP results when the collection of discrete decision variables is non-empty in several stages, including the first two.

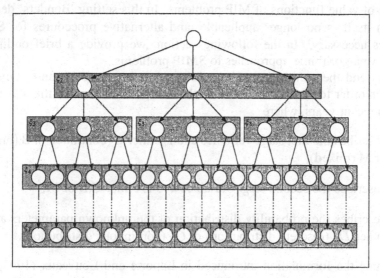

Figure 8.1. A scenario tree

1.3 Classification of stochastic mixed integer programs

In describing an instance of a multi-stage SMIP, it is important to list the time stages in which the integer decision variables appear. We let V_B, V_C, V_D denote the sets of time indices (t) for which the sets B_t, C_t, D_t are nonempty, respectively. By using these sets (V_B, V_C, V_D) we can characterize the extent to which discrete decision variables appear in any SMIP instance. At one end of the spectrum we have a multi-stage SLP problem where $V_B = V_D = \{\phi\}$ (the empty set), whereas, a purely discrete SMIP is one in which $V_C = \phi$ (empty set) while V_C and V_D together include every time-stage of the decision problem. Note that a purely binary SMIP has the property that $V_C = V_D = \phi$. Another common class of problems is one for which $V_B = V_D = \{1\}$, that is, the discrete variables only appear in the first stage. Moreover, the traditional Benders' decomposition algorithm is intended for those classes of instances for which $V_B = V_D = \{1\}$. Thus, these sets (V_B, V_C, V_D) can be used to characterize the class of instances for which an algorithm is applicable. For SLP problems (i.e. $V_B = V_D = \phi$) in which $\tilde{\omega}$ is a discrete random variable, the objective function in (8.1) is a weighted average of LP value functions, and consequently, problem (8.1) results in a convex program. In fact, the same is true in a two stage SMIP in which the integer variables are restricted to the first stage (i.e. $V_B = V_D = \{1\}$). In these cases, the basic structure of Benders' decomposition remains applicable. However, when integer variables appear in stages beyond the first stage, the objective function becomes a weighted average of value functions of MIP problems. In this setting, Benders' decomposition itself is no longer applicable, and alternative procedures for SMIP becomes necessary. In the following section, we provide a brief outline of alternative algorithmic approaches to SMIP problems.

We extend the compact notation **a/b/c** adopted by Laporte and Louveaux (1993) in order to summarise some of the computational work that has taken place in recent years, where

a describes the first stage variables and is either C (continuous), B (binary) or M (mixed);

b describes the second stage variables and is either C,B or M;

c describes the probability distribution of the unknown parameters and is either D_d (discrete distribution) or C_d (continuous distribution).

Combining the formulation introduced in Laporte and Louveaux (1993) and our notation in the section 8.1.3, it is easy to represent multi-stage SIPS compactly.

For instance

1. A 3-stage SMIP model with binary variables in the first-stage, mixed integer in the second stage and continuous in the third stage and having the random variables represented as discrete scenarios can be written as follows:

$$\{V_B \cup V_C \cup V_D, \quad \text{discrete}\}$$

$$V_B = \{1, 2\} \quad ; \quad V_C = \{2\} \quad ; \quad V_D = \{2\}$$

2. A 4-stage SP model with continuous variables in the first-stage, binary in the second stage, continuous in the third, mixed integer in the fourth

stage and having random variables that are represented as continuously distribution can be written as follows:

$$V_B \cup V_C \cup V_D, \ \text{continuous}\}$$

$$V_B = \{2,4\}; \quad V_C = \{1,3,4\}; \quad V_D = \{4\}$$

Table 8.1 represents some of the existing applications in which the models are formulated as SMIP models.

SMIPs	Scale	Parallel	References
$V_B \cup V_C \cup V_D, discrete$ $V_B = \{1,2\}; V_C = \{1,2\}; V_D = \{1,2\}$	Small	No	Bienstock and Shapiro 1984,
$V_B \cup V_C \cup V_D, continuous$ $V_B = \{1,2\} V_C = \{1,2\} V_D = \{1,2\}$	Small	No	Laporte et al. 1994
$V_B \cup V_C \cup V_D, discrete$ $V_B = \{1..T\}, V_C = \{1..T\}, V_D = \{1..T\}$	Small	No	Ahmed et al 2003
$V_B \cup V_C \cup V_D, discrete$ $V_B = \{1..T\}, V_C = \{1..T\}, V_D = \{1..T\}$	Large	No	Dert 1995, Drijver et al 2003
$V_B \cup V_C \cup V_D, continuous$ $V_B = \{1..T\}, V_C = \{1..T\}, V_D = \{1..T\}$	Large	No	Bitran et al.1986
$\{V_B \cup V_C \cup V_D, discrete\}$ $V_B = \{1\}; V_C = \{2\}; V_D = \{\phi\}$	Small	No	Lageweg et al. 1985,
$\{V_B \cup V_C \cup V_D, continuous\}$ $V_B = \{1\}; V_C = \{2\}; V_C = \{\phi\}$	Small	No	Laporte and Louveaux 1993,
$\{V_B \cup V_C \cup V_D, discrete\}$ $V_B = \{1,2\}; V_C = \{1\}; V_D = \{1\}$	Large	No	Leopoldinho et al. 1994,
$\{V_B \cup V_C \cup V_D, discrete\}$ $V_B = \{1\}; V_C = \{1,2\}; V_D = \{1\}$	Large	Yes	Mulvey and Valdimirou 1992
$V_B \cup V_C \cup V_D, discrete$ $V_B = \{1,2\}, V_C = \{2\}, V_D = \{1,2\}$	Small	No	Laporte et al 1992, Caroe et al 1997, Lulli and Sen 2002,Nowak and Roemisch 2000, Takriti et al. 1996, Escudero et al 1996, Carpentier et al 1996
$V_B \cup V_C \cup V_D, discrete$ $V_B = \{1\}, V_C = \{1,2\}, V_D = \{1\}$	Large	Yes	Mirhassani et al. 2000
$V_B \cup V_C \cup V_D, discrete$ $V_B = \{1\}, V_C = \{1,2\}, V_D = \{1\}$	Large	No	Escudero and Kamesan 1993 , Escudero et al 1999.
$V_B \cup V_C \cup V_D, discrete$ $V_B = \{1,2\}, V_C = \{1,2\}, V_D = \{\phi\}$	Large	No	Alonso-Ayuso et al 2000, Alonso-Ayuso et al. 2003
$V_B \cup V_C \cup V_D, discrete$ $V_B = \{1,2\}, V_C = \{\phi\}, V_D = \{\phi\}$	Large	No	Norkin et al 1998,Sen and Ntaimo 2003

Table 8.1. Applications of SMIPs

1.4   Outline of the chapter

The rest of this chapter is organised as follows. In section 8.2 we discuss the computational approaches, in particular the decomposition based solution algorithms, for processing SMIPs. In section 8.3 we consider the generic approach of applying mathematical programming to supply chain planning and management. In section 8.3.1 we introduce the structure of strategic planning models and in section 8.3.2 we consider the salient aspects of inventory control problems and explain how these are typical examples of tactical decision-making. In section 8.3.3 the interaction of Strategic and Tactical planning models are discussed. In section 8.4 we present a case study of a strategic planning model which was developed for a consumer goods manufacturer. After introducing the model in section 8.4.1 we explain in section 8.4.2 how a column generation approach captures the key strategic decisions. In section 8.4.3 we introduce our Lagrangean relaxation based solution method for the resulting SMIP model; the computational results are summarised and discussed in section 8.4.4. In section 8.4.5 we provide a description of how the heuristics can be extended to an optimum seeking method. In section 8.5 we discuss future research directions and our conclusions.

2.   Algorithms for stochastic mixed integer programs

In this section, we provide a brief survey of algorithmic procedures for SMIP. The main focus here is on two stage problems, although we do provide some remarks on multi-stage problems towards the end of this section. For a more detailed exposition of these methods, the reader is referred to Schultz (2003), and Sen (2003).

The fact that there are numerous potential applications for SMIP is not in question. However, due the lack of formal, and effective methodology for their solution, it is fair to suggest that prior to the 1990's, very few attempts had been made to understand and characterize the structure of SMIP problems. Even fewer attempts were made to develop solution procedures for these problems. Among the exceptions, we cite the early paper Wollmer (1980). For the case in which $V_B = V_D = \{1\}$, this paper applied Benders' decomposition because the structure of the second-stage problem remained a linear program. For the same class of problems Norkin, Ermoliev and Ruszczynski (1995) combine a branch-and-bound algorithm with sample-based function evaluation for the second stage expectation. As in stochastic decomposition methods (Higle and Sen, 1991), this strategy allows the inclusion of both discrete as well as continuous random variables, and overcomes some of the difficulties associated with evaluating multi-dimensional integrals in calculating the expected recourse function. Since errors arise due to errors from statistical estimates, care must be exercised in fathoming nodes (of the B&B tree) for which the lower bound (a statistical estimate) is not significantly different from the upper bound (also a statistical estimate). However, under certain fathoming rules, Norkin, Ermoliev, and Ruszczynski (1998) show that their method does provide an optimal solution with probability one. More recently, Alonso-Ayuso et al (2003) have proposed the use of a Branch-and-Fix algorithm for this class of problems (i.e. $V_B = V_D = \{1\}$). To be precise, their development is stated for the case in which $V_D = \{\phi\}$, although there are no conceptual barriers to using the method even if this assumption does not hold. In proposing their method, Alonso-Ayuso et al (2003) assume that there are finitely many scenarios. Their algorithm begins by solving the LP relaxation of each scenario subproblem, and the non-anticipativity restrictions for the scenario solutions are coordinated during the B&B process. Although their exposition suggests recording as many B&B trees as there are scenarios in the problem, we believe that the method can be streamlined using one B&B tree that is common to all the scenario subproblems. Such an implementation will allow the solution of problems with many more scenarios.

For cases in which integer variables appear in stages beyond the first, the simple integer recourse (SIR) is, perhaps, the most well studied problem (Klein Haneveld et al. 1995, 1996). Here, the second stage integer program is the integer analog of the continuous simple recourse problem (Birge and Louveaux 1997), and is applicable in situations where the recourse decision involves a penalty for straying from the forecast. Just as the continuous simple recourse models arise in the newsvendor problem, one can envision the SIR problem arising in planning for "large ticket" items such as aircrafts, ships etc., or in planning with severe penalties for over/under production. Under certain assumptions (e.g. random variables with finite support, and properties of penalty costs), it can be shown that a solution to SIR problems can be obtained by mapping the data to a different collection of random variables that can be used within a continuous simple recourse problem. A survey of results in the area is provided in Klein Haneveld et al. (1999). Recently, van der Vlerk (2003) has extended this idea to simple, mixed-integer recourse problems too.

Another special class of SMIP problems arises when the first-stage variables are strictly binary, whereas the second-stage variables may be mixed-integer. In fact, the method suggested in Laporte and Louveaux (1993) does not even require the second stage problem to include linear functions. In essence, the method is based on deriving cuts based on the value of the second-stage objective function, under the requirement that the first-stage variables are binary. Since each iteration requires that the second-stage integer subproblem be solved to optimality, the scalability of this method is suspect. In its defense though, this procedure does separate decisions between the two stages, thus creating smaller IP instances in each stage.

If in addition to the binary requirements of stage 1, we have mixed-integer 0-1 second stage decisions (i.e. $1 \notin V_C, V_D = \phi$), one can also apply a cutting plane method referred to as the Disjunctive Decomposition (D^2) method proposed in Sen and Higle (2000). In recent computational tests with a stochastic server location problem (with $V_C = V_D = \{\phi\}$), Sen and Ntaimo (2003) have reported the solution of instances in which the deterministic equivalent contains over a million binary variables. While these experiments do not address the case of general SMIP problems, it is the first known case of solving (to optimality), truly large-scale SIP problems. The D^2 has also been extended to allow a branch-and-cut approach which accommodates the presence of integer variables (Sen and Sherali 2002). We should note that these methods also allow randomness in the tenders matrix T.

In connection with branch-and-bound methods for two stage SMIP problems, we should also mention the method proposed in Tawarmalani et al. (2000). They observe that when the tenders matrix is fixed, a branch-and-bound search in the space of tender-variables (i.e. Tx) can be carried out by branching along the "coordinate axes" of the tender variables. This approach also allows the presence of general integer variables in both stages.

Another approach that addresses purely integer stochastic programs is the method studied in Schultz et al. (1998). Their method, based on the notion of test sets, is not very sensitive to the number of scenarios in the SIP, provided certain calculations for deriving the test set can be performed efficiently. Geometrically, a test set provides a finite set of directions, such that for any integer feasible point, one need only scan these directions to either obtain a better (integer) solution or declare the feasible solution optimal. One of the key observations that makes this approach attractive for SIP is that the characterization of a test set depends only on the cost and technology matrix of the second stage. Consequently, identification of a test set immediately opens the door to solutions of an entire family of problems that differ only in the right hand side. While the ideas underlying this approach are elegant, the computational scale for such methods remains unclear at this time. Moreover, the literature for this approach has thus far been restricted to pure integer problems.

Most of the methods discussed above decompose the integer program into problems where the decision variables from both stages do not appear in one integer program. We refer to this property as temporal decomposition of IPs. One of the early cutting plane methods for SMIP is based on viewing the deterministic equivalent SMIP as a giant integer programming problem, and deriving cutting planes for this problem (Caroe 1998). In his method, Caroe (1998) uses lift-and-project cuts of Balas et al. (1993). More recently, Sherali and Fraticelli (2002) provide cutting planes based on a reformulation-linearization technique. Due to the block-angular structure of SMIP, cuts can be derived in such a way as to maintain the block-angular structure of the updated deterministic equivalent. Hence the LP relaxation of these updated approximations remain amenable for Benders' decomposition.

The state-of-the-art for multi-stage SMIP is relatively uncharted territory. Prior to Lokketangen and Woodruff (1996), efforts to solve multi-stage SMIP problems were restricted to special applications, especially for power generations (Takriti et al. 1996). This application continues to draw a fair amount of attention (Nowak and Roemisch 2000). Caroe and Schultz (1999) proposed a more general purpose algorithm in which a B&B search was used, with lower bounds being obtained using a Lagrangian relaxation in which the non-anticipativity conditions were relaxed. The advantage of this approach is that lower bounds are obtained by solving each scenario problem independently of the others. Moreover, any special structure of these problems can be used within each scenario. The dualization required by Caroe and Schultz (1999) requires a solver for the non-differentiable optimization problems. Their computational results are based on two-stage power generation problems. In contrast, Sen and Lulli (2003) have provided a branch-and-price algorithm in which the dual multiplier estimates are provided by a simplex method, and moreover, the branching strategy is guided by a solution obtained from linear programming. They report solving multi-stage SMIPs arising in batch-sizing problems over many stages, and their computational results suggest that as the number of stages increase, decomposition algorithms gain in efficiency over direct methods that solve the deterministic equivalent.

3. Supply chain planning and management

The manufacturing supply chain can be viewed as a complex process, which sequentially transforms raw materials into intermediate or work-in-progress (WIP) products; these are then converted into finished goods inventory (FGI), which are delivered to the customers.

The different levels of planning in the supply chain can be classified in a hierarchical structure as follows:

Strategic planning: decisions about major resource acquisitions and investments and the manufacture and distribution of new and existing products over the coming years. These decisions determine the most effective long-term organisation of the company's supply chain network, and aim at maximising return on investment or net revenues.

Tactical planning : decisions about how the facilities will be used, and how market demands will be met over the next few time period(s). These decisions, which have a medium-term scope, aim at minimising manufacturing, transportation, and inventory holding costs while meeting the demand.

Operational planning: decisions about the production, schedule and distribution of the products manufactured. These decisions, which typically af-

fect activities over a short-term planning horizon, determine the most effective resource utilisation within the supply chain network, and aim at minimising short-term production cost while meeting demand, and generating the best
goods delivery schedule. As in the case of tactical planning operational plans
involve making best use of available resources; the relationship of these problems are illustrated in Figure 8.2. Optimisation models set out to make best

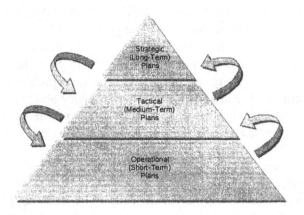

Figure 8.2.   Hierarchy of the supply chain planning

use of available (limited) resources. This makes optimisation an attractive tool
for decision support, often proposed by economists and operational research
practitioners.

Mathematical programming models have been used to represent the different activities that are involved in a manufacturing supply chain, such as facilities location, production capacity, production scheduling, and management of
WIP and FGI inventory levels.

Our work in SCHUMANN (SCHUMANN, 2000) provides a recent example of a prototype supply chain optimisation-based DSS. The software tool
was built to aid supply chain planning and management in the automobile
sector (Escudero *et al.*, 1999) as well as in the pharmaceutical sector. Two
models establish the core of the system, a strategic model, which focuses on
the optimisation of the network design over a long-term, and a tactical model,
which assists in the optimisation of resources across the supply chain network.
Moreover, these two models were developed to represent supply chain capacity and inventory decisions taking into account the uncertainty inherent in the
production costs, procurement costs, and demand. Demspter et al. (2000) report application of SP to the planning of strategic and tactical level logistics
operations in the oil industry.

3.1   Strategic planning models

In making strategic plans, a company is often concerned with acquiring assets it needs to survive and prosper over the long-term. As a part of this planning process the company must analyse its options and evaluate the potential benefits of new assets and their utilization in conjunction with existing assets. The company must also evaluate the long-term effect that new assets would have on its resource utilization.

In these strategic planning decisions, time and uncertainty play important roles (Bienstock and Shapiro 1988, Eppen et al. 1989, Escudero et al. 1993). It is therefore necessary to construct a decision model which enables the decision maker to adopt a strategic policy that can hedge against uncertainty and respond to events as they unfold (Malcolm and Zenios 1994). We are concerned with effective ways of representing uncertainty due to forecast demands, cost estimates and equipment behaviour. In particular, the potential impact of important uncertainties due to the interaction with the external environment has to be identified and assessed. The deterministic mathematical programming model for allocating assets has to be combined with a procedure for evaluating risk which in turn provides hedging strategies.

We discuss a strategic model that represents the entire manufacturing chain, from the acquisition of raw material to the delivery of final products. In order to capture the interactions existing among different stages of the manufacturing process (Davies 1994), at each time period $t \in T$ the chain is simplified to four stages related to production, packing, distribution and customer zones.

In Figure 8.3 PR, PC, DC and CZ denote the sets of production plants, packing plants, distribution centres and customer zones. Production and packing plants consist of several lines, with distribution responsible for deliveries between sites and customer zones. Different types of a product can be produced and distributed. The result is a multi-period multistage model which captures

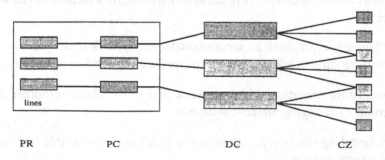

| PR | PC | DC | CZ |

Figure 8.3.   A strategic supply chain network

the main strategic characteristics of the problem as, site location decisions, choices of production and packing lines and capacity increment or decrement policies.

3.2   Tactical planning models

The primary concerns of strategic planning is to consider snapshots of the future and to make asset allocation decisions. In manufacturing this translates to processing capacities for product (competitiveness, and volume) and in logistics which is customer service focused this translates to determining warehousing and distribution (assets) capacities. Determining human resource levels are also part of strategic plans.

In contrast tactical plans are concerned with the best way of deploying assets and the overwhelming focus is on the bottom line that is either to maximise profit or minimise cost.

If projected demand is treated as fixed and given, then net revenue maximisation is achieved by minimising the total cost of meeting it. This approach to tactical planning is undesirable because most companies have flexibility over the medium term about how much of each product to sell. Because supply chain models provide meaningful product costs and margins, the company can benefit from insights they provide about adjusting sales plans to achieve greater profit. The issue to be addressed is describing how product mix options should be modelled. In the simplest case, they may be described by lower and/or upper bounds on sales of each product (product family), reflecting quantities that should or can be sold. The selling price per unit in such ranges may be constant or decreasing to reflect price elasticities.

Tactical planning models are developed within the context of a strategic plan (their interaction are considered in section 8.3.3).

Instead of relying only on snapshots of the future, tactical planning models are temporal that is multi-period with linking between the periods. According to Shapiro (2000) the dynamic effects to be captured by a tactical model should include:

- Planning inventories to accommodate seasonal demand patterns or to achieve smooth adjustments to unexpected ebbs and flows in demand

- Adjusting operating shifts per week and other labour resources as the result of changing market conditions

- Scheduling yearly plant maintenance to minimise avoidable costs across the supply chain

- Exercising vendor contracts on a monthly basis to reduce inventory holding costs while avoiding contractual penalty costs for taking insufficient volumes

An effective approach is to link the single period models by interperiod inventory flows, smoothing constraints, and other constraints and variables that cut across periods in describing dynamic effects.

In our SCHUMANN project we had developed both a tactical model (see, Escudero et al (1999)) and a strategic model (web: www.supply-chain-optimise.com). The importance of interacting between these two model classes are discussed in section 8.3.3.

Organisationally, the use of tactical models is quite different from that of strategic models. Tactical models are meant to provide plans for managing the company's supply chain on a repetitive and routine basis. Strategic planning studies using models are easier to undertake because they are preformed by an ad hoc team whose activities do not interfere with execution of the company's business. Thus far, few companies have implemented tactical planning processes driven by analysis with an optimisation model, or any other tool for that matter. Given the top-down and bottom-up pressures to use data to improve supply chain management, it seems to be only a matter of time before the development and use of modelling system for tactical planning will proliferate and the importance of using SP, in particular SMIP, models are to be noted.

3.3   Interaction of strategic and tactical planning models

Supply chain management is defined as *"the integration of business processes from end user through original suppliers that provide products, services and information that add value for customers"* (Cooper *et* al., 1997).

Consequently, the hierarchy in the supply chain activities is reflected in the management hierarchy in an organisation: the senior managers take the strategic decisions, whereas the junior managers take the tactical decisions. Preferably, decisions are assigned jointly when they are strongly related, or interact in a way that notably affects the overall performance.

It has long been recognised that a successful implementation of a strategic plan requires careful coordination of all the activities within the company at an operational level. In fact, long-term strategic planning and medium-term tactical planning, such as scheduling decisions, closely interact in defining robust solutions. This is often known as *Integrated Planning* (or *Integrated Supply Chain Management*), which is based on the integration of the different supply chain activities from three perspectives: functional, spatial, and inter-temporal integrations (Porter, 1985; Shapiro, 2001).

This need to coordinate strategic, tactical, and operational production plans is very important. Such coordination is achieved through a pragmatic approach of linking and overlapping the different modelling systems for the various supply chain activities with supply chain decision databases (Shapiro, 2001). Detailed operational models are linked with more aggregate tactical models, and in the same way, strategic models about investments and divestments over the future years are linked with aggregate tactical models that decide how these facilities will be used. Figure 8.4 illustrates the supply chain systems hierarchy. SCHUMANN (Schumann, 2000) puts into practice the integrated planning of

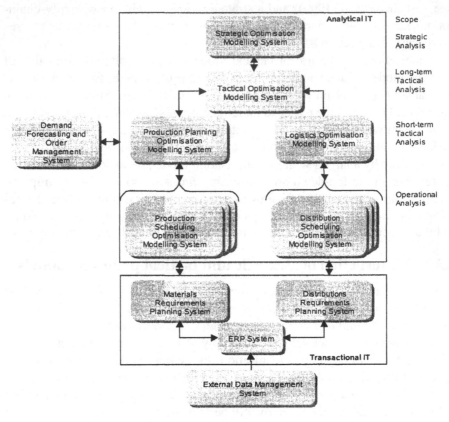

Figure 8.4.   Supply chain systems hierarchy (source- Shapiro, 1998)

the strategic and tactical decisions of an automobile manufacturing, and a pharmaceutical goods distribution networks. It is normal practice to re-examine a supply chain-planning network after a significant change has occurred in the operational environment. Typical examples are changes in regulation requirements (emission levels), a new market entrant, or the introduction of a new production technology type. The strategic model is used to determine the best strategy for asset (plant capacity, technology type) allocation. The tactical model complements the strategic model by focusing on the logistics, optimising scheduling, inventory control, and distribution of goods, which make best use of the allocated assets. A single model, which combines both the strategic and tactical decisions, is clearly desirable (Bradley and Arntzen 1999; Shapiro,

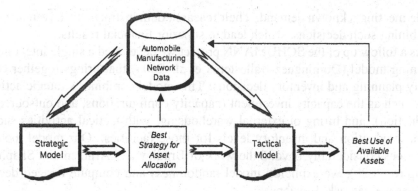

Figure 8.5. SCHUMANN Models/Data Flow

2000). An important aspect of creating such an integrated model is either the availability or the implementation of a supporting data mart. The various model parameters used in this integrated model must first reside in such an analytic database (data mart, or decision database). The results are also stored as decision data, and the database is explored using suitable Relational OLAP, that is, (R)OLAP tools; (see Figure 8.5).

A number of researchers have focused their attention on two groups of decisions faced by the supply chain logistics planners: capacity planning, and inventory planning.

Capacity planning is typically concerned with decisions of location, retooling, size, timing, and acquisition or selling of capacity in order to optimise the profits over a given planning horizon.

Inventory planning is concerned mainly with the holding levels of work-in-progress (WIP) and finished goods inventory (FGI), and equally with decisions of location, acquisition or rental of warehouses. These decisions are extremely important in operational conditions of uncertain supply or demand, seasonal demand, or in planning promotional sales.

These two types of activities are strongly related, and the combination of capacity and inventory decisions within the same modelling framework leads to a trade-off between the capacity and inventory levels. Finding the appropriate balance involves identifying the combination of capacity and inventory that allows performance criteria to be met while optimising a firm's financial criteria.

Bradley and Arntzen describe a practical approach of simultaneous planning of production capacity and inventory for an electronics manufacturer and for a manufacturer of office supplies. They developed a model that decides the optimal production schedule, capacity levels of number of machines and people, and inventory levels of raw materials, work-in-progress, and finished goods, while meeting a known demand. Their research demonstrates the advantage of combining such decisions, which lead to superior financial results.

As a follow up of the SCHUMANN project we developed a single integrated planning model (Dominguez-Ballesteros et al. 2004) that brings together capacity planning and inventory decisions. This model combines strategic activities, such as the capacity investment (capacity configurations, and outsourced production), and hiring of external warehousing, with tactical activities such as the production and inventory levels for those facilities. Our model, however, goes considerably beyond the work of Bradley and Arntzen and Shapiro (Shapiro, 2000), we explicitly model randomness and compute hedged decisions as well as risk decisions.

4.   Strategic supply chain planning: a case study

In the strategic planning model we are mainly concerned with decisions on the capacity configuration over the time horizon, i.e. the number of lines the company must open/close each year in order to minimise investment costs and yet best satisfy customer demand.

4.1   The problem description

$l \in L_i$ denotes the type of line where L_i is the set of line types at site i;

M_i is the maximum number of lines allowed at site i;

$k \in DC$ denotes distribution centres;

$h \in CZ$ denotes customer zones;

$p \in P$ and $b \in B$ denotes final and basic product types respectively; these are defined as mutually exclusive sets, i.e. $p=1,2,\ldots,|P|$ and $b=(|P|+1),\ldots,(|P|+|B|)$;

$t \in T$ denotes time periods;

The decision variables at each time period $t \in T$ are the following:

$z_{ti} \in \{0,1\}$ indicates if site i is open;

$r_{til} \in \{0,1\}$ indicates if a line of type l is open at site i;

$C_{til} \in Z^+$ represents the number of lines of type l open at site i;

$x_{til} \in Z^+$ is the number of lines of type l opened at site i;

$O_{til} \in Z^+$ is the number of lines of type l taken out before the end of their life time at site i;

$Q_{tkp} \in Z^+$ represents the ordering quantity of product p at the distribution centre k;

$F_{tkhp} \in Z^+$ is the number of product units of type p shipped from k to h;

$q_{tilp} \in Z^+$ is the number of final product units of type p packed by a line of type l at site i;

$q_{tilp} \in Z^+$ is the number of basic product units of type b produced by a line of type l at site i;

$\mu_{thp} \in Z^{+}$ represents the shortage of product p at customer zone h.

In many cases it has been shown that is important to take into account uncertainty in the inventory management of manufacturing systems (Cohen et al.). In particular, distribution centres must be managed so that they achieve minimum inventory costs while meeting expected customer demand.

Let D_{thp} be the customer demand in time period t for product p at the customer zone h. Then the following constraint must be satisfied:

$$\mu_{thp} \geq D_{thp} - \sum_{k} F_{tkhp} \qquad (8.4)$$

where shortage of product μ_{thp} is minimised by including μ_{thp} in the objective function. Distribution centres must make a decision about its ordering quantity Q_{tkp} such that:

$$\sum_{h} F_{tkhp} = Q_{tkp} \qquad (8.5)$$

Note that constraints (8.4)-(8.5) represent the link between demand in a customer zone and quantity ordered from different distribution centres.

Another source of uncertainty can be represented by the transportation times. In the context of strategic decision planning problems this level of detail and the corresponding uncertainty is not relevant and is therefore omitted.

We assume that products are shipped to distribution centres continuously; otherwise excessive quantities are stockpiled at distribution centres. Demands for such products, however, would depend on the market characteristics, with some delivered continuously in the long run while others once in a while. We assume generally that products are sent to customers at a fixed time interval. We then refer to stock that satisfies normal or expected customer demands as operational stock.

However, additional stocks are needed to hedge against uncertain demand, to make up for occasions when demand exceeds capacity, to accommodate seasonal demands, to cover demand while other products are being produced and to reduce the impact of major facility failures. Besides, peak stock levels are required for promotional sales. We refer to the sum of these stocks used to hedge against uncertainty as safety stock.

We next focus our attention on promotion and packing stages and assume that both production and packing lines are located in the same area whereby these two stages can be considered as a single one.

At this stage the demand for a given product p is represented by $\sum_{k} D_{thp}$ so that the quantity q_{tilp} of product p packed by line l at site i must satisfy:

$$\sum_{i,l} q_{tilp} = \sum_{h} D_{thp} \quad \forall t, p \qquad (8.6)$$

and

$$\sum_{i,l} q_{tilb} = \sum_{i,l,p} \beta_{bp} q_{tilp} \quad \forall t, p \tag{8.7}$$

where β_{bp} is the quantity of basic product b needed to obtain a final product p.

Thus, denoting by f_{tilp} the coefficient, expressed in *lines product units,* indicating the usage of line of type l needed to produce one unit of final product p and by f_{tilb} the coefficient, expressed in *lines/product units,* indicating the usage of line of type l needed to produce one unit of basic product b then the planned plant capacity must be at least equal to $\sum_p f_{tilp} q_{tilp}$ and $\sum_b f_{tilp} q_{tilp}$, expressed in *lines/product units* $\times$ *product units = lines.*

In order to determine the capacity C_{til}, of each plant, note that a line of type l can be operating in a site if and only if the site is open, i.e.

$$z_{ti} \le \sum_l r_{til} \le M_i z_{ti} \quad \forall t, i \tag{8.8}$$

If line l is not operating then C_{til} is also forced to be 0:

$$r_{til} \le C_{til} \le M_i r_{til} \quad \forall t, i, l \tag{8.9}$$

Then the site capacity C_{til} is given by

$$C_{til} = C_{t-1,tl} + x_{til} - \theta_{til} \quad \forall t, i, l \tag{8.10}$$

However, in order to obtain C_{til} we must keep track of t^1, i.e. the time when the expiring investment started. Note that t^1 depends both on the time τ_{il} required for the capacity of the new investment to be ready for production and on the total operation time T_{il} of such investment. In real systems such values depends on the technological behaviour of the investment such as set up time and life time of the technology. In the case it is necessary to adjust the capacity C_{til} by introducing a capacity adjustment variable C'_{til} such that:

$$C_{til} = C_{t-1,il} - C'_{til} + x_{til} - 0_{tl} \quad \forall t, i, l \tag{8.11}$$

where C'_{til} indicates the number of lines reaching the end of their life time at the beginning of the current period t plus the number of lines on which the investment has already been made but that are not yet operating. C'_{til} is a particular feature of our model, it allows us to consider investment with a limited life time investment already done but not yet in operation. We consider the problem of representing capacity configurations in the next section.

Now we can formulate the constraint which links production capacity and ordering quantity as:

$$\sum_p f_{tilp} q_{tilp} \le e_{til} \left(C_{til} - \sum_{k-t-\tau_{il}}^{t} x_{kil} \right) \quad \forall t, i, l \tag{8.12}$$

$$\sum_b f_{tilp} q_{tilp} \leqslant e_{til} \left(C_{til} - \sum_{k-t-\tau_{il}}^{t} x_{kil} \right) \quad \forall t, i, l \qquad (8.13)$$

where e_{til} is the efficiency rate of line of type l at site i.

Because the technology used is not perfectly reliable the efficiency rate e_{til} is another source of uncertainty. In fact, in managing production and packing plants the uncertain behaviour of the technological equipment may be taken into account. Machines can be subject to random failures and random repair times. Often efficiency rates e_{til} depend also on the cost of the technological equipment and on the costs allocated for maintenance, but we will not discuss the specific case here.

Finally, another source of uncertainty that must be taken into account when considering long term planning is that due to the economical and political situation introduced in the model by different type of costs.

When a production or packing site is opened for the first time, it is reasonable to assume that there is one-off site opening cost as a result of installing either production or packing lines at the site. Correspondingly, there are site closing down cost savings.

Unit capital costs of lines are assumed fixed regardless of initial or incremental investment.

This assumption could be generalised to situations where unit investment costs are different for initial and incremental investment or where unit investment costs vary depending on the number of lines due to price discounts.

The minimum number of units of capacity requirement for initial investment is not represented explicitly to avoid creating a model of large dimensions. On the other hand, the optimal solution would provide some degree of balance between minimum capacity requirement and capital investment cost.

Given the side opening cost v_{ti} of a new site and the saving s_{ti} made by closing a site, the total investment cost on production lines in time t is

$$\sum_{t,i} [v_{ti} \max\{0, z_{ti} - z_{t-1,i}\} + s_{ti} \max\{0, z_{t-1,i} - z_{ti}\}] + \sum_{t,i,l} v'_{til} x_{til} \quad (8.14)$$

where v'_{til} is the unit capital cost of production line l at site i in time period t. Costs included in (8.14) are not the only ones; in order to find the optimal strategy the operation costs must be considered as well. Some of the operation costs are fixed and volume independent as those due to administration and maintenance; some others, such as depreciation, depend on the investment strategy.

Depreciation costs for existing production lines can be assumed to be constants while those related to new investments can be calculated as a percentage of the capital cost. Depreciation costs for new investments start to be charged when the investments are ready for production and the charge is made over the depreciation period. Both the depreciation rate and the depreciation period can be modelled as stochastic variables which affect both constraints and the objective function. Other costs are considered in the model some of them are strongly related to the capacity, like repair and maintenance, or have a stepwise nature, like fixed labour, or are not only capacity related but also product type and volume related, like set up costs, raw material costs.

4.2   Representing strategic decisions through column generation

In the strategic planning model we are mainly concerned with decisions on the capacity configuration over the time horizon, i.e. the number of lines the company must open/close each year in order to minimise investment costs and yet best satisfy customer demand. In modelling capacity requirements we need to take into account the capacity loss due to set up time and limited operational time of lines. Due to the delay introduced by the set up time the number of operating lines may be less than the number of lines actually installed at the plant at any instant. Furthermore, in order to have an operating line starting from time t the investment decision must be taken at time $(t - \tau_s)$ where τ_s is the set up time (see Figure 8.6).

The exact computations of investment costs is further complicated by the limited life span of the technological equipment together with the possibility of premature closure. Moreover, investment options must be analysed and

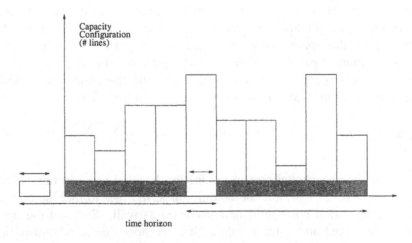

Figure 8.6.   Influence of time on the strategic decisions

evaluated in conjunction with the existing production capacity, i.e. future decisions depend on the number of existing lines and their remaining life time.

In order to be able to consider the realistic case of existing lines with different initial conditions, i.e. different age, the only possibility is to represent each line individually.

We accomplish this task by invoking an external procedure which automatically generates a matrix A, called the configuration matrix where each column A_v represents a possible configuration of a single line n over the whole planning horizon.

Note that the configuration matrix $A = \{A_v\}$ can be viewed as a multidimensional array where each element $a_{vt} = a[i, l, n, v, t]$ identifies the state that a line n of type l at site i assumes under configuration v at time t. Table 8.2 shows an example of a generated configuration for a given site i and a given line type l where V_n is the set of possible configurations that line v can assume and N is the maximum number of different lines considered.

Denoting $m_1 = |V_1|, m_2 = m_1 + |V_2|, ..., m_N = m_{N-1} + |V_N|$ Table 8.2 shows that line $n = 1$ under configuration $v = 1$ is operating from time t=1 to time t=8, while under configuration $v = |V_1|$ it is operating from time t=9 to time t=T.

A strategy decision is then identified by a vector

$$Y(i, l) = (y_{11}, ..., y_1 |v_1|, y_{21}, ..., y_2 |v_2|, ..., y_{N1}, ..., y_N |v_N|) \qquad (8.15)$$

where $y_n |v_n| \in \{0, 1\}$ and $\sum_{v-1}^{|V_n|} y_{nv} = 1, \forall n$, i.e. only one configuration is selected for each line. In using an external procedure for "column generation"

| | n=1 | | | n=2 | | ... | n=N | | |
	v=1	...	v=$|V_1|$	v=1	...	v=$|V_2|$	...	v=1	...	v=$|V_n|$
t=1	1	...	0		...		...	1	...	0
t=2	1	...	0		...		...	1	...	0
t=3	1	...	0		...		...	1	...	0
t=4	1	...	0		...		...	1	...	0
t=5	1	...	0		...		...	1	...	0
t=6	1	...	0		...		...	1	...	0
t=7	1	...	0		...		...	1	...	0
t=8	1	...	0		...		...	1	...	0
t=9	0	...	0		...		...	0	...	0
⋮	⋮	⋮	⋮	⋮	⋮	⋮	⋮	⋮	⋮	⋮
t=10	0	...	1		...		...	0	...	1

Table 8.2. Configuration generation

several "generation rules" can be introduced such that configurations of existing lines can be taken into account and only "reasonable" configurations are generated.

Consider the situation where after a line is opened the set-up period τ_s must be taken into account and further operation time τ_m must be introduced. Similarly, after a line has been closed it must remain closed until the end of the planning horizon.

These and other rules enable the decision maker to construct refined models using domain knowledge and at the same time greatly reduce the total number of configurations generated.

Using the proposed approach constraints (8.9)-(8.10) are implicitly satisfied by defining:

$$C_{til} = \sum_n \sum_{v=1}^{|Vn|} a_{nvt} y_{nv} \quad \forall t, i, l \tag{8.16}$$

Further constraints are defined to represent site capacity limits. These can be stated as: the maximum number M_1 of lines of the same type operating at each time period at the same site:

$$\sum_l \sum_n \sum_{v=1}^{|V_n|} a_{nvt} y_{nv} \leqslant M_1 \quad \forall t, i \tag{8.17}$$

the maximum number M_2 of lines that can be simultaneously operating at the same site

$$\sum_n \sum_{v=1}^{|V_n|} a_{nvt} y_{nv} \leqslant M_2 \quad \forall t, i \tag{8.18}$$

In this approach the problem of computing the exact cost of each investment strategy is also overcome. In fact, during the generation procedure the cost of each configuration at each time period can be computed. This cost vector includes the opening cost (if the line is not initially operational at time t=0) operational costs for each year the line has been operating and a closing cost (or saving) if the line is closed before the end of its life time.

Taking into consideration the details described in sections 8.4.1 and 8.4.2 and using the matching data set supplied by the manufacturer, SMIP models were generated to study the problem. The model size is substantial even without considering the scenarios (see Table 8.3).

4.3   Lagrangian Decomposition based solution method

The model discussed is formulated as a 2-stage Stochastic integer programming problem with the integer variables in the first-stage. The model can be presented in an algebraic summary form as follows:

Network Dimensions		Planning Problem	
The number of Sites, I :		8	
The types of packing line technology, Y_C:		4	
The types of production line technology, Y_R:		2	
The number of distribution centres , J :		15	
The types of DC line technology , Y_D:		2	
The number of Customer Zones , H :		30	
The number of Products , P :		13	
The number of time periods, T :		6	
Discrete Decision Variables: Sites, DCs, Production lines, Packing lines, DC lines.		$n_1 = 2096$	
Continuous Variables: Production, Packing, Ordering, Transportation, and Shortage quantities.		$n_2 = 54400$	56496
Logical Constraints: Sites, DCs opening and closing, Limit on number of Sites, DCs, and Lines.		$m_1 = 968$	
Other Constraints: Production, Packing, Ordering, Transportation, Balance, Demand, and also Production and Packing Capacities.	Mixed	$m_2 = 850$	6768
	Continuous	$m_3 = 4950$	
Non-zeros		1154034	
Scenarios		100	

Table 8.3. Dimension of the strategic supply chain model

$$\min \ Z = cx + \sum_s p_s f y_s$$
$$subject\ to$$
$$Ax = b$$
$$P_{2SIP} \quad Bx + Dy_s \leqslant h \quad \forall s \in \{1, ..., S\} \qquad (8.19)$$
$$Ey_s = d_s \quad \forall s \in \{1, ..., S\}$$
$$x \in \{0, 1\}^{n_1}$$
$$y_s \geqslant 0$$

For $s = 1..S$ scenarios we obtain the corresponding scenario related deterministic models P_{WS} which represent $|S|$ deterministic problems.

$$\min \ Z = cx + fy$$
$$subject\ to$$
$$Ax = b \quad \text{Logical Constraints,}$$
$$P_{WS} \quad Bx + Dy \leqslant h \quad \text{Capacity Constraints,} \qquad (8.20)$$
$$Ey = d_s \quad \text{Demand Constraints,}$$
$$x \in \{0, 1\}^{n_1} \quad \text{Discrete Variables}$$
$$y \geqslant 0 \quad \text{Continuous Variables}$$

For the data instances under consideration, each P_{WS} problem is a difficult MIP model in its own right and cannot be solved to optimality. For cases in which the wait-and-see problems can be solved in reasonable time, Lulli and Sen 2003 present a branch-and-price algorithm which solves a collection of P_{WS} together with a master problem that yields a solution to P_{2SIP}. However, our earlier experience of applying a branch and bound technique to obtain an first integer solution to P_{WS} was computationally difficult and time consuming. Given that there are often ($|S|$) 100 or more such problems to solve, this leads to very long processing time (MirHassani et al. 2000).

We have developed a heuristic approach based on Lagrangian relaxation which improves the computational performance of finding good feasible integer solutions for each P_{WS}. We relax a group of constraints and include them in the objective function with appropriate penalties (Lagrange multipliers). At each iteration of Lagrangian relaxation we produce a lower and upper bound for the scenario related deterministic model. The modified problem is solved to generate a set of integer feasible first-stage solutions (configurations) for a given P_{WS}. Thus by applying Lagrangian relaxation we are able to (i) generate a set of integer feasible solutions for each scenario related deterministic problem, and (ii) decrease the gap between the upper and lower bound.

We perform these iterative Lagrangian relaxation steps on $|S|$ instances of the scenario related deterministic models which results in a large set of integer feasible solutions (configurations), which we call the 'Approximated Configuration Set' (ACS). We use the ACS solutions to compute an approximate solution to the original two-stage model, P_{2SIP}.

4.3.1 Exploitation of the problem structure .

Consider the wait-and-see model, P_{WS} of the strategic supply chain model. It has a structure that can be exploited during the Lagrangian relaxation. In principle, Lagrangian relaxation can be applied to all the second stage constraints for each scenario. However, we exploit the matrix partition in the second-stage constraints so that we relax only the capacity constraints.

$$production\_quantity_{t,i,rtype} - production\_capacity_{t,i,rtype} \leqslant h' \quad \forall t, i, rtype \quad (8.21)$$

$$packing\_quantity_{t,i,ctype} - packing\_capacity_{t,i,ctype} \leqslant h'' \quad \forall t, i, ctype \quad (8.22)$$

$$order\_quantity_{t,j,dtype} - distribution\_capacity_{t,j,dtype} \leqslant h''' \quad \forall t, j, dtype \quad (8.23)$$

where rtype = product type, ctype= customer type, and dtype= distribution zone.

These are 'mixed integer' which ensure that we do not produce, pack or distribute more goods than the available capacity. These constraints combine strategic decisions concerning production, packing and distribution capacities with corresponding continuous decision variables about appropriate production, packing and distribution quantities.

Having decided to dualize the constraints $Bx + Dy \leqslant h$, we add these constraints to the objective function.

Therefore the relaxation problem for each realisation $s=1 \ldots S$ is:

$$\min \ Z = cx + fy + \lambda(Bx + Dy - h)$$

OR

$$\min \ Z = (c + \lambda B)x + (f + \lambda D)y - \lambda h$$

P_{LR} *subject to*

$$Ax = b$$
$$Ey = d_s$$
$$x \in \{0,1\}, y \geq 0, \lambda \geq 0, \lambda \in \mathbf{R}^{m_2}$$

By examining the structure of problem P_{WS}, we are able to decompose the problem into two smaller independent problems. The first problem shown in P_{PIP} is a pure integer problem for a fixed value of λ while P_{LP} is a larger LP problem.

$$\min \ Z = (c + \lambda B)x$$

subject to

P_{PIP}
$$Ax = b$$
$$x \in \{0,1\}, \lambda \geqslant 0$$

$$\min \ Z = (f + \lambda D)y$$

subject to

P_{LP}
$$Ey = d_s$$
$$y \geqslant 0, \lambda \geqslant 0$$

4.3.2 Generation of Configurations and Lower Bounds. Let $\bar{x}$ be a solution to the problem P_{PIP}. The group of constraints in P_{WS} have shortage variables which lead to "one sided" goal programming constraints. As a consequence for any integer feasible solution $\bar{x}$ to problem P_{PIP}, it is always possible to solve P_{WS} for continuous variables $\bar{y}$, such that $(\bar{x}, \bar{y})$ is a feasible solution to P_{WS}. So solving problem P_{WS} in which all discrete variables x are fixed to $\bar{x}$, leads to a feasible solution $(\bar{x}, \bar{y})$ and an upper bound ub_0 for problem P_{LR}. Thus we use the Lagrangian relaxation solution $\bar{x}$ to obtain a solution to the original problem involving all the decisions variables. By solving the LP problem, P_{LP}, while λ is fixed to λ_0 a relaxed solution to the continuous variables is found. Adding the objective values of problems P_{PIP} and P_{LP} together and subtracting the constant term λh leads to a lower bound lb_0 to the problem P_{LR}.

Having obtained an initial lower bound and an upper bound to the original problem P_{LR}, we then apply the sub-gradient method to update the Lagrangian multipliers to obtain better bounds in successive iterations. By repeating this process for different Lagrangian multipliers, λ, and different scenarios we are able to create a set of attractive configurations at the end of phase one of scenario analysis.

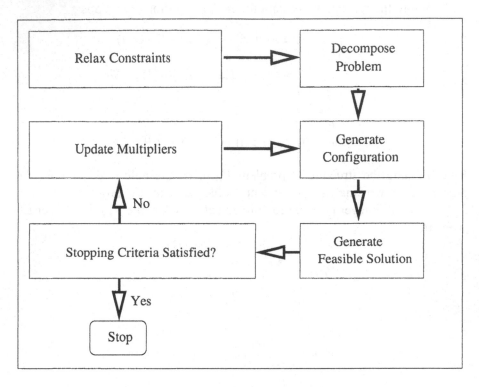

Figure 8.7. The Lagrangian algorithm

The Lagrangian Iterations:
Consider the following notations
BUB: Best Upper Bound.
BLB: Best Lower Bound.
λ :Lagrangian multiplier.
π :The step length parameter for the subgradient procedure.
T: The step size.
Gi : The gradient of the ith relaxed row.
The Lagrangian relaxation iterations are described in Pseudo Code 1 in Figure 8.8 and each step of the Pseudo Code are annotated below.
In step 0 we initialise *BUB* and *BLB* to ∞ and $-\infty$ respectively, and λ to zero. Also we set the pass number (number of times the process of solving problem P_{PIP}, P_{LP}, and P_{WS}(with x fixed) to zero.
In step 1 the relaxed problem is solved and a new configuration $\bar{x}$ is found.

In step 2 we find a feasible solution to the original problem after fixing x to $\bar{x}$. Because of the existence of shortage variables in the original problem a feasible solution to the problem P_{WS} is guaranteed for any given configuration $\bar{x}$.

In step 3 the stopping criteria is applied to terminate the algorithm as appropriate. A good termination criterion is when all the side constraints are satisfied. However, this rarely happens and sometimes it results in a poor solution. The other criterion used is the relative gap between the *BUB* and the *BLB*. We also apply an ultimate stopping criteria whereby we limit the number of passes of the decomposition algorithm.

In step 4 our aim is to prepare a new updated vector of Lagrangian multipliers in order to create tighter bounds. For this type of discrete programming problem the Lagrangian relaxation method converges without much difficulty. There is a trade off between using a small step size to create a better bound but with more passes versus using a large step size with less number of passes. Also if the step size is considered too small then an excessive number of iterations may be needed to get a new bound. This is practically important when we have to solve an IP or MIP problem at each pass, as is the case in our investigation. The manipulation of the step size is assessed by changing the parameter π and testing its performance.

Configuration Evaluation We compute a set of N distinct configurations renumbered as n=$1,\ldots,N$, where $N > S$. We then determine the best configuration for each scenario by solving problem P_{WS} for all configurations n=$1,\ldots,N$. This usually results in a substantially smaller number of good configurations than we originally considered.

4.4 Computational results

In order to compare the performance of the two approaches, computational tests were carried out. A Pentium PC 500 MHz with 128-MB memory is used for solving the large model. The phases of Scenario Analysis have been carried out on the real planning model with 100 scenarios in the direct approach and using Lagrangian relaxation. Overall, by applying 25 Lagrangian iterations for 100 scenarios, 2500 configurations are generated of which 2050 are unique. The time taken for generating the integer solutions through Lagrangian relaxation is 26 hrs and 23 mins. The performance of all configurations has been verified and finally a lower bound, an upper bound and a feasible integer solution for each scenario is calculated. The results show that for all scenarios the lower bound calculated by Lagrangian relaxation is better than the LP relaxation. In order to find a feasible integer solution for each scenario the Lagrangian solution is expanded to a complete solution to the original problem.

Step 0: Initialise
$\qquad BUB = \infty, BLB = -\infty$, Pass $= 0, \lambda = 0$
Step 1: Solve relaxation problem
$\qquad$ Solve IP Problem P_{PIP} use B& B Algorithm
$\qquad$ Save new configuration
$\qquad$ Solve LP Problem P_{LP}
Step 2: Create a feasible solution
$\qquad$ Solve LP Problem P_{WS} while x is fixed to $\bar{x}$
Step 3: Solution quality
$\qquad BLB = \text{Max } \{BLB, \text{ Obj Value } P_{PIP} + \text{ Obj Value } P_{LP} - \lambda\, h\}$
$\qquad BUB = \text{Min } \{BUB, \text{ Obj Value } P_{WS} \text{ with } x \text{ fixed }\}$
$\qquad$ **if** (Stopping Criteria Satisfied) **then**
$\qquad\qquad\qquad s = s + 1$
$\qquad\qquad\qquad$ **if** ($s > $ Max No. of Scenario) **then** Stop
$\qquad\qquad\qquad$ Go To Step 0
$\qquad$ **else**
$\qquad\qquad\qquad$ Pass = Pass + 1
$\qquad\qquad\qquad$ Go To Step 4
$\qquad$ **endif**
Step 4: Update Lagrangean multiplier
$\qquad$ Compute step size $T = \dfrac{\pi(BUB - BLB)}{\|G\|^2}$
Update the multiplier values
$\qquad \lambda_i = \max\{0, \lambda_i + TG_i\}$
$\qquad$ Go To Step 1

Figure 8.8. Pseudo Code 1

Our result shows that this solution is achieved with a better objective value than the direct approach when using Branch and Bound.

Because we are not able to achieve an optimum solution for each scenario, the second phase of Scenario Analysis is carried out over all scenarios. This leads to an improvement on the upper bound of some scenarios and an increase in the number of scenarios with relatively better objective values. The time taken during this procedure is 447 hrs and 36 minutes.

We treat Lagrangian relaxation as a column generator that is able to generate more configurations in each run. The only disadvantage of this approach is that we have to evaluate a large number of potentially good configurations. To overcome this problem we analyse objective function values. We find that a configuration that corresponds to high capacity leads to good objective values. Consequently good configurations are characterised by having a high strategic cost. Therefore in a simplification of the evaluation, we concentrate on configurations whose strategic cost is greater than the average strategic cost. This results in 70% less computational effort in evaluating the configurations. Also it allows us to consider more configurations that lead to better objective values for each scenario and improves the average upper bound. It takes 26 hours and 23 minutes to generate the 2050 configurations, whereas it takes 447 hours and 37 minutes to evaluate these 2050 configurations against the 100 scenarios.

We rank all the $x \in$ ACS by calculating their hedged *here-and-now* objective value over all 100 scenarios (see figure 8.9). The solution with the least "here-and-now" objective value is then chosen as the best solution for the problem. Next we analyse each scenario related deterministic model. We calculate the number of scenarios for which a given configuration, $\bar{x}$, provides the best solution (see figure 8.10). This figure shows that only 8 of out of the 2050 configurations provide the best solution. It is therefore natural to conclude that some, if not all, of them are good choices for the first stage decisions. Looking from the perspective of P_{2SIP} model this is definitely not the case! None of the configurations in figure 8.9 appear in figure 8.10. Although configuration number 1512 provided the best performance for 60 scenarios, however, its hedged objective of 3.2×10^6 is much higher that 2.64×10^6- of the *best hedged* solutions of configurations 857 and 1762. Comparing figure 8.9 and figure 8.11 it is apparent that a solution that performs well on a single scenario or on a subset of scenarios need not be the best hedged decision when compared with all the scenarios.

Z_{WS}	Z_{HN}	Z_{EEV}	$EVPI = Z_{HN} - Z_{WS}$	$VSS = Z_{EEV} - Z_{HN}$
2.52×10^6	2.64×10^6	3.59×10^6	1.2×10^5	9.5×10^5

Table 8.4. The stochastic metrics

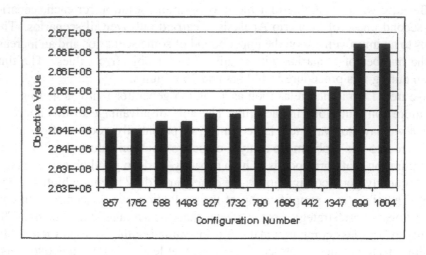

Figure 8.9.   Best hedged-value of the configuration

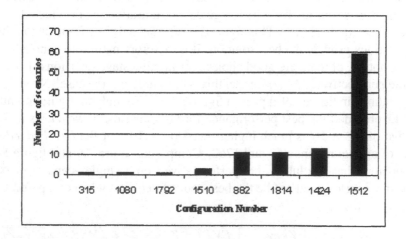

Figure 8.10.   The frequency of the configuration selected

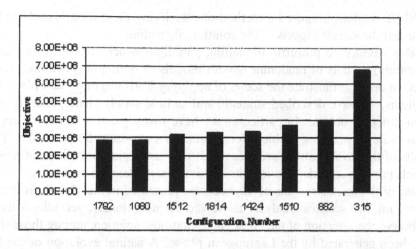

Figure 8.11. The probability weighted objective value of the configuration

4.5 Generalizing the Heuristic to an "Optimum-Seeking" Method

The heuristic described in section 8.4.4 can be generalized to an optimum-seeking method by adopting a branch-and-price approach. In order to do this, we consider a master problem in which we use the columns generated during the Lagrangean phase are used as columns of the master program. The rows of the master program should enforce the constraints that were relaxed in forming the Lagrangean relaxation Moreover, the master program is also required to enforce the condition that all scenarios share the same decision vector x, which must satisfy binary restrictions. Using this master program as the starting point of a branch-and-price algorithm, we can follow a procedure similar to that suggested in Lulli and Sen [2003]. Essentially, this generalization would alternate between a branch-and-bound algorithm for the master, and the Lagrangean relaxation, with the Lagrange multiplier estimates being generated within the branch-and-bound algorithm. Thus, for instances in which an optimal solution is essential, our heuristic presents a natural starting point for an optimum-seeking algorithm.

5. Discussion and conclusions

The success of Linear Programming and Mixed Integer Programming has in turn fuelled considerable interest in the study of Stochastic Programming and more recently Stochastic Mixed Integer Programming. (SMIP). SP has wide ranging applications in situations where uncertainty and risk are taken into consideration in the planning process. In this chapter we have introduced the SMIP models, proposed a method of classifying these models and briefly discussed the salient aspects of the solution algorithms.

SP is based on a powerful modelling paradigm which combines (scenario generation) models of randomness with models of optimum (planning) decisions. In order to illustrate the scope of applying SMIPs to important planning problems, we have described strategic and tactical supply chain planning and management problem. In particular we have investigated a strategic supply chain strategic planning problem formulated as a two-stage SMIP model. The results of the investigations are presented; they underpin our belief that SMIP models provide considerable insight into important planning problems.

One of the main strengths of the heuristic presented in this chapter is that it closely mimics scenario analysis as viewed by practitioners, yet, allows them to observe the selection of the "best" here-and-now solution, among those that have been generated by the Lagrangean phase. A natural evolution of the SP and SMIP models are to bring together ex ante optimum decision making with ex post simulation evaluation. Our future research focus is therefore to develop modelling and solution environments (Valente et al 2003, Di Domenica et al 2004, Poojari et al 2004) which enable the problem owners to seriously consider 'business analytics' with which they can study hedged decisions and risk decisions under conditions of uncertainty.

References

S. Ahmed, A. J. King, and G. Parija. (2003) A Multi-Stage Stochastic Integer Programming Approach for Capacity Expansion under Uncertainty," *Journal of Global Optimization*, vol. 26, pp.3-24.

A. Alonso-Ayuso, L.F. Escudero, and M.T. Ortuno (2000), A stochastic 0-1 program based approach for air traffic management, European Journal of Operations Research 120, pp. 47-62, 2000.

A. Alonso-Ayuso, L.F. Escudero, A. Garin, M.T. Ortuno, and G. Perez (2003), An approach for strategic supply chain planning under uncertainty based on stochastic 0-1 programming, Journal of Global Optimization, 26, pp. 97-124.

E. Balas and S. Ceria and G. Cornuejols (1993), A lift-and-project cutting plane algorithm for mixed 0-1 programs, Mathematical programming, Vol. 58,pp. 295-324

M. O. Ball, R. Hoffman, A. Odoni, R. Rifkin (2003), A stochastic integer program with dual network structure and its application to the ground holding problem, working paper, Robert. H. Smith, School of Business, University of Maryland.

D. Bienstock and J.F. Shapiro (1988), Optimizing resource acquisition decisions by stochastic programming, Management Science, Vol. 34, pp 215-229.

G.R. Bitran and E.A. Haas and H.Matsuo (1986), Production planning of style goods with high setup costs and forecast revisions, Operations Research, Vol. 34, pp 226-236.

J. R. Birge, F Louveaux (1997), Introduction to stochastic programming, Springer-Verlag, New York, NY.

J.R. Birge (1988), Large-scale linear programming techniques in Stochastic Programming, Numerical techniques for stochastic optimisation (Y. Ermoliev and R. Wets eds.) Springer, Berlin

C. C. Caroe, R. Schultz (1999), Dual composition in stochastic integer programming, Operations Research Letters, 24, pp. 37-45.

C. C. Caroe (1998), Decomposition in Stochastic Integer Programming, PhD. Thesis, Institute of Mathematical Sciences, Dept. of Operations Research, University of Copenhagen, Denmark.

C.C. Caroe and A. Ruszczynski and R. Schultz (1997), Unit commitment under uncertainty via two-stage stochastic programming, NOAS, C.C. Caroe and D. Pisinger Eds., Department of Computer Science, University of Copenhagen, pp 21-30.

P. Carpentier, G. Cohen, J.C. Culioli and A. Renaud(1996), Stochastic optimization of unit commitment, a new decomposition framework, IEEE Transactions on Power Systems, 11. pp. 1067-1073.

M.A. Cohen, P.V. Kamesam, P. Kleindorfer, H.L. Lee, and A. Tekerian (1990), "Optimizer: IBM's Multi-Echelon Inventory System for Managing Service Logistics," Interfaces, 20, 1, pp65-82.

M.C. Cooper, Douglas M. Lambert and Janus D. Pugh (1997),"Supply Chain Management: More Than a New Name for Logistics," Vol. 8, No. 1, pp. 1-14.

T. Davies. (1993), "Effective supply chain management", Sloan Management Review, summer, pp 35-45.

M.A.H Dempster(ed) (1980), Stochastic Programming, Academic Press.

M.A.H Dempster, N Hicks Pedron, E a Medova, J E Scott and A Sembos (2000), Planning logistics operations in the oil industry: 2. Stochastic modelling, Journal of the Operational Research Society. Vol 51, No 11,pp"1271-1288.

C.L. Dert (1995), Asset Liability Management for Pension Funds: A Multistage Chance Constrained Programming Approach, Erasmus University, Rotterdam, The Netherlands.

N. Di Domenica, G. Mitra, G. Birbilis, P Valente (2004), Stochastic programming and scenario generation within a simulation framework an information systems perspective, working paper, CARISMA.

B. Domniguez-Ballesteros, G Mitra, C Lucas and C Poojari (2004), " An integrated Strategic and Tactical supply chain model under uncertainty", working paper, Centre for the analysis of risk and optimisation modelling applications.

S. Drijver, W. Klein Haneveld and M. van der Vlerk (2003), Asset Liability Management modeling using multistage mixed-integer Stochastic Programming in Asset and Liability Management Tools, A handbook for Best Practice, Risk Books.

G.D. Eppen and R.K. Martin and L. Schrage (1989), A Scenario Approach to capacity planning, Operational Research, Vol. 37, No. 4, pp. 517-525

L.F. Escudero and P.V. Kamesam and A.J. King and R-J.B. Wets (1993), Production planning via scenario modelling, Annals of Operations Research, 43, pp 311-335.

. L.F. Escudero, E. Galindo, G. Garcia, E. Gomez and V. Sabau (1999), "SCHUMANN. A modelling framework for supply chain management under uncertainty," European Journal of Operational Research, 1999, Vol. 119, pp. 14-34.

J.L. Higle and S. Sen, Stochastic Decomposition (1996): A Statistical Method for large-scale Stochastic Linear Programming, Kluwer Academic Publishers, and University of Arizona.

J. L. Higle and S. Sen (1991), Stochastic Decomposition: An algorithm for two-stage linear programs with recourse, Math. Of Operations Research, 16. pp. 650-669.

W. K. Klein Haneveld, L Stougie, and M. H. van der Vlerk (1995), On the convex hull of the simple integer recourse objective function, Annals of Operations Research, 56, pp. 209-224.

W. K. Klein Haneveld, L Stougie, and M. H. van der Vlerk (1996), An algorithm for the construction of convex hulls in simple integer recourse programming, Annals of Operations Research, 64, pp. 67-81.

W. K. Klein Haneveld, L Stougie, and M. H. van der Vlerk (1999), Stochastic integer programming: general models and algorithms, Annals of Operations Research, 85, pp. 39-57.

B.J. Lageweg and J.K. Lenstra and A.H.G. Rinnooy Kan and L. Stougie (1985), Stochastic integer programming by dynamic programming, Statistica Neerlandica, 39, 97113.

G. Laporte, F.V. Louveaux and L. Van Hamme (1994), Exact solution to a location problem with stochastic demands, Transportation Science, Vol. 28, pages 95-103.

G. Laporte, F. V. Louveaux (1993), The integer L-shaped methods for stochastic integer programs with complete recourse, Operations Research Letters, 13, pp. 133-142.

G. Laporte, F.V. Louveaux and H. Mercure (1992), The vehicle routing problem with stochastic travel times, Transportation Science, Vol. 26, pp. 161-170.

C.M.A Leopoldino, M.V.F Pereira, L.M.V Pinto and C.C Ribeiro (1994), A constraint generation scheme to probabilistic linear problems with an application to power system expansion planning, Annals of Operational Research, Vol. 50, pp367-385

A. Lokketangen and D.L. Woodruff (1996), Progressive hedging and Tabu search applied to mixed integer (0,1) multi-stage stochastic programming, Journal of Heuristics, Vol. 2, pp. 111-128.

G. Lulli, S. Sen (2002), Stochastic batch-sizing problems, Network Interdiction and Stochastic Integer Programming (D. L. Woodruff, ed.), pp. 85-103, Kluwer Academic Press.

G. Lulli, S. Sen (2003), A branch and price algorithm for multi-stage stochastic integer programs with applications to stochastic batch-sizing problems, to appear in Management Science.

S.A. Malcolm, and S.A. Zenios, (1994). Robust optimization of power systems capacity expansion under uncertainty, Journal of the Operational Research Society 45: 1040–1049.

S.A MirHassani, C Lucas, G Mitra, E. Messina, C.A Poojari (2000), Computational solutions to capacity planning under uncertainty, in Parallel Computing Journal (26), pages-511-538, 2000.

J.M. Mulvey and H. Vladimirou (1992), Stochastic Network Programming for Financial Planning Problems, Management Science, Vol. 38, 11, pp 1642-1664

V. I. Norkin, Y. M. Ermoliev and A Ruszczynski (1998), On optimal allocation of indivisibles under uncertainty, Operations Research, 46, pp. 381-395.

M. P. Nowak, W. Roemisch (2000), Lagrangian relaxation applied to power scheduling in a hydro-thermal system under uncertainty, Annals of Operations Research, 100, pp. 251-272.

C.A. Poojari, C.A Lucas and G. Mitra (2003), A heuristic technique for solving stochastic integer programming models – a supply chain application, submitted to Journal of Heuristics, 2003.

C.A. Poojari, G. Mitra, E.F.D Ellison, S. Sen (2004) "An investigation into Stochastic programming algorithms", working paper, CARISMA.

M. Porter, (1985), The value chain and competitive advantage, Chapter 2 in Competitive Advantage: Creating and Sustaining Superior Performance, Free Press, and New York, 33-61.

M. Riis, A. J. V. Schriver, and J. Lodahi (2002), Network planning in telecommunications: a stochastic programming approach, working paper, Department of Operations Research, University of Aarhus, Aarhus, Denmark, 2002.

L. Schrage (2000), An optimization modelling system, International Thomson Publishing, USA.

R. Schultz (2003), Integers in stochastic programming, working paper, Institute of Mathematics Gerhard-Mercator University, Duisberg, Germany, to appear in Mathematical Programming.

R. Schultz, L. Stougie, and M. H. van der Vlerk (1998), Solving stochastic programs with integer recourse by enumeration: a framework using Grobner basis reduction, Mathematical Programming, 83, pp. 71-94.

S. Sen (2003), Algorithms for stochastic mixed-integer programming models, to appear in Handbook of Discrete Optimization, IK. Aardal, G. L. Nemhauser and R. Weismantel eds.) Elsevier.

S. Sen and J.L. Higle (2000), The C^3 theorem and a D^2 algorithm for large scale stochastic integer programming: Set convexification, working paper, MORE Institute, SIE Department, University of Arizona, Tucson, AZ 85721 (also Stochastic Programming E-Print Series 2000-26.

S. Sen, L. Ntaimo (2003), The million variable "march" for stochastic combinatorial optimization, working paper, MORE Institute, SIE Department, University of Arizona, Tucson, AZ 85721, submitted for publication, 2003.

S. Sen and H. D. Sherali (2003), Decomposition with Branch-and-Cut Approaches for Two-Stage Stochastic Integer Programming, working paper, MORE Institute, SIE Department, University of Arizona, Tucson, AZ 85721, submitted for publication.

J. Shapiro (2000), "Modelling supply chain", Duxbury Press.

H.D. Sherali and B.M.P Fraticelli (2002), A modified Benders 'partitioning approach for discrete subproblems: An approach for stochastic programs with integer recourse, Journal of Global Optimization Vol. 22 pp. 319 –342.

S. Takriti, J.R. Birge and E. Long (1996), A Stochastic Model for the Unit Commitment Problem, IEEE Transactions on Power Systems, Vol. 11, pp 1497-1508.

M. Tawarmalani, S. Ahmed, and N.V. Sahinidis (2000), A finite branch and bound algorithm for two-stage stochastic integer programs, working paper, Department. Of Chemical Engineering, Univ. of Illinois, Urbana-Champagne, IL.

M. H. van der Vlerk (2003), Six-pack recourse: a mixed-integer simple recourse problem, working paper, Department of Econometrics and Operations Research, University of Groningen, The Netherlands.

P. Valente, G. Mitra, and C.A. Poojari (2003), Software tools for stochastic programming: A stochastic programming integrated environment (spine), accepted in Applications of Stochastic Programming, MPS-SIAM- Series in Optimisation, Springer-Verlag.

H. Wagner (1969), Principles of Operations Research with Application to Managerial Decisions, Prentice Hall, USA.

H. P. Williams (1999), Model Building in Mathematical Programming, John Wiley.

R.D. Wollmer (1980), Two-stage linear programming under uncertainty with 0-1 integer first-stage variables, Mathematical Programming, Vol. 19, 279-288, 1980.

Chapter 9

LOGIC INFERENCE AND A DECOMPOSITION ALGORITHM FOR THE RESOURCE-CONSTRAINED SCHEDULING OF TESTING TASKS IN THE DEVELOPMENT OF NEW PHARMACEUTICAL AND AGROCHEMICAL PRODUCTS

Christos T. Maravelias and Ignacio E. Grossmann*

Department of Chemical Engineering
Carnegie Mellon University, Pittsburgh
PA 15213, U.S.A.

ig0c@andrew.cmu.edu

Abstract   In highly regulated industries, such as agrochemical and pharmaceutical, new products have to pass a number of regulatory tests related to safety, efficacy and environmental impact, to gain FDA approval. If a product fails one of these tests it cannot enter the market place and the investment in previous tests is wasted. Depending on the nature of the products, testing may last up to 10 years, and the scheduling of the tests should be made with the goal of minimizing the time to market and the cost of testing. Maravelias and Grossmann (2001) proposed a mixed-integer linear program (MILP) that considers a set of candidate products for which the cost, duration and probability of success of their tests is given, as well as the potential income if the products are successfully launched. Furthermore, there are limited resources in terms of laboratories and number of technicians. If needed, a test may be outsourced at a higher cost. The major decisions in the model are: (i) the decision to perform in-house or outsource a testing task, (ii) the assignment of resources to testing tasks, and (iii) the sequencing and timing of tests. The objective is to maximize the net present value of multiple projects. The mixed-integer linear program can become very expensive for solving real world problems (2-10 products and 50-200 tests). In order to improve the linear programming relaxation, we propose the use of logic cuts that are derived from implied precedences that arise in the graphs of the cor-

---

*Author to whom correspondence should be addressed

responding schedules. The solution of a single large-scale problem is avoided with a heuristic decomposition algorithm that relies on solving a reduced MILP model that embeds the optimal schedules obtained for the individual products. It is shown that a tight upper bound can be easily determined for this decomposition algorithm. On a set of test problems the proposed algorithm is shown to be one to two orders of magnitude faster than the full space method, yielding solutions that are optimal or near optimal.

Keywords: New Product Development (NPD), Food and Drug Administration (FDA), Environmental Protection Agency (EPA), Clinical Trials, Scheduling, Mixed Integer Linear Programming (MILP), Logic Cuts, Preprocessing, Implied Precedences, Decomposition Heuristics.

1.   Introduction

The problem of selecting, testing and launching new agrochemical and pharmaceutical products (Robbins-Roth, 2001) has been studied by several authors. Schmidt and Grossmann (1996) proposed various MILP optimization models for the case where no resource constraints are considered. The basic idea in this model is to use a discretization scheme in order to induce linearity in the cost of testing. Jain and Grossmann (1999) extended these models to account for resource constraints. Honkomp et. al. (1997) addressed the problem of scheduling R&D projects, and Subramanian et. al. (2001) proposed a simulation-optimization framework that takes into account uncertainty in duration, cost and resource requirements. Maravelias and Grossmann (2001) proposed an MILP model that integrates the scheduling of tests with the design and production planning decisions. Schmidt et. al. (1998) solved an industrial scale problem based on a subset of DowElanco's testing processes, with one agrochemical product that must undergo 65 tests, without taking into account resource constraints. It was shown that the optimization of the testing process resulted in very substantial savings. In pharmaceutical industry the savings for one product can be of the order of million dollars. Although real-world problems are better structured than randomly generated problems, they are also hard to solve if resource constraints are taken into account.

In this chapter, we first discuss logic based approaches that reduce the combinatorics of the problem, and then propose a decomposition algorithm for the solution of the resource-constrained scheduling problem for which a tight upper bound is derived. We use the scheduling MILP model developed by Jain and Grossmann (1999) and refined by Maravelias and Grossmann (2001).

2.   Motivating Example

To illustrate the trade-offs in the scheduling of testing tasks for new products, consider the example of Figure 9.1 (from Schmidt and Grossmann, 1996).

A new agrochemical product must undergo three tests: ground water studies, an acute toxicology test and formulation chemistry studies. For each test k, the cost c_k, duration d_k and probability of success p_k are assumed to be known and are shown in Figure 9.1. If one of the tests fails the product cannot enter the market and the investment in the previous tests is wasted. The probability of conducting a test k is equal to the product of the probabilities of the tests that are scheduled before test k. In the schedule shown on the left of Figure 9.1,

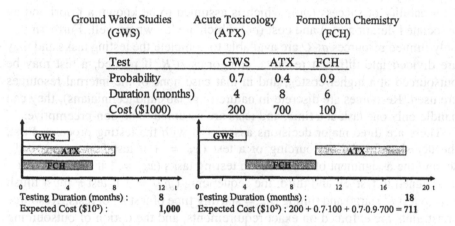

	Ground Water Studies (GWS)	Acute Toxicology (ATX)	Formulation Chemistry (FCH)
Test	GWS	ATX	FCH
Probability	0.7	0.4	0.9
Duration (months)	4	8	6
Cost ($1000)	200	700	100

Testing Duration (months) :	8	Testing Duration (months) :
Expected Cost ($10³) :	1,000	Expected Cost ($10³) : 200 + 0.7·100 + 0.7·0.9·700 = 711

Figure 9.1. Motivating example

all tests are scheduled to start at time $t = 0$. If all tests are successful then the product can be launched at $t = 8$. Assuming that the cost for each test is paid when the test starts, the expected cost for the first schedule is $1,000,000, since the probability of conducting all tests is 1. In the second schedule, the formulation chemistry test is scheduled to begin after the ground water test is finished, at $t = 4$, and the toxicology test is scheduled after the formulation chemistry test is finished, at $t = 10$. This implies that the formulation chemistry test will be conducted only if the ground water test is successful, and that the toxicology test will be conducted only if the other two tests are successful. Thus, the probability of conducting the three tests are 1, 0.9*0.7=0.63 and 0.7 respectively, which, in turn, means that the expected cost of the second schedule is $711,000 and that, if all tests are successful, the new product can be launched after 18 months, i.e. 10 months later than in the first schedule. This means that in the second case, the expected cost of testing is smaller but the income from sales, which is a decreasing function of time, is also smaller. The problem becomes more difficult when resource constraints and technological precedences are also taken into account. The MILP scheduling model of

Maravelias and Grossman (2001) determines the schedule that maximizes the net present value of several products. In the present chapter we develop an algorithm for the solution of this model.

3.   Model

The problem addressed in this chapter can be stated as follows. Given are a set of potential products that are in various stages of the company's R&D pipeline. Each potential product $j \in J$ is required to pass a series of tests. Failure to pass any of these tests implies termination of the project. Each test $k \in K$ has a probability of success (p_k), which is assumed to be known a priori and an associated duration (d_k) and cost (c_k) which are known as well. Furthermore, only limited resources $q \in Q$ are available to complete the testing tasks and they are divided into different resource categories $r \in R$. If needed, a test may be outsourced at a higher cost $\hat{c}_k$, and in that case none of the internal resources are used. Resources are discrete in nature (e.g. labs and technicians), they can handle only one task at a time, and tests are assumed to be non-preemptive.

There are three major decisions associated with the testing process. First, the decision for the outsourcing of a test ($x_k = 1$ if test k is outsourced), second the assignment of resources to testing tasks ($\hat{x}_{kq} = 1$ if resource unit q is assigned to test k), and third, the sequencing ($y_{kk'} = 1$ if test k must finish before test k' starts) and timing ($s_k =$ starting time of test k) of tests. Resource constraints are enforced on exact requirements, and the option of outsourcing is used when existing resources are not sufficient. Thus, the schedule obtained is always feasible, and rescheduling is needed only when a product fails a test.

The proposed scheduling model **(M)** consists of equations (9.1) to (9.16). The nomenclature is given at the end of the chapter.

$$\max NPV = \sum_j (INC_j - TC_j) \tag{9.1}$$

$$u_{jm} \geqslant T_j - b_{jm} \quad \forall j \in J, \forall m \tag{9.2}$$

$$INC_j = F_j - \sum_m f_{jm} u_{jm} \quad \forall j \in J \tag{9.3}$$

$$TC_j = \sum_{k \in K(j)} \left\{ c_k \left(\sum_n e^{\alpha_n} \lambda_{kn} \right) + \hat{c}_k \left(\sum_n e^{\alpha_n} \Lambda_{kn} \right) \right\} \quad \forall j \in J \tag{9.4}$$

$$y_{kk'} = 1 \wedge y_{k'k} = 0 \quad \forall (k, k') \in A \tag{9.5}$$

$$s_k + d_k \leqslant T_j \quad \forall j \in J, \forall k \in K(j) \tag{9.6}$$

$$s_k + d_k - s_{k'} - U(1 - y_{kk'}) \leqslant 0 \quad \forall j \in J, \forall k \in K(j), \forall k' \in K(j) \quad (9.7)$$

$$w_k = -\rho s_k + \sum_{k' \in KK(k)|k' \neq k} \ln(p_{k'}) y_{k'k} \quad \forall k \in K \qquad (9.8)$$

$$w_k = \sum_n \alpha_n (\lambda_{kn} + \Lambda_{kn}) \quad \forall k \in K \qquad (9.9)$$

$$\sum_n \lambda_{kn} = 1 - x_k \quad \forall k \in K \qquad (9.10)$$

$$\sum_n \Lambda_{kn} = x_k \quad \forall k \in K \qquad (9.11)$$

$$\sum_{q \in (QT(k) \cap QC(r))} \hat{x}_{kq} = N_{kr}(1 - x_k) \quad \forall k \in K, \forall r \in R \qquad (9.12)$$

$$\hat{x}_{kq} + \hat{x}_{k'q} - y_{kk'} - y_{k'k} \leqslant 1$$
$$\forall q \in Q, \forall k \in K(q), \forall k' \in (K(q) \cap KK(k))|k < k' \qquad (9.13)$$

$$\hat{x}_{kq} + \hat{x}_{k'q} - \hat{y}_{kk'} - \hat{y}_{k'k} \leqslant 1$$
$$\forall q \in Q, \forall k \in K(q), \forall k' \in (K(q) \backslash KK(k))|k < k' \qquad (9.14)$$

$$s_k + d_k - s_{k'} - U(1 - \hat{y}_{kk'}) \leqslant 0$$
$$\forall q \in Q, \forall k \in K(q), \forall k' \in K(q) \backslash KK(k) \qquad (9.15)$$

$$x_k, \hat{x}_{kq}, y_{kk'}, \hat{y}_{kk'} \in \{0, 1\}, \quad T_j, s_k, u_{jm} \geqslant 0, \quad \lambda_{kn}, \Lambda_{kn} \in [0, 1] \quad (9.16)$$

The objective in (9.1) is to maximize the NPV of multiple projects, i.e. the income from sales minus the cost of testing. The income is assumed to be a piecewise linear decreasing function of completion time [equations (9.2) and (9.3)]. To avoid nonlinearities, the discounted, expected cost of testing is approximated as a piecewise linear function [eq. (9.4)]. Technological precedences are enforced through constraints (9.5). Constraint (9.6) ensures that the completion time T_j of a product j should be greater than the completion time of any test $k \in K(j)$. Constraint (9.7) enforces that if test k is scheduled before test k' ($y_{kk'} = 1$), the starting time of k' should be greater than the completion time of k. Constraints (9.8)-(9.11) are used for the linearization of the discounted cost of each test k (see Schmidt and Grossmann, 1996).

In order to calculate the discounted, expected cost C_k of test k, assuming that test k can only be performed in-house, we should multiply the nominal cost c_k with a discounting factor $e^{-\rho s_k}$, and the probability of conducting test k, i.e. the product of probabilities of success of the tests scheduled before test k, $\prod_{k' \neq k} p_{k'} y_{k'k}$:

$$C_k = c_k e^{-\rho s_k} \prod_{k' \neq k} p_{k'} y_{k'k} = c_k e^{-\rho s_k + \sum_{k' \neq k} \ln(p_k) y_{k'k}} = c_k e^{w_k}$$

which is a non-linear expression. To avoid solving a large-scale MINLP we express the exponent as a linear combination of fixed grid points α_n and we approximate the exponential factor e^{w_k} as a piecewise linear function:

$$-\rho s_k + \sum_{k \in KK(k)|k' \neq k} \ln(p_k) y_{k'k} = w_k = \sum_n \lambda_{kn} \alpha_n$$

$$C_k = c_k e^{w_k} = c_k e^{\sum_n \alpha_n \lambda_{kn}} \approx c_k \sum_n e^{\alpha_n} \lambda_{kn}$$

where λ_{kn} are the weights for the expression of the exponent w_k as linear combination of grid points α_n ($\lambda_{kn} \in [0,1]$). If outsourcing is considered, a similar linearization is introduced, with weights $\Lambda_{kn} \in [0,1]$. Thus, w_k is given by equations (9.8) and (9.9), the linearization weights are "activated" through equations (9.10) and (9.11), and the approximated cost is calculated in equation (9.4).

Constraint (9.12) ensures that if a test is not outsourced ($x_k = 0$) the required number of resources from each resource category is assigned to that test. Constraints (9.13) and (9.14) are the integer linear transformations of the following logical relationship: *if a test k is assigned to resource q ($\hat{x}_{kq}=1$), then for resource constraints to hold, any other test k' is either not assigned to resource q ($\hat{x}_{k'q}=0$), or there is a precedence relation between test k and k' ($y_{kk'}=1$ or $y_{k'k} = 1$).* Constraint (9.13) is expressed for tests k and k' that can be scheduled in the same resource q ($k \in K(q)$, $k' \in K(q)$) and belong to the same product j (i.e. $k' \in KK(k)$), while constraint (9.14) is expressed for tests that can be scheduled in the same resource but belong to different products ($k' \notin KK(j)$). Constraint (9.15) is similar to constraint (9.7) and it is used to enforce sequencing between tests of different products that can be assigned to the same resource unit. Maravelias and Grossmann (2004) extended model (M) to account for resource installation during the course of testing and variable resource allocation.

4. Logic Cuts

4.1 Cycle Breaking Cuts

In order to enhance the solution of the MILP given by (9.1)–(9.16), we consider two major types of logic cuts (Hooker et al., 1994). These cuts are redundant in the sense that they are not necessary for the formulation and they do not cut-off any integer feasible solution. First consider the cuts,

$$y_{kk'} + y_{k'k} \leqslant 1 \quad \forall k, \forall k' \in KK(k) | k < k' \tag{9.17}$$

$$y_{kk'} + y_{k'k''} + y_{k''k} \leqslant 2 \quad \forall k, \forall k' \in KK(k), \forall k'' \in KK(k) | k < k' < k'' \tag{9.18}$$

$$y_{k'k} + y_{kk''} + y_{k''k'} \leqslant 2 \quad \forall k, \forall k' \in KK(k), \forall k'' \in KK(k) | k < k' < k'' \tag{9.19}$$

These logic cuts were proposed by Schmidt and Grossmann (1996), and they forbid 2-cycles ($k \rightarrow k' \rightarrow k$) and 3-cycles ($k \rightarrow k' \rightarrow k'' \rightarrow k$); i.e. they do not allow both variables $y_{kk'}$ and $y_{k'k}$ to be 1, or more than two of the $y_{kk'}$, $y_{k'k''}$, $y_{k''k}$ to be 1. In that work the authors proposed a model in which all these cuts are added a priori. For larger problems, though, the number of these cuts is very large (e.g. for a problem with two products, 90 total tests and ten resource units the number of these cuts is 78,765, whereas the necessary equations are only 10,156). Moreover, only a few of these cuts are violated, and thus not all of them are needed. An obvious improvement is to initially add at the root node only the cuts that are violated, and then apply a branch and cut scheme (Johnson et al., 2001), where in each node we add only the violated cuts. Additional improvements can be achieved by adding cuts that prevent 4-, 5-, 6-, etc., cycles. In order to explore the effectiveness of these ideas we have solved three test problems with two to four products and 90 to 150 tests, and for which the data is given in Appendix A. The numbers of products, tests, resource categories and units, as well as the number of binary and continuous variables of the three examples are given in Table 9.1. First, we explored the effect of the cycle-breaking cuts by solving four different cases (Table 9.2).The first case corresponds to model (M) with no cycle-breaking cuts; in the second case we added all 2- and 3-cycle breaking cuts (Schmidt and Grossmann, 1996); in the third one we added only the 2- and 3-cycle cuts that are violated at the root node (RNV cuts); and in the fourth case we added 2-, 3-, 4- 5-cycle breaking cuts that are violated at the root node. As shown, by adding only the violated 2- and 3-cycle breaking cuts, we can get an LP relaxation that is almost as tight as the one obtained by adding all 2- and 3-cycle breaking cuts. Furthermore,

	Example 1	Example 2	Example 3
Products	2: A & B	3:A, B & C	4:A, B, C & D
Tests	60,30	60,30,40	60,30,40,20
Resource categories/units	4/10	4/10	4/10
Binary variables	5,628	10,569	14,060
Continuous variables	1,991	2,875	3,319

Table 9.1. Model statistics of test problems

by adding the violated 2-, 3-, 4- and 5-cycle breaking cuts, we obtain LP relaxations that are tighter than the relaxations obtained by 2- and 3- cycle-breaking cuts, and with no significant increase in the number of constraints. Thus, the last approach will be further explored.

	Example 1		Example 2		Example 3	
Optimal Objective	4,701.8		8,002.3		12,428.0	
	Cons.	LP Rel.	Cons.	LP Rel.	Cons.	LP Rel.
(M): (9.1) – (9.16)	10,156	6,176.8	19,931	9,929.3	26,069	14,415.9
w/ all 2-, 3- cuts	88,921	5,698.7	119,236	9,253.5	127,844	13,706.0
w/ RNV 2-,3- cuts	10,447	5,698.7	20,524	9,265.1	26,725	13,718.0
w/ RNV 2-,3-,4-,5- cuts	14,408	5,084.5	29,711	8,462.1	36,354	12,909.3

Table 9.2. Addition of cycle-breaking cuts: Number of constraints and LP relaxation

Note that for n tests $(k_1, k_2, \ldots k_n)$ there are $n!/n = (n-1)!$ possible cycles, and thus $(n-1)!$ cycle-breaking cuts. Here, we add only two of these cuts; i.e. the ones that prevent the forward $k_1 \rightarrow k_2 \rightarrow \ldots \rightarrow k_n \rightarrow k_1$, and backward $k_n \rightarrow k_{n-1} \rightarrow \ldots \rightarrow k_1 \rightarrow k_n$ cycles. For the case of 4-cycles, for instance, we add the cuts that correspond to cycles (a) and (b) of Figure 9.2.

$$y_{kk'} + y_{k'k"} + y_{k"k"'} + y_{k"'k} \leqslant 3 \quad \forall k, \forall k' \in KK(k),$$
$$\forall k" \in KK(k), \forall k"' \in KK(k) | k < k' < k" < k"' \tag{9.20}$$

$$y_{k"'k"} + y_{k"k'} + y_{k'k} + y_{kk"'} \leqslant 3 \quad \forall k, \forall k' \in KK(k),$$
$$\forall k" \in KK(k), \forall k"' \in KK(k) | k < k' < k" < k"' \tag{9.21}$$

4.2   Implicit Precedence Cuts

A second class of logic cuts can be derived from the sequencing of tests for a given new product. Consider the case where a product X has to pass the three tests of Table 9.3. Due to technological precedences test 2 must be performed after the completion of test 1, and test 3 must be performed after the

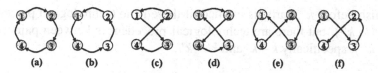

Figure 9.2. Different cycles for four tests

completion of test 2. This implies that binaries y_{12} and y_{23} are fixed to 1 and binaries y_{21} and y_{32} are fixed to 0 [constraint (9.5)].

Test k	1	2	3
p_k	1.0	1.0	0.5
c_k	20	10	5
d_k	5	5	5

Table 9.3. Testing data for the motivating example

No of intervals for piecewise linear income function: $m = 3$		
$F_X = 100$		
$b_{X1} = 0$	$b_{X2} = 12$	$b_{X3} = 24$
$f_{X1} = 2$	$f_{X2} = 2$	$f_{X3} = 2$

Table 9.4. Income data for the motivating example.

The solution of the LP relaxation, for the income data of Table 9.4, is the following (using $U = 60$):
Starting times: $s_1 = 0, s_2 = 5, s_3 = 10$.
Sequencing binaries: $y_{13} = 0.25, y_{31} = 0.75$.
Cost of testing = 37.7.
This means that we should next branch on binary y_{13} (or y_{31}) to find a feasible solution. As shown in Figure 9.3 the branch-and-bound tree, for the model with no logic cuts, consists of 7 nodes, and the optimal solution is 29.7. If, on the other hand, we had noticed that the precedence between tests 1 and 3 is implied by the other two precedences, we could have fixed binaries y_{13} and y_{31} as well. In this case the LP relaxation at the root node could have been tightened yielding a feasible solution with $y_{12} = y_{23} = y_{13} = 1, y_{21} = y_{32} = y_{31} = 0$, same starting times as before, and a cost of testing equal to 29.7; i.e. the optimal integer solution is found at the root node by solving the corresponding LP. In larger instances it is unlikely, of course, to find a feasible solution at the root node, but the example shows that by fixing a priori all the implied precedences we reduce the number of free binaries, and thus reduce the size of the branch-and-bound tree. In general, the precedence $k \rightarrow k''$ (k

must finish before k'' starts) is implied, if there is no technological precedence for pair (k, k''), but there are technological precedences between pairs (k, k') and (k', k''), specifically $k \rightarrow k'$ and $k' \rightarrow k''$.

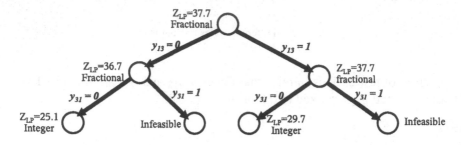

Figure 9.3.   Branch & bound tree of motivating example

In order to derive all implied precedences, we construct the acyclic activity-on-node directed graph, D, of technological precedences. Note that the nodes of an acyclic digraph can always be labeled in a way that the heads of the arcs are higher numbered than the tails of the arcs. Such a labeling is called topological ordering (Ahuja et al., 1993). A topological ordering of an acyclic directed graph can be found in time linear to the number of nodes. Let $n(v)$ be the label of node v, and assume that such a labeling is available for digraph D. Then, for any implied precedence $k \rightarrow k''$ we will have $n(k) < n(k'')$, because the existing technological precedences $k \rightarrow k'$ and $k' \rightarrow k''$ imply $n(k) < n(k')$ and $n(k') < n(k'')$. Thus, if a topological ordering is available, we need only examine ordered triplets (k, k', k'') with $n(k) < n(k') < n(k'')$. The preprocessing algorithm exploits the topological ordering and derives implied precedences in one iteration. If A is the set of technological precedences (i.e. $k \rightarrow k'' \Leftrightarrow (k,k') \in A$), and $|K_j|$ is the cardinality of K_j, the preprocessing algorithm is as follows:

For all products $j \in J$,
  For $k | n(k) = 1 \dots |K_j| - 2$
    For $k' | n(k') = n(k) + 1 \dots |K_j| - 1$
      For $k'' | n(k'') = n(k') + 1 \dots |K_j|$
        If $(k, k') \in A$ AND $(k', k'') \in A$, then $(k, k'') \in A$

Table 9.5.   Preprocessing algorithm (PPROCALG)

The idea of implied precedences can be further exploited during the branch-and-bound search. Consider tests k, k', k'' with no fixed technological precedence among them, i.e. no implied precedences can be deduced for binaries $y_{kk'}$, $y_{k'k''}$, $y_{kk''}$, $y_{k'k}$, $y_{k''k'}$, and $y_{k''k}$ during preprocessing. However, when binaries $y_{kk'}$ and $y_{k'k''}$, for example, are fixed to 1 at some node of the branch-and-bound tree, binary $y_{kk''}$ should also be equal to 1, and binary $y_{k''k}$ should be equal to 0. In the existing formulation, however, there are no constraints forcing binaries $y_{kk''}$ and $y_{k''k}$ to be equal to 1 and 0, respectively. Thus, we can add a new class of cuts that force binary $y_{kk''}$ to be equal to 1 whenever $y_{kk'}$ and $y_{k'k''}$ are equal to 1 at some node of the branch-and-bound tree. In general, for tests k, k', k'', this can be achieved by adding the following six cuts:

$$y_{kk''} \geqslant y_{kk'} + y_{k'k''} - 1 \quad \forall k, \forall k' \in KK(k), \forall k'' \in KK(k) | k < k' < k'' \tag{9.22}$$

$$y_{k''k} \geqslant y_{k''k'} + y_{k'k} - 1 \quad \forall k, \forall k' \in KK(k), \forall k'' \in KK(k) | k < k' < k'' \tag{9.23}$$

$$y_{kk'} \geqslant y_{kk''} + y_{k''k'} - 1 \quad \forall k, \forall k' \in KK(k), \forall k'' \in KK(k) | k < k' < k'' \tag{9.24}$$

$$y_{k'k} \geqslant y_{k'k''} + y_{k''k} - 1 \quad \forall k, \forall k' \in KK(k), \forall k'' \in KK(k) | k < k' < k'' \tag{9.25}$$

$$y_{k'k''} \geqslant y_{k'k} + y_{kk''} - 1 \quad \forall k, \forall k' \in KK(k), \forall k'' \in KK(k) | k < k' < k'' \tag{9.26}$$

$$y_{k''k'} \geqslant y_{k''k} + y_{kk'} - 1 \quad \forall k, \forall k' \in KK(k), \forall k'' \in KK(k) | k < k' < k'' \tag{9.27}$$

Similar cuts that force binary $y_{k''k}$ to be equal to 0 when binaries $y_{kk'}$ and $y_{k'k''}$ are equal to 1 can also be added. The cut, for example, that forces $y_{k''k}$ to be equal to 0 when $y_{kk'} = y_{k'k''} = 1$ [i.e. corresponds to (9.22)] is the following:

$$y_{k''k} \leqslant 2 - y_{kk'} - y_{k'k''} \quad \forall k, \forall k' \in KK(k), \forall k'' \in KK(k) | k < k' < k'' \tag{9.28}$$

The number of these cuts in (9.22)-(9.27) is large ($O(n^3)$), but since only a few of them are violated, we can, again, add only the ones that are violated at the root node. Table 9.6 contains the number of constraints and the LP relaxation of the three test problems that were considered for the cycle breaking logic cuts. Four different cases of the second class of logic implication are shown: the first row corresponds to model (M); the second row corresponds to the case where all the precedences identified by PPROCALC are added a priori; the third corresponds to the case where all cuts (equations (9.22) - (9.27)) are added, and the last one to the case where only the violated cuts are added at the root node.

	Example 1		Example 2		Example 3	
Objective	4,701.8		8,002.3		12,428.0	
	Cons.	LP Rel.	Cons.	LP Rel.	Cons.	LP Rel.
(M): (9.1) – (9.16)	10,156	6,176.8	19,931	9,929.3	26,069	14,415.9
w/ fixed implied precedences	14,356	5,079.3	25,904	8,470.8	32,360	12,918.5
w/ all cuts in (9.22)-(9.27)	239,836	5,001.3	396,461	8,378.6	509,019	12,825.9
w/ RNV cuts in (9.22)-(9.27)	18,817	5,001.3	50,144	8,378.6	60,909	12,825.9

Table 9.6. Precedence implications and cuts: No. of constraints and LP relaxation

As shown, preprocessing has a very large impact in the LP relaxation without increasing the number of constraints (all of the 14,356-10,156=4,200 new constraints are of the form $y_{kk'} = 1$ or $y_{kk'} = 0$). The addition of all logic cuts of the second class increases the number of constraints and yields the same LP relaxation as the fourth case, i.e. the addition of the violated precedence (PR) cuts only. This means that the fourth case dominates the third. We will, therefore, further examine the preprocessing of precedences, and the addition of the violated precedence cuts at the root node. We will also examine the formulation with the violated 2-, 3-, 4- and 5-cycle breaking (CB) cuts at the root node.

The solution statistics of these three cases are given in Table 9.7. A relative optimality tolerance of 0.5% has been used as termination criterion. In all cases reported in this chapter we have used GAMS 20.0/CPLEX 7.0 on a Pentium III at 1000 MHz.

The addition of the violated cycle-breaking cuts does not appear to be effective. For the smallest problem (example 1) it gives shorter computational times, but as the problem size grows, computational times do not improve. The addition of the violated precedence cuts seems to be more effective. The computational time required for example 1 is almost one order of magnitude shorter than the one required by model (M). For examples 2 and 3 computational times are comparable with the ones required by (M), but note that the solutions found have a higher objective value. Preprocessing is not very effective for small problems (example 1), but it gets better for bigger problems; note that for examples 2 and 3 both computational times and the quality of the solution are better than the ones of model (M). Note also that for all examples

	Example 1			Example 2			Example 3		
	Nodes	CPUs	OBJ	Nodes	CPUs	OBJ	Nodes	CPUs	OBJ
(M): (9.1) – (9.16)	130,034	4,029	4,699	201,526	10,917	7,971*	95,668	6,886	12,317*
w/ RNV 2-5 CB cuts	31,486	1,459	4,701	130,852	12,021	7,976*	81,036	9,290	12,374*
w/ RNV PR cuts	8,259	375	4,701	116,268	9,222	7,979*	93,537	9,753	12,382*
w/ fix implied precedences	140,505	3,677	4,695	114,120	5,738	7,990*	73,719	4,797	12,428

*Terminated due to tree size limit

Table 9.7. Solution statistics

the addition of cuts leads to improved solutions, even if the number of explored nodes is smaller. Combinations of the above approaches were also examined, with similar results. This implies that the addition of logic implication (as cuts or as fixing of technological precedences) yields improvement only for small problems. For larger instances it does not improve computational times to a point that they can be solved to optimality in reasonable times. It should be pointed out, though, that the logic cuts presented above are most suitable for a branch-and-cut framework, where in each node of the tree we add only the cuts that are violated, or we fix the value of a binary variable by fixing its lower bound to 1 or its upper bound to 0. We are currently working towards the development of an integrated tool that uses logic implication for the solution of general problems.

5. Decomposition Heuristic

The incidence matrix of the constraints of model (M) exhibits a block triangular form with a few linking constraints. This led us to develop a decomposition heuristic that exploits this special structure. In particular, each one of the constraints in (9.2)-(9.13) include variables that refer to only one product, and only constraints in (9.14) and (9.15) include variables that refer to tests of two different products. Constraints (9.14) and (9.15) include, also, the "mixed" binaries $\hat{y}_{kk'}$ where tests k and k' belong to different products. The incidence matrix of the constraints, therefore, exhibits the structure shown in Figure 9.4, where there is one block for each product [constraints (9.2)-(9.13)] and the two linking constraints (9.14) and (9.15).

If x^j is the set of variables that refer to product j, then the scheduling model can be written as follows:

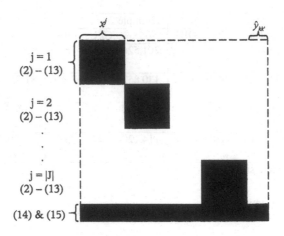

Figure 9.4.   Incidence matrix of constraints

$$\max \ Z = \sum_{j} \left[c^j\right]^T x^j$$
$$s.t. \ A^j x^j \leqslant b^j \quad \forall j \tag{M}$$

$$\hat{x}_{kq} + \hat{x}_{k'q} - \hat{y}_{kk'} - \hat{y}_{k'k} \leqslant 1$$
$$\forall q, \forall k \in K(q), \forall k' \in (K(q)\backslash KK(k))|k < k' \tag{9.29}$$

$$s_k + d_k - s_{k'} - U(1 - \hat{y}_{kk'}) \leqslant 0 \quad \forall q, \forall k \in K(q), \forall k' \in K(q)\backslash KK(k) \tag{9.30}$$

where $A^j x^j \leqslant b^j$ is the block of constraints that corresponds to product j [constraints (9.2) – (9.13)].

By removing constraints (9.14) and (9.15) and placing them in the objective multiplied with Lagrangean multipliers, we get the Lagrangean relaxation (LR) of problem (M). Problem (LR) consists of $|J|$ independent blocks of constraints, but it cannot be decomposed into $|J|$ independent subproblems because the variables $\hat{y}_{kk'}$ refer to two tests of different products. Furthermore, even if variables $\hat{y}_{kk'}$ are included in only one subproblem and (LR) is decomposable, the upper bounds obtained from the solution of (LR) are very poor due to the fact that the relaxed constraint (9.15) is a big-M constraint.

By completely removing constraints (9.14) and (9.15), we get problem (M') which is decomposable into $|J|$ independent subproblems. Each subproblem corresponds to the resource-constrained scheduling problem for only one product (note that variables $\hat{y}_{kk'}$ are also removed).

$$\max \ Z = \sum_{j} \left[c^j\right]^T x^j$$
$$s.t. \ A^j x^j \leqslant b^j \quad \forall j \tag{M'}$$

For each product j, the resource-constrained scheduling subproblem (M^j) is simply as follows:

$$\max \ Z^j = \left[c^j\right]^T x^j$$
$$\text{s.t. } A^j x^j \leqslant b^j \qquad\qquad (M^j)$$

The basic idea of the proposed heuristic is to solve each subproblem (M^j) separately and then combine all the individual solutions to get a feasible global solution. More precisely, we solve each subproblem (M^j) separately assuming that all resources are available for the testing of product j. Since all resources are used for the tests of only one product, the optimal schedule of a subproblem will in general be shorter than the schedule of the same product when tested simultaneously with other products, because resources are shared among different products. This means that by solving $|J|$ subproblems we get a set of $|J|$ schedules that cannot be implemented simultaneously. The sequences found in the solution of the subproblems, however, are very useful and in most cases are present in the global optimal solution.

To illustrate this, consider the example shown in Figure 9.5, where products 1 and 2 are tested. When tested separately, the completion times for products 1 and 2 are 17 and 9 months, respectively. When tested simultaneously the optimal schedules are longer: 22 and 10 months, respectively. In terms of the graph of the sequences, this means that the graph that depicts the solution of a subproblem is "stretched" to accommodate more tests allocated on the same resources. This "stretching" is done by the inclusion of some new sequences (thick lines of Figure 9.5). If only product 1 is tested, tests 4 and 5 of product 1, for example are performed simultaneously, and thus, there is no precedence between them. When both products are tested simultaneously, however, test 5 is delayed (because resource 1 is used for tests 15 and 11 of product 2 as well) and thus there is a new arc between tests 4 and 5. Note, also, that all the precedences defined in the subproblems, are preserved in the global solution (for presentation reasons only the non-redundant arcs are shown in Figure 9.5b).

In general, we expect that any precedence present in the optimal solution of a subproblem will be preserved in the optimal solution of the entire problem, and, in addition, some new precedences will be determined. Thus, in the proposed heuristic we first solve all subproblems separately, we then fix the sequencing variables $y_{kk'}$ and $y_{k'k}$ for which $y_{kk'} = 1$ in the optimal solution of a subproblem, and we leave free the remaining binaries. Next, we perform a new round of preprocessing and we, finally, solve the simultaneous testing problem with fixed precedences.

The proposed decomposition scheme is as follows:

As explained above, when a new product is tested separately, its completion time is shorter than its actual completion time. This means that the income from sales for a particular new product will be smaller in reality than the income predicted by a subproblem. Likewise, the actual testing cost of a new

(a) Gantt Charts

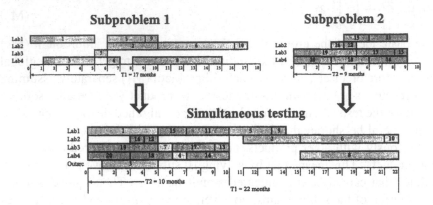

(b) Graphs of Sequence of Tests

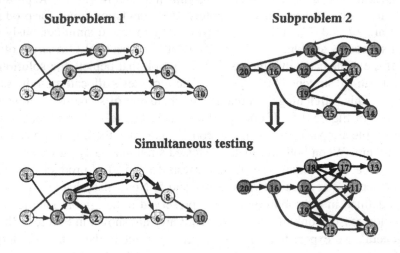

Figure 9.5.   Decomposition heuristic

Apply Preprocessing algorithm (PPROCALG)
For each product $j \in J$, solve the corresponding subproblem (M^j).
Fix $y_{kk'} = 1 \wedge y_{k'k} = 0 \; \forall k, \forall k' | y_{kk'} = 1$ in the optimal solution of (M^j).
Apply Preprocessing algorithm (PPROCALG).
Solve (M) with fixed sequencing binaries from steps 1, 3 and 4.

Table 9.8.   Decomposition algorithm (DECALG)

product will be higher than the testing cost predicted by a subproblem, since, in the simultaneous testing resources are scarcer and more outsourcing is needed. The sum of the objective values $\Sigma_j Z^j$ of the subproblems, therefore, is an upper bound on the optimal value Z^* of the objective of the entire problem. This upper bound provides an estimate of how good the heuristic solution Z^H is, and as shown in Table 9.9, it is very tight. The values of the objective functions of the subproblems, the upper bound, the optimal solution (found by the full space method using zero tolerance as termination criterion), and the solution found by the proposed heuristic are presented in Table 9.7. Note that the solution of the decomposition algorithm is always within 0.6% of the optimal solution, and that the gap between the upper bound and the heuristic solution is always less than 0.8%.

	Example 1	Example 2	Example 3
Subproblem: Product A (Z_A)	3,896.3	4,150.3	4,150.3
Subproblem: Product B (Z_B)	824.7	1,251.5	1,251.5
Subproblem: Product C (Z_C)	-	2,619.5	2,619.5
Subproblem: Product D (Z_D)	-	-	4,430.4
Upper Bound ($UB=\Sigma_j Z^j$)	4,721.0	8,021.3	12,450.6
Optimal Solution (Z^*)	4,701.8	8,002.3	12,428.0
Heuristic Solution (Z^H)	4,688.0	7,969.4	12,362.3
% Optimality Gap [$(Z^* - Z^H)/Z^*$]	0.29	0.41	0.53
% Bound Gap [$(UB-Z^H)/Z^H$]	0.70	0.65	0.71

Table 9.9. Solution statistics of the examples

6. Computational Results

The computational results for examples 1, 2 and 3 are given in Table 9.10. An optimality tolerance of 0.5% has been used as termination criterion for all cases. Compared to the formulation of Jain and Grossmann (1999), the proposed algorithm provides solutions of almost the same quality, and it is one to two orders of magnitude faster. If preprocessing is used for the implied precedences, the solutions of the full space method improve and the computational times are reduced, but the proposed decomposition algorithm is still almost one order of magnitude faster.

7. Example

Detailed results for Example 1 are presented in this section. In Example 1 two new products P1 and P2 should undergo 60 and 30 tests respectively. There are four categories of resources (I, II, III, IV) and ten resource units (1, 2,... 10). Units 1, 5, and 9 are of type I, units 2, 6, and 10 are of type II, units 3 and 7 are of type III, and units 4 and 8 are of type IV. Each test

	Example 1		Example 2		Example 3	
	OBJ	CPUs	OBJ	CPUs	OBJ	CPUs
Full Space – No implication	4,699	4,029	7,971*	10,917	12,317*	6,886
Full Space – Preprocessing	4,695	3,677	7,990*	5,738	12,428	4,797
Decomposition Algorithm	4,688	230	7,969	224	12,362	520

*Terminated due to tree size limit

Table 9.10. Solution statistics of the examples

requires one unit of one type of resource. Cost, duration, probability of success, resource requirement and technological precedences for each test are given in Appendix A. The solution found from the full space method using zero tolerance has a NPV of $4,701,800 and is given in Table 9.11. The solution found from the heuristic has a NPV of $4,688,000 (0.29% suboptimal) and is given in Table 9.12. The resource assignment of the heuristic solution is given in Table 9.13.

Product	Completion Time (days)	Income (10^3)	Testing Cost (10^3)
P1	1,480	6,382	2,496.3
P2	1,070	3,944	3,127.9

Table 9.11. Optimal solution

Product	Completion Time (days)	Income (10^3)	Testing Cost (10^3)
P1	1,490	6,366	2,485.5
P2	1,080	3,926	3,118.5

Table 9.12. Heuristic solution

8. Conclusions

Two classes of logic cuts and a decomposition algorithm were considered for the scheduling of testing tasks in new product development. The logic cuts, cycle breaking and implied precedences, tighten the LP relaxation and reduce the search space, improving the computational performance of full space methods. The proposed decomposition algorithm solves a series of subproblems, one for each product, and a final reduced model that embeds the individual optimal schedules obtained from the subproblems. The proposed heuristic yields near optimal solutions (typically within 1% of the optimum) and it is one to two orders of magnitude faster than the full space method.

Resource type/unit	Tests of P1	Tests of P2
I/1	9, 17, 53, 57	69, 73, 77
II/2	2, 6, 14, 38, 42	62, 66, 74, 78, 82, 86, 90
III/3	7, 11, 15, 19, 39, 51, 55, 59	63, 67, 71, 75, 83
IV/4	4, 8, 12, 16, 28, 36, 44, 56, 60	64, 76, 80, 84
I/5	5, 21, 25, 33, 45	81, 85, 89
II/6	10, 26, 34, 46, 50, 58	
III/7	5, 23, 27, 35, 43, 47	79, 87
IV/8	20, 48, 52	68, 72, 88
I/9	37, 41, 49	61
II/10	18, 22, 30	70
Outsourced	1,13, 24, 29, 31, 32, 40, 54	65

Table 9.13. Resource assignment of heuristic solution

9.   Nomenclature

Indices:

j	New products
k, k', k''	Tests
r	Resource categories
q	Resource units
n	Grid points for linearization of testing cost
m	Grid points for the calculation of income

Sets:

J	New products
K	Tests
$K(q)$	Tests that can be scheduled on unit q
$K(j)$	Tests of potential product $j \in J$ ($K = \cup_{j \in JP} K(j)$ and $\cap_{j \in JP} K(j) = \emptyset$)
$KK(k)$	Tests corresponding to the product that has k as one of its tests
Q	Resource units
R	Resource categories
$QC(r)$	Resources units of category r ($Q = \cup_{r \in R} QC(r)$)
$QT(k)$	Resource units that can be used for test k
A	Technological precedences: $(k, k') \in A$ if test k should finish before test k' starts

Parameters:

d_k	Duration of test k
p_k	Probability of success of test k
c_k	Costs of test k when performed in-house
$\hat{c}_k$	Costs of test k when outsourced
N_{rk}	Resource units of category r needed for test k
a_n	Grid points for linear approximation of testing cost

ρ Discount factor
F_j Income from product j at $t = 0$
b_{mj} Slope of income decrease for product j
U Upper bound on the completion time of testing

Binary Variables:

$y_{kk'}$ 1 if test k must finish before k' starts and $k' \in KK(k)$
$\hat{y}_{kk'}$ 1 if test k must finish before k' starts and $k' \notin KK(k)$
x_k 1 if test k is outsourced
$\hat{x}_{kq}$ 1 if resource q is assigned to test k

Continuous Variables:

INC_j Income from launching product j
TC_j Total expected, discounted testing cost for product j
C_k Expected, discounted cost of test k
T_j Completion time of testing of potential product j
s_k Starting time of test k
w_k Exponent for the calculation of the expected, discounted cost of test k
λ_{kn} Weight factors for linear approximation of cost of test k when performed in-house
Λ_{kn} Weight factors for linear approximation of cost of test k when outsourced

10.   Acknowledgment

The authors would like to gratefully acknowledge financial support from the National Science Foundation under Grant ACI-0121497.

References

Ahuja, R.K.; Magnanti, T.L.; Orlin, J.B. Network Flows: Theory, Algorithms and Applications, Prentice Hall, **1993**.

Honkomp, S.J.; Reklaitis, G.V.; Pekny, G.F. Robust Planning and Scheduling of Process Development Projects under Stochastic Conditions. AIChE Annual Meeting, Los Angeles, CA, **1997**.

Hooker, J.N.; Yan, H.; Grossmann, I.E.; Raman, R. Logic Cuts for Processing Networks with Fixed Charges, *Computers and Operations Research*, **1994**, 21 (3), 265-279.

Jain, V; Grossmann, I. E. Resource-constrained Scheduling of Tests in New Product Development. *Ind. Eng. Chem. Res,*. **1999**, 38, 3013-3026.

Johnson E.L.; Nemhauser, G.L.; Savelsbergh, M.W.P.; Progress in Linear Programming Based Branch-and-Bound Algorithms: Exposition, *INFORMS Journal of Computing*, **2000** (12).

Maravelias, C.; Grossmann, I.E. Simultaneous Planning for New Product Development and Batch Manufacturing Facilities. *Ind. Eng. Chem. Res.*, **2001**, 40, 6147-6164.

Maravelias, C.T.; Grossmann, I.E. Optimal Resource Investment and Scheduling of Tests for New Product Development. To appear in *Computers and Chemical Engineering*.

Robbins-Roth, C. From Alchemy to IPO: The Business of Biotechnology, Perseus Books, **2001**

Schmidt, C.W.; Grossmann, I.E. Optimization Models for the Scheduling of Testing Tasks in New Product Development. *Ind. Eng. Chem. Res.*, **1996**, 35, 3498-3510.

Schmidt, C. W.; Grossmann, I. E.; Blau, G. E. Optimization of Industrial Scale Scheduling Problems in New Product Development. *Computers Chem. Engng.*, **1998**, 22, S1027-S1030.

Subramanian, D.; Pekny, J.F.; Reklaitis, G.V. A Simulation-Optimization Framework for Research and Development Pipeline Management. *Journal of AIChE.* **2001**, 47 (10), 2226-2242.

Appendix: Example Data

The testing data for products A, B, C and D, that are included in Examples 1, 2, and 3 are given in Tables 9.A.1, 9.A.2, 9.A.3, 9.A.4 and 9.A.5, respectively. The cost of outsourcing is twice as much as the in-house cost, c_k (2nd column, $\$10^3$). The duration, d_k, is in days (3rd column). The probability of success, p_k, is given in the 4th column, and the resource requirement in the 5th column. There are four different resource categories (denoted as I, II, III, IV) and there are three units of category I, three units of II, two units of III, and two units of IV. All tests require one unit of a resource category. The tests that should be performed before test k are listed in the 6th column. The data are similar to the data of (Schmidt et. al., 1998); i.e. only few tests have probability of success lower than 1, there are many technological precedences, and there is large variation in the duration and the cost of different tests. Income regions (b_{jm}) and income decrease coefficients (f_{jm}) for all products are given in Table 9.A.5.

k	c_k	d_k	p_k	N_{rk}	Precedences	k	c_k	d_k	p_k	N_{rk}	Precedences
1	8	80	1	I		31	0	0	0.82	III	29,30
2	8	80	1	II		32	0	0	0.95	IV	29
3	5	60	1	III	1	33	10	10	1	I	31
4	1	10	0.84	IV		34	2	20	1	II	32
5	49	60	1	I		35	369	90	1	III	30,33
6	111	100	1	II	2	36	9	60	1	IV	
7	6	30	1	III		37	119	10	1	I	31,34
8	174	100	1	IV	4,7	38	4	10	1	II	35
9	62	130	1	I		39	1014	40	0.97	III	33
10	1	20	1	II	2,5	40	1	60	1	IV	30,31,35,37
11	16	80	1	III	3	41	3	0	1	I	34,36,38
12	113	12	0.87	IV	10	42	4	1060	1	II	
13	0	0	0.91	I	8,9,12	43	576	20	0.91	III	31,37,38
14	13	60	1	II	7	44	2	50	1	IV	35,39
15	53	30	1	III	11	45	14	50	1	I	37
16	9	30	1	IV		46	274	70	1	II	35,40,44
17	117	90	1	I	9,11,14	47	241	50	1	III	42,45
18	40	90	1	II	13	48	667	180	1	IV	46,47
19	57	50	1	III	11,12	49	3	20	1	I	41,44,46,47
20	23	90	1	IV	9,13	50	23	80	0.98	II	47,48
21	10	30	1	I	16,19	51	98	60	1	III	42,43,46, 49,50
22	15	20	1	II		52	146	120	1	IV	48,50
23	6	60	1	III	14,17,18	53	130	120	1	I	41,51
24	0	0	0.84	IV	21	54	0	0	0.68	II	48,52
25	92	80	1	I		55	2421	50	1	III	50,53
26	46	100	1	II		56	26	90	1	IV	52,53
27	7	10	1	III	18,21,23,24, 25	57	29	30	1	I	54
28	38	30	1	IV	26	58	45	70	1	II	51,53,55
29	0	0	0.94	I	19,23,25,26	59	206	180	0.95	III	54,55,56, 57,58
30	42	40	1	II	20,27,28	60	364	80	1	IV	55,57,58,59

Table 9.A.1.   Testing data of product A

k	c_k	d_k	p_k	N_{rk}	Precedences	k	c_k	d_k	p_k	N_{rk}	Precedences
61	473	180	1	I		76	1111	110	1	IV	70,73,74
62	7	30	1	II	61	77	121	40	1	I	68,72,75
63	180	120	1	III	62	78	211	20	1	II	76,77
64	10	30	1	IV	63	79	311	30	1	III	71,73,76
65	0	0	0.94	I	62,64	80	401	20	1	IV	77,78
66	40	80	1	II	61	81	110	20	1	I	76,79,80
67	80	60	0.87	III		82	87	60	0.87	II	78,80
68	10	80	0.95	IV	63,65	83	91	10	0.91	III	78,80
69	50	60	1	I	66	84	211	60	1	IV	81
70	80	40	1	II	64,67	85	315	70	1	I	78,82
71	100	120	1	III	68	86	615	110	1	II	82
72	110	80	1	IV		87	51	60	1	III	84
73	1	90	1	I	67,69	88	101	70	1	IV	83,85
74	84	50	0.84	II	69,70	89	100	60	1	I	84,86,88
75	210	90	1	III	70,71	90	210	50	1	II	85,87,89

Table 9.A.2. Testing data of product B

k	c_k	d_k	p_k	N_{rk}	Precedences	k	c_k	d_k	p_k	N_{rk}	Precedences
91	135	20	1	III		111	120	130	1	III	107,108,109
92	10	30	0.95	IV	91	112	123	120	0.92	IV	109,111
93	15	40	1	I		113	134	110	1	I	109,110,111 112
94	25	50	1	II	91,92	114	145	100	1	II	110,112
95	20	60	0.98	III	92,93,94	115	111	90	1	III	109,111,112 113
96	40	70	1	IV	92,93,95	116	132	80	0.9	IV	112,114
97	80	90	1	I	91,94	117	213	70	1	I	113
98	0	100	0.99	II	93,97	118	234	60	1	II	114,115
99	100	120	1	III	94,95,96	119	101	50	1	III	115
100	120	150	1	IV	93,95,97,98	120	123	50	1	IV	113,114,117 118,119
101	120	150	1	I	97,98	121	90	40	0.85	I	114,116,118
102	150	160	1	II	96,98,99,100	122	123	30	1	II	117,119,120
103	145	180	0.9	III	98,99,100	123	147	30	1	III	18,120,121
104	164	200	1	IV	100,102,103	124	89	20	1	IV	120,122
105	172	190	1	I	101	125	56	20	0.9	I	119,121,123 124
106	161	180	1	II	101,103,104	126	23	20	1	II	124,125
107	192	170	0.8	III	102	127	185	10	1	III	124,125
108	154	160	1	IV	101,106	128	89	10	1	IV	123,124,125 126
109	184	150	0.85	I	103,104	129	90	50	0.88	I	125,127,128
110	120	140	1	II	106,107,108	130	120	100	1	II	125,126,128 129

Table 9.A.3. Testing data of product C

k	c_k	d_k	p_k	N_{rk}	Precedences	k	c_k	d_k	p_k	N_{rk}	Precedences
131	24	10	1	III		141	35	30	1	I	134,135,137 1138,139
132	56	20	1	IV	131	142	76	60	1	II	137,141
133	45	40	0.98	I	131	143	85	80	0.78	III	139,140
134	105	20	1	II		144	98	100	1	IV	136,138,141
135	200	80	0.95	III	131,132,133	145	125	120	1	I	140,142,144
136	89	60	1	IV	.	146	142	50	0.85	II	139,141,142 143
137	45	120	1	I	134,136	147	53	80	1	III	143,145
138	76	150	1	II	134	148	98	180	1	IV	143,144,145 146
139	85	100	0.8	III	134,135,136	149	108	110	0.95	I	147
140	95	40	1	IV	134,137	150	38	160	1	III	148,149

Table 9.A.4. Testing data of product D

Product	A	B	C	D
F_j (10_3)	8,000	5,000	6,000	5,000
f_{j1} (10^3/day)	0.2	0.2	0.2	0.2
f_{j2} (10^3/day)	0.6	0.6	0.6	0.6
f_{j3} (10^3/day)	1.0	1.0	1.0	1.0
b_{j1} (days)	0	0	0	0
b_{j2} (days) of Example 1	300	300		
b_{j3} (days) of Example 1	600	600		
b_{j2} (days) of Example 2	500	500	500	
b_{j3} (days) of Example 2	1000	1000	1000	
b_{j2} (days) of Example 3	500	500	500	500
b_{j3} (days) of Example 3	1000	1000	1000	1000

Table 9.A.5. Income data for products

Chapter 10

A MIXED-INTEGER NONLINEAR PROGRAMMING APPROACH TO THE OPTIMAL PLANNING OF OFFSHORE OILFIELD INFRASTRUCTURES

Susara A. van den Heever and Ignacio E. Grossmann*

Department of Chemical Engineering
Carnegie Mellon University, Pittsburgh
PA 15213, U.S.A.

{ susara,ig0c } @andrew.cmu.edu

Abstract   A multiperiod Mixed Integer Nonlinear Programming (MINLP) model for off-shore oilfield infrastructure planning is presented where nonlinear reservoir behavior is incorporated directly into the formulation. Discrete decisions include the selection of production platforms, well platforms and wells to be installed/drilled, as well as the drilling schedule for the wells over the planning horizon. Continuous decisions include the capacities of the platforms, as well as the production profile for each well in each time period.

For the solution of this model, an iterative aggregation/disaggregation algorithm is proposed in which logic-based methods, a bilevel decomposition technique, the use of convex envelopes and aggregation of time periods are integrated. Furthermore, a novel dynamic programming sub-problem is proposed to improve the aggregation scheme at each iteration in order to obtain an aggregate problem that resembles the disaggregate problem more closely. A number of examples are presented to illustrate the performance of the proposed method.

Keywords: Oilfield planning, MINLP, aggregation, decomposition

1.   Introduction

The work presented here is extracted from a paper first published in I&ECR[1], and has been condensed here so as to share with the operations research com-

*Author to whom correspondence should be addressed

munity some of the advances achieved in design and planning optimization for the oil industry. Offshore oilfield infrastructure planning is a challenging problem encompassing both complex physical constraints and intricate economical specifications. An offshore oilfield infrastructure consists of Production Platforms (PP), Well Platforms (WP), wells and connecting pipelines (see Figure 10.1), and is constructed for the purpose of producing oil and/or gas from one or more oilfields. Each oilfield (F) consists of a number of reservoirs (R), while each reservoir in turn contains a number of potential locations for wells (W) to be drilled.

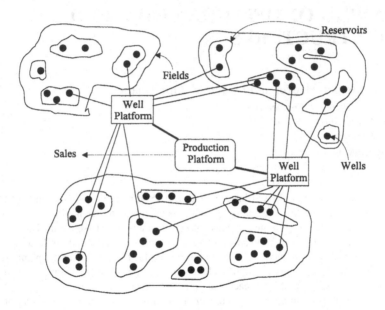

Figure 10.1. Configuration of fields, well platforms and production platforms (Iyer *et al.*, 1998)

Offshore oilfield facilities are often in operation over several decades and it is therefore important to take future conditions into consideration when designing an initial infrastructure. This can be incorporated by dividing the operating horizon into a number of time periods and allowing planning decisions in each period, while design decisions are made for the horizon as a whole. The corresponding model can be run both at the start of the project to facilitate decisions regarding initial investments, and periodically during the life of the project to update planning decisions based on the most recent information regarding oilfield production capabilities and costs.

Uncertainty in the capacities of oil reserves and oil prices is an important consideration in oilfield facility planning. Even though running the model periodically with updated data mitigates some of this uncertainty, it may be necessary to explicitly incorporate the uncertainties into the optimization model to better capture their effect. While the explicit handling of uncertainty is not the subject of this paper, we refer the reader to the work of Goel and Grossmann[2], who extended our work to address uncertainty by using a stochastic programming approach. Another important consideration is addressing the effects of taxes, tariffs and royalties on investment and planning decisions. This is the subject of two follow-up papers, where we explicitly model government imposed tax structures, tariffs and royalties[3] and propose a decomposition scheme[4] to handle the increase in computational complexity resulting from this inclusion.

Design decisions typically involve the capacities of the PPs and WPs, as well as decisions regarding *which* PPs, WPs and wells to install over the whole operating horizon. Planning decisions involve the production profiles in each period, as well as decisions regarding *when* to install PPs, WPs, and wells included in the design. Decision variables can also be grouped into discrete variables, for example those representing the installation of PPs, WPs and wells in each period, and continuous variables, for example those representing the production profiles and pressures in each period. This discrete/continuous mixture, together with the nonlinear reservoir behavior, requires a Mixed Integer Nonlinear Programming (MINLP) model, where the objective is to maximize or minimize a specific value function. MINLPs are known to be NP-complete [5], and when the models are multiperiod, their solution becomes even more intractable. Specialized techniques are therefore needed to solve these models. In the case of oilfield infrastructure planning, MINLP models have traditionally been avoided in favor of Mixed Integer Linear Programs (MILPs) or Linear Programs (LPs), due to the difficulties associated with dealing directly with nonlinearities and, in the latter case, discrete decisions. It is important to note that specialized algorithms may be necessary even when nonlinearities are avoided, due to the large magnitude of the multiperiod optimization problem.

In the past, decisions regarding capacities of platforms, drilling schedules and production profiles have often been made separately with certain assumptions to ease the computational burden [6–9]. The unfortunate consequence of not including all the design and planning decisions in one model is that important interactions between these variables are not taken into account, leading to infeasible or suboptimal solutions. Simultaneous models have only emerged recently. Some of these were solved up to a limited size using commercial MILP solvers [10–13], while others were solved with specialized solution algorithms [14–15]. In all these models, however, the nonlinear reservoir behavior was approximated by one or more linear constraints. It is important to deal with nonlinearities directly, not only to avoid the large number of integer variables associated with for example piecewise linear interpolation as an approximation, but also to reduce the loss of accuracy and improve the quality of the solution.

In this chapter we address the formulation and solution of a discrete non-linear optimization model for oilfield planning. Our goal is to develop an algorithm capable of (1) dealing with nonlinearities directly, and (2) solving realistic instances of nonlinear offshore oilfield infrastructure design and planning models in reasonable time. The model we consider to this end, uses as a basis the one proposed by Iyer et al.[15], and includes decisions regarding investment of PPs, WP, and wells, the well drilling schedule and the well production profiles in each time period. While Iyer et al.[15] used linear interpolation to approximate nonlinearities, we include these nonlinearities directly into the model. Specifically, these are the reservoir pressures, gas to oil ratio, and cumulative gas produced expressed as nonlinear functions of the cumulative oil produced.

We propose an iterative aggregation/disaggregation algorithm, where the time periods are aggregated in the design problem, and subsequently disaggregated when the planning problem is solved for the fixed infrastructure obtained from the design problem. The solution from the planning problem is used to update the aggregation scheme after each iteration. This is done through a dynamic programming subproblem which determines how time periods should be aggregated to yield an aggregate problem that resembles the disaggregate problem as close as possible. Convex envelopes are used to deal with non-convexities arising from non-linearities. A large-scale example is presented to show the effectiveness of the proposed algorithm and potential economic benefits.

2. Problem Statement

In this chapter, we consider the design and planning of an offshore oilfield infrastructure (refer to Figure 10.1) over a planning horizon of Y years, divided into T time periods (e.g. quarterly time periods). An oilfield layout consists of a number of fields, each containing one or more reservoirs and each reservoir contains one or more wellsites. After the decision has been made to produce oil from a given wellsite, it is drilled from a WP using drilling rigs. A network of pipelines connects the wells to the WPs and the WPs to the PPs. For our purposes, we assume that the location/allocation problem has been solved, i.e. the possible locations of the PPs and WPs, as well as the assignment of wells to WPs and WPs to PPs, are fixed. In the model, one can easily relax the assumption of fixed allocation of each well to a WP and consider that each well may be allocated to two or more WPs. However, this will clearly increase the computational effort significantly. A more practical option might be to consider allocating each well to up to two WPs. In this case, the two different allocations are treated as two choices of identical wells of which only one can be selected.

The design decisions we consider are valid for the whole planning horizon and are:

a) whether or not to include each PP, WP, and well in the infrastructure over the planning horizon

b) the capacities of the PPs and WPs

The planning decisions are made in each time period and are:

a) whether or not to install each PP and WP

b) whether or not to drill each well

c) the production profile of each well

These decisions are made under the assumption that operating conditions are constant during a given time period.

3.   Model

The following is a complete mathematical model of the offshore oilfield infrastructure planning problem. The model is based on the one proposed by Iyer *et al.* [15], with the main differences being that we include non-linearities directly into the model instead of using linear interpolation, do not consider the constraints for the movement of drilling rigs explicitly, and formulate the problem in disjunctive form. For ease of comparison, we use the same nomenclature and make the same assumptions as Iyer *et al.* [15]. Please refer to the list of nomenclature at the end of this chapter.

3.1   Objective function:

The objective function is to maximize the Net Present Value (NPV) which includes sales revenues, investment costs and depreciation.

$$\max \Psi = \sum_{t=1}^{T} \{ Rev_t - \sum_{p \in PP} [CI_t^p + \sum_{\pi \in WP(p)} \{ CI_t^{\pi,p} + \sum_{w \in W_{WP}(\pi)} CI_t^{w,\pi,p} \}] \}$$

(10.1)

The cost of exploration and appraisal drilling is not included in the objective, seeing that this model will be used after the results from exploration and appraisal have become available. Costs related to taxes, royalties and tariffs are not included in this study, seeing that they add additional computational complexity and detract from the main focus of this work, namely inclusion of nonlinear reservoir behavior and development of a decomposition algorithm. We do, however, cover these complex economic objectives in two follow-up papers[3,4]. Costs that we have not addressed include the costs of decommissioning and well maintenance. The modeling of these costs requires additional discrete variables that result in an increase in the computational complexity above and beyond that of the taxes, tariffs and royalties. This is therefore a challenge we leave for future work. Some other economic considerations one may include in variations of the objective function are different depreciation rates for different types of expenses, or different exchange rates in the case of multi-national investments. These latter considerations can be captured by simply changing the input data and do not affect the complexity of the model or the validity of the results discussed in this work.

3.2   Constraints valid for the whole infrastructure:

In (10.2) the sales revenue in each time period is calculated from the total oil produced, which is in turn calculated in (10.3) as the sum of the oil produced from all production platforms. (10.4) calculates the amount of oil flow from each reservoir in each time period to be the sum of all oil flowrates from wells associated with that reservoir times the duration of the time period. The cumulative flow of oil from each reservoir is calculated in (10.5). Note that (10.5) is one of the linking constraints that links the time periods together and thus prevents a solution procedure where each period is solved individually. The cumulative flow of oil is used in (10.6) to calculate the reservoir pressure through the exponential function which is obtained by fitting a nonlinear curve to the linear interpolation data used by Iyer *et al.*[15].

$$Rev_t = c_{1t}x_t^{total} \tag{10.2}$$

$$\sum_{p \in PP} x_t^p = x_t^{total} \tag{10.3}$$

$$l_t^{r,f} = \Delta t \sum_{(w,\pi,p) \in W_{F,R}(f,r)} x_t^{w,\pi,p} \forall r \in R(f), f \in F \tag{10.4}$$

$$xc_\theta^{r,f} = \sum_{t=1}^{\theta-1} l_t^{r,f} \forall r \in R(f), f \in F \tag{10.5}$$

$$v_t^{r,f} = \gamma_{p1}^{r,f} \exp(\gamma_{p2}^{r,f} xc_t^{r,f}) \forall r \in R(f), f \in F \tag{10.6}$$

for $t = 1 \ldots T$.

3.3   Disjunction for each PP,WP and well:

We exploit the hierarchical structure of the oilfield to formulate a disjunctive model. The production platforms are at the highest level of the hierarchy, and the disjunction includes all constraints valid for that PP, as well as the disjunction for the next hierarchical level, i.e. for all WPs associated with that PP. In turn, the disjunction for each WP, which is located within the disjunction of a PP, contains all constraints valid for that WP, as well as the disjunctions for all wells associated with that WP. We present the disjunctions here with numbers indicating the constraints present, and follow with the explanation of the individual constraints contained in each disjunction:

$$
\left[
\begin{array}{c}
\bigvee_{\theta=1}^{t} z_\theta^p \\[4pt]
\left[\begin{array}{c} z_t^p \\ (10.7),(10.8) \end{array}\right] \vee \left[\begin{array}{c} \neg z_t^p \\ CI_t^p, e_t^p = 0 \end{array}\right] \\
(10.9),(10.10),(10.11),(10.12) \\[6pt]
\left[
\begin{array}{c}
\bigvee_{\theta=1}^{t} z_\theta^{\pi,p} \\[4pt]
\left[\begin{array}{c} z_t^{\pi,p} \\ (10.13),(10.14) \end{array}\right] \vee \left[\begin{array}{c} \neg z_t^{\pi,p} \\ CI_t^{\pi,p}, e_t^{\pi,p} = 0 \end{array}\right] \\
(10.15),(10.16),(10.17),(10.18),(10.19) \\[6pt]
\left[
\left[\begin{array}{c}
\bigvee_{\theta=1}^{t} z_\theta^{w,\pi,p} \\[4pt]
\left[\begin{array}{c} z_t^{w,\pi,p} \\ (10.20) \end{array}\right] \vee \left[\begin{array}{c} \neg z_t^{w,\pi,p} \\ CI_t^{w,\pi,p} = 0 \end{array}\right] \\
\begin{array}{c}(10.21),(10.22),(10.23),(10.24),(10.25),\\(10.26),(10.27),(10.28),(10.29)\end{array}
\end{array}\right]
\vee
\left[\begin{array}{c}
\neg \bigvee_{\theta=1}^{t} z_\theta^{w,\pi,p} \\[2pt]
x_t^{w,\pi,p} = 0 \\
g_t^{w,\pi,p} = 0 \\
xc_t^{w,\pi,p} = 0 \\
gc_t^{w,\pi,p} = 0
\end{array}\right]
\right] \\
\forall w \in W_{WP}(\pi)
\end{array}
\right]
\vee
\left[\begin{array}{c}
\neg \bigvee_{\theta=1}^{t} z_\theta^{\pi,p} \\[2pt]
x_t^{\pi,p} = 0 \\
g_t^{\pi,p} = 0
\end{array}\right] \\
\forall \pi \in WP(p)
\end{array}
\right]
\vee
\left[\begin{array}{c}
\neg \bigvee_{\theta=1}^{t} z_\theta^p \\[2pt]
x_t^p = 0 \\
g_t^p = 0
\end{array}\right] \\
\forall p \in PP, t \in T
\end{array}
\right]
$$

The outer disjunction is valid for each PP in each time period and can be interpreted as follows: If production platform p has been installed during or before period t (discrete expression $\bigvee_{\theta=1}^{t} z_\theta^p$ = true), then all constraints in the largest bracket are applied:

$$CI_t^p = c_{2t}^p + c_{3t}^p e_t^p \qquad (10.7)$$

$$e_t^p \leqslant U \qquad (10.8)$$

$$
\left[\begin{array}{c}
z_t^p = 1 \\
CI_t^p = c_{2t}^p + c_{3t}^p e_t^p \\
e_t^p \leqslant U
\end{array}\right]
\vee
\left[\begin{array}{c}
z_t^p = 0 \\
CI_t^p, e_t^p = 0
\end{array}\right]
$$

$$x_t^p \leqslant d_t^p \tag{10.9}$$

$$d_t^p = d_{t-1}^p + e_t^p \tag{10.10}$$

$$\sum_{\pi \in WP_{(p)}} x_t^{\pi,p} = x_t^p \tag{10.11}$$

$$\sum_{\pi \in WP_{(p)}} g_t^{\pi,p} = g_t^p \tag{10.12}$$

First, the smaller nested disjunction is used to calculate the discounted investment cost (including depreciation) of the production platform in each time period. This cost is calculated if production platform p is installed in period t ($z_t^p = True$), otherwise it is set to zero. (10.7) relates the cost as a function of the expansion capacity, which is set to zero if the production platform is not installed ($z_t^p = False$), while (10.8) sets an upper bound on the expansion. (10.9) determines the design capacity to be the maximum flow among all time periods, and this is modeled linearly by defining the expansion variable which can take a non-zero value in only one time period. (10.11) and (10.12) are mass balances calculating the oil/gas flow from the PP as the sum of the flow from all WPs associated with that PP. If the production platform has not been installed yet (discrete expression $\bigvee_{\theta=1}^{t} z_\theta^p$ = false), the oil/gas flows, as well as investment cost, are set to zero.

The middle disjunction is valid for all well platforms associated with production platform p and is only applied if the discrete expression $\bigvee_{\theta=1}^{t} z_\theta^p$ is true. This disjunction states that if well platform π has been installed before or during period t (discrete expression $\bigvee_{\theta=1}^{t} z_\theta^{\pi,p}$ = true), then the constraints present in that disjunction are applied:

$$CI_t^{\pi,p} = c_{2t}^{\pi,p} + c_{3t}^{\pi,p} e_t^{\pi,p} \tag{10.13}$$

$$e_t^{\pi,p} \leqslant U \tag{10.14}$$

$$\begin{bmatrix} & z_t^{\pi,p} & \\ CI_t^{\pi,p} = c_{2t}^{\pi,p} + c_{3t}^{\pi,p} e_t^{\pi,p} \\ e_t^{\pi,p} \leqslant U \end{bmatrix} \vee \begin{bmatrix} & \neg z_t^{\pi,p} & \\ CI_t^{\pi,p}, e_t^{\pi,p} = 0 \end{bmatrix}$$

$$x_t^{\pi,p} \leqslant d_t^{\pi,p} \qquad (10.15)$$

$$d_t^{\pi,p} = d_{t-1}^{\pi,p} + e_t^{\pi,p} \qquad (10.16)$$

$$\sum_{\pi \in W_{WP}(\pi)} x_t^{w,\pi,p} = x_t^{\pi,p} \qquad (10.17)$$

$$\sum_{\pi \in W_{WP}(\pi)} g_t^{w,\pi,p} = g_t^{\pi,p} \qquad (10.18)$$

$$v_t^p = v_t^{\pi,p} - \alpha x_t^{\pi,p} - \beta g_t^{\pi,p} - \delta_t^{\pi,p} \qquad (10.19)$$

Again, the smaller nested disjunction is used to calculate the discounted investment cost (including depreciation) of the well platform in each time period. This cost is calculated if well platform π is installed in period t ($z_t^{\pi,p} = True$), otherwise it is set to zero. (10.13) relates the cost as a function of the expansion capacity, which is set to zero if the well platform is not installed ($z_t^{\pi,p} = False$), while (10.14) sets and upper bound on the expansion. (10.15) and (10.16) determine the design capacity as described in the case of the production platform. (10.17) and (10.18) are mass balances calculating the oil/gas flow from the WP as the sum of the flow from all wells associated with that WP. (10.19) relates the pressure at the WP to the pressure at the PP it is associated with. The pressure at the PP is the pressure at the WP minus the pressure drop in the corresponding pipeline, which is given by the remaining terms in (10.19). If the production platform has not been installed yet (discrete expression $\bigvee_{\theta=1}^{t} z_\theta^{\pi,p} = $ false), the oil/gas flows, as well as investment cost, are set to zero.

The innermost disjunction is valid for each well w associated with well platform π, and is only included if well platform π has already been installed (discrete expression $\bigvee_{\theta=1}^{t} z_\theta^{\pi,p} = $ true). If well w has been drilled during or before period t (discrete expression $\bigvee_{\theta=1}^{t} z_\theta^{w,\pi,p} = $ true), then the following constraints are applied:

$$CI_t^{w,\pi,p} = c_{2t}^{w,\pi,p} \qquad (10.20)$$

$$\begin{bmatrix} z_t^{w,\pi,p} \\ CI_t^{w,\pi,p} = c_{2t}^{w,\pi,p} \end{bmatrix} \vee \begin{bmatrix} \neg z_t^{w,\pi,p} \\ CI_t^{w,\pi,p} = 0 \end{bmatrix}$$

$$v_t^{\pi,p} = v_t^{w,\pi,p} - \alpha x_t^{w,\pi,p} - \beta g_t^{w,\pi,p} - \delta_t^{w,\pi,p} \qquad (10.21)$$

$$x_t^{w,\pi,p} = \rho^{w,\pi,p}(v_t^{r,f} - v_t^{w,\pi,p}) \qquad (10.22)$$

$$g_t^{w,\pi,p} \leqslant x_t^{w,\pi,p} GOR_{\max} \qquad (10.23)$$

$$x_t^{w,\pi,p} \leqslant \rho^{w,\pi,p} P_{\max} \qquad (10.24)$$

$$xc_t^{w,\pi,p} = \sum_{\theta=1}^{t-1} x_t^{w,\pi,p} \Delta t \qquad (10.25)$$

$$gc_t^{w,\pi,p} = \sum_{\theta=1}^{t-1} x_t^{w,\pi,p} \Delta t \qquad (10.26)$$

$$x_t^{w,\pi,p} \geqslant x_{t+1}^{w,\pi,p} \qquad (10.27)$$

$$gc_t^{w,\pi,p} = \gamma_{g1}^{r,f} + \gamma_{g2}^{r,f} xc_t^{w,\pi,p} + \gamma_{g3}^{r,f}(xc_t^{w,\pi,p})^2 \qquad (10.28)$$

$$\forall (w,\pi,p) \in W_{F,R}(f,r)$$

$$GOR_t^{w,\pi,p} = \gamma_{gor1}^{r,f} + \gamma_{gor2}^{r,f} xc_t^{w,\pi,p} + \gamma_{gor3}^{r,f}(xc_t^{w,\pi,p})^2 \qquad (10.29)$$

$$\forall (w,\pi,p) \in W_{F,R}(f,r)$$

The smaller nested disjunction is used to calculate the discounted investment cost (including depreciation) of the well in each time period. This cost is calculated in (10.20) if well w is drilled in period t ($z_t^{w,\pi,p} = True$), otherwise it is set to zero. (10.21) relates the pressure at the well to the pressure at the WP it is associated with. The pressure at the WP is the pressure at the well minus the pressure drop in the corresponding pipeline, which is given by the remaining terms. (10.22) states that the oil flowrate equals the productivity index times the pressure differential between reservoir and well bore. (10.23) restricts the gas flowrate to be the oil flow times the GOR, while (10.24) restricts the maximum oil flow to equal the productivity index times the maximum allowable pressure drop. The productivity index and the reservoir pressure determine the oil production rate from a well in a given time period. The well is usually capped when the GOR (gas to oil ratio) exceeds a certain threshold limit or when the pressure of the reservoir is lower than a minimum pressure. (10.25) and (10.26) calculate the cumulative flow to be the sum of flows over all periods up to the current one. Note that (10.25) and (10.26) are linking constraints that link the time periods together and prevent a solution procedure where every time period is solved separately. (10.27) denotes a specification by the oil company which restricts the flow profile to be non-increasing. While this company does not require a lower bound on the quantity of oil flowing through a pipeline, one could consider adding such a lower bound in the form of a threshold constraint where the flow is either zero or above some minimum, in order to address concerns that the pipeline may seize up if the flow rate were to drop below a certain level. The linear interpolation to calculate cumulative gas and

GOR as functions of cumulative oil, are replaced by the nonlinear constraints (10.28) and (10.29). These quadratic equations are obtained from a curve fit of the linear interpolation data from Iyer *et al.*[15]. If the well has not been drilled yet (discrete expression $\bigvee_{\theta=1}^{t} z_\theta^{w,\pi,p} =$ false), the oil/gas flows, cumulative flows, as well as investment cost, are set to zero.

3.4   Logical constraints:

These represent logical relationships between the discrete decisions. (10.30) – (10.32) specify that each well, WP and PP can be drilled/installed in only one period. (10.33) states that if a WP has not been installed by t, then any well w associated with that WP cannot be drilled in t. Likewise, (10.34) states that if a PP has not been installed by period t, then any WP associated with that PP cannot be installed in t. The restriction that only M_w wells can be drilled in any given time period, is given by (10.35).

$$\bigvee_{t=1}^{T} z_\theta^{w,\pi,p} \forall w \in W_{WP}(\pi), \pi \in WP(p), p \in PP \qquad (10.30)$$

$$\bigvee_{t=1}^{T} z_\theta^{\pi,p} \forall \pi \in WP(p), p \in PP \qquad (10.31)$$

$$\bigvee_{t=1}^{T} z_\theta^{p} \forall p \in PP \qquad (10.32)$$

$$\neg \bigvee_{\theta=1}^{t} z_\theta^{\pi,p} \Rightarrow \neg z_t^{w,\pi,p} \forall w \in W_{WP}(\pi), \pi \in WP(p), p \in PP \qquad (10.33)$$

$$\neg \bigvee_{\theta=1}^{t} z_\theta^{p} \Rightarrow \neg z_t^{\pi,p} \forall \pi \in WP(p), p \in PP \qquad (10.34)$$

$$\bigvee_{(w,\pi,p)} z_t^{w,\pi,p} \leqslant M_w \qquad (10.35)$$

4.   Solution Strategy

Even though the elimination of linear interpolation binary variables as used by Iyer *et al.* reduces the problem size significantly, the resulting model (P) still becomes very large as the number of time periods increases. In addition to this difficulty, the newly included nonlinear constraints introduce non-convexities into the model, thereby increasing the chance of finding sub-optimal solutions. A specialized solution strategy capable of dealing with both the problem size and the non-convexities is therefore needed. We propose such a strategy based on logic-based methods, the use of convex envelopes, bilevel decomposition, and the aggregation of time periods. In the following four sections we discuss these aspects and show how we integrate them to form an iterative aggregation/disaggregation algorithm for oilfield infrastructure planning.

4.1   Logic-Based Methods

Model (P) is logic-based since it is in disjunctive form, and a logic-based method is required for the solution thereof. Turkay and Grossmann [19] proposed a logic-based Outer Approximation (OA) algorithm for MINLPs based on the OA method by Duran and Grossmann [20]. The latter involves iteration between an NLP subproblem where all binary variables are fixed, and an MILP master problem where the nonlinear equations are relaxed and linearized at the NLP solution points. This algorithm can be initialized with either the relaxed solution, or by specifying an initial feasible set of fixed binaries for the first NLP. Variables of subsequent NLPs are initialized from the solution of the preceding MILP. In the logic-based method, linearizations are only added to the MILP if the disjunction has true value, and the NLP subproblems only include constraints for existing units (i.e. constraints of disjunctions with true value). All disjunctions are converted to mixed-integer form through the convex hull formulation [19,21]. The advantage of this formulation is that it reduces the dimensionality of the nonlinear sub-problem by only considering disjunctions for which the Boolean variable is true, thereby avoiding singularities due to linearizations at zero flows, and eliminating non-convexities of non-existing processes.

In this paper, we modify the logic-based OA algorithm to deal with non-convexities at the level of the master problem, by using convex envelopes where non-convexities occur, instead of linearizations, as discussed in the next paragraph (see Figure 10.2).

4.2   Dealing with non-convexities

When non-convexities are present in an optimization model, the OA algorithm cannot guarantee a global optimum. Standard NLP solvers such as MINOS and CONOPT cannot guarantee a global optimum for the NLP subproblem, and the NLP solution often depends on the starting point given by the modeler. For the MILP master problems, parts of the feasible region may be cut off by linearizations of non-convex constraints. When this happens, it is possible that the optimal solution is eliminated from the optimization procedure and never found. The crucial drawback therefore lies in the MILP, since a large number of feasible solutions, including the optimal solution, might be missed completely. We address this problem by replacing the linearizations of the non-convex nonlinear functions in model (P) with convex envelopes [23]. For concave inequalities we need to use a convex envelope for a valid relaxation over the domain $x^j \in [x_L^j, x_U^j]$, where L and U indicate the upper and lower bounds respectively:

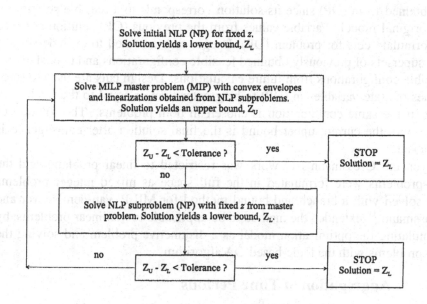

Figure 10.2. Logic-based OA algorithm

$$x^i \leqslant g(x_L^j) + \left(\frac{g(x_U^j) - g(x_L^j)}{x_U^j - x_L^j} \right) \left(x^j - x_L^j \right)$$

In the case of equalities, the multiplier of $x^i = g(x^j)$ from the NLP sub-problem indicates in which way the equality relaxes, and determines whether a linearization or convex envelope will be used in the MILP master problem.

4.3 Bilevel Decomposition

The bilevel decomposition approach, first introduced by Iyer and Grossmann [24], reduces the problem size by decomposing the original design and planning model into an upper level design problem and a lower level planning problem. The upper level design problem (DP) includes only the discrete variables representing design decisions, in our context whether or not to include a well or platform during the whole planning horizon. The lower level planning problem (OP) is solved for a fixed design or configuration and determines the operational plan in each time period, for example when to drill a well. (DP) is a relaxation of the original model, since only design decisions are considered, and thus yields an upper bound. (OP) is in a reduced solution space, since a subset of fixed discrete variables are used as obtained from (DP), making it possible to ignore a large number of equations and variables. A lower bound is obtained from (OP) since its solution corresponds to a feasible solution of the original model. Variable values from the previous (OP) solution are used to formulate cuts for problem (DP). Integer cuts are used to exclude subsets and supersets of previously obtained feasible configurations and to exclude infeasible configurations from future calculations. Design cuts are used to force values of state variables in (DP) to be greater than or equal to their values in (OP) if the same configuration is chosen in both problems. The solution of (OP) with the current upper bound is the final solution after convergence is achieved.

Iyer and Grossmann's [24] work was restricted to linear problems and the sub-problems were formulated in the full space as mixed integer problems and solved with a branch and bound method for MILP. Van den Heever and Grossmann [18] extended the method to be applicable to nonlinear problems, by formulating the optimization model as a disjunctive problem and solving the sub-problems with the logic-based OA algorithm.

4.4 Aggregation of Time Periods

Due to the size of the oilfield infrastructure planning problem, it cannot be solved in reasonable time with the disjunctive bilevel decomposition algorithm. To address this difficulty, we propose the aggregation of time periods for the upper level problem. We propose an iterative scheme in order to try and improve on the first feasible solution by using information obtained from previous iterations. For this purpose, we integrate the aggregation of time periods with the logic-based bilevel decomposition algorithm, while incorporating the use of convex envelopes.

In order to insure that an upper bound is obtained from the aggregate design problem, the discounting factors in the objective function need to be chosen in such a way as to overestimate the objective and the production also needs to be overestimated. Iyer *et al.* [15] present a proof that this can be accomplished by specifying that the flow profile is non-increasing for each well, and that investments are discounted as if they occur at the end of an aggregate period, while sales and depreciation are discounted as if they occur at the beginning of an aggregate period. We incorporate this property in our aggregate model to insure an upper bound.

The solution from the aggregate design problem is also used to curb the investment horizon for each well in the planning problem. This is done by specifying that well w can only be drilled in disaggregate period t if $\bigvee_{\theta=1}^{\tau} z_\theta^{w,\pi,p} = True$ and $t \leqslant \sum_{\theta=1}^{\tau} m_\tau$, where m_τ is the length of aggregate period τ.

A difficulty that may be encountered here is that the aggregate problem may give poor upper bounds depending on the aggregation scheme, i.e. which time periods are grouped together. To address this, we propose an intermediate optimization problem where the basic idea is to determine the lengths of the aggregate time periods that minimize the NPV, subject to the constraint that the NPV is an upper bound to the original problem. We formulate and solve this intermediate problem (PIA) with dynamic programming [25] in a recursive scheme as follows (please refer to the list of nomenclature at the end of the chapter):

(PIA):

$$h_\tau(s_{\tau-1}) = \min_{m_\tau} \left[g_\tau(s_{\tau-1}, m_\tau) + h_{\tau+1}(s_\tau) \right], \quad \tau = TA \ldots 1 \quad (10.36)$$

$$s_\tau = s_{\tau-1} - m_\tau, \quad \tau = TA \ldots 1 \quad (10.37)$$

$$g_\tau(s_{\tau-1}, m_\tau) =$$

$$\left[f_{inv,(T-s_{\tau-1}+m_\tau)}(1 - f_{dpr,(T-s_{\tau-1}+1)}) \sum_{t=T-s_{\tau-1}+1}^{t=T-s_{\tau-1}+m_\tau} I_t \right] + \quad (10.38)$$

$$f_{rev,(T-s_{\tau-1}+1)} \sum_{t=T-s_{\tau-1}+1}^{t=T-s_{\tau-1}+m_\tau} R_t, \quad \tau = TA \ldots 1$$

$$h_{TA+1}(s_{TA}) = 0 \quad (10.39)$$

The objective function (10.36) is to minimize the NPV. This is counter-intuitive, since one would normally expect to maximize NPV. However, by chosing the discounting factors appropriately, we ensure that the optimal value is an upper bound to the original disaggregate problem. The optimization has the effect of determining the length of aggregate time periods such that we obtain the lowest possible upper bound, i.e. the upper bound that is closest to the previous lower bound. The investments, I_t, and revenues, R_t, in each time period are obtained from the solution of the previous disaggregate problem. The objective function is calculated recursively for each stage/aggregate time period from *TA* down to 1. Constraint (10.37) relates the current state to the previous state through the length of the current aggregate time period. The number of disaggregate periods available for assignment to remaining aggregate periods equals the number of periods that were available at the previous stage minus the number that are assigned to the current aggregate period. Constraint (10.38) calculates the actual objective value at each stage of recursion for all possible values of the state at that stage, while constraint (10.39) gives the initial condition for the start of recursion. The final solution is given by $h_1(s_0)$. We programmed this model in C++ and were able to solve it in less than 1 CPU second for all instances.

The solution of (PAI) yields an objective value which is an indication of how close the upper bound can get to the lower bound, as well as an aggregation scheme to obtain an improved upper bound. The short solution time allows the possibility of experimenting with different numbers of total aggregate periods to determine if we can have an even smaller aggregate problem without losing too much accuracy. We thus determined to fix the number of aggregate periods to 6, since this allows a significant speed-up in solution time, while keeping an appropriate level of accuracy.

The final form of the iterative aggregation/disaggregation algorithm is shown in Figure 10.3.

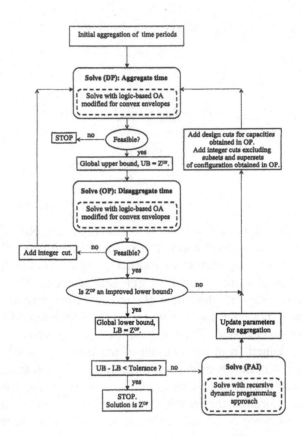

Figure 10.3.   Iterative aggregation/disaggregation algorithm

5.   Example

In this section we solve the oilfield infrastructure planning problem for a range of problem sizes up to the largest instance consisting of 1 PP, 2 WPs, and 10 reservoirs containing a total of 25 well sites. All problems are solved for the complete horizon of 24 time periods and the results are presented in Figure 10.4. The first four columns show the various MINLP problem sizes in terms of number of constraints, total number of variables, and number of binary variables (treated as SOS1 variables) for the given number of wells. The next two columns show a comparison of the best lower bounds obtained by the OA algorithm (as used in DICOPT++ [22]) and the proposed algorithm respectively. Note that the OA algorithm is coded in GAMS to include convex envelopes in order to obtain reasonable lower bounds. From these values it can be seen that the bounds obtained by the proposed method are very similar to, or slightly better than, the ones obtained when the problems are solved in the full space by the OA algorithm. This indicates that accuracy is maintained in the proposed decomposition and aggregation/disaggregation.

The next two colums (7th and 8th) show the solution times in CPU seconds, obtained on a HP9000/C110 workstation, for the fullspace OA algorithm and the proposed algorithm respectively. All NLP sub-problems are solved with CONOPT2, while all MILP master problems are solved with CPLEX 6.5. The results show an order of magnitude reduction in solution time for the proposed method compared to the OA algorithm as used in DICOPT++ [22]. For the largest instance the OA algorithm solved in 19385.9 CPU seconds, while the proposed method solved the problem in 1423.4 CPU seconds.

The last two columns compare the gap between the upper bound obtained from the aggregate design problem and the lower bound obtained from the disaggregate planning problem for the cases when sub-problem (PAI) is and is not included in the algorithm. For the case where it is not included, groups of four quarterly periods are aggregated into years. The gap when (PAI) is included ranges between 3.7% and 8.7%, while the gap when (PAI) is not included ranges between 8.5% and 22.6%. A better indication of the quality of the solution is therefore obtained by the inclusion of (PAI) into the algorithm.

Figures 10.5 and 10.7, together with Figure 10.6 show the final solution obtained for the largest problem instance of 25 wells for the planning horizon of 24 periods. The final configuration is shown in Figure 10.5. Note that only 9 of the potential 25 wells are drilled over the 24 periods. Of these, 3 are drilled in the first period, 1 in the second, 1 in the third, 1 in the fourth, and 3 in the fifth period as shown in Figure 10.6.

Figure 10.7 shows the production profile for the whole infrastructure over the 24 time periods encompassing the six years from January 1999 up to December 2004. The net present value obtained for the profit is \$ 67.99 million. This final solution is found in less than 25 minutes by the proposed algorithm, whereas a traditional solution approach such as the OA algorithm needs more than 5 hours to find the solution. Due to the short solution time, the model can quickly be updated and resolved periodically as more information about the future becomes available. Also, different instances of the same problem can

Problem size (24 time periods)				Best LB* ($mil.)		Solution time* (CPU sec.)		Gap w/o	Gap w/
# wells	# constraints	# variables	# 0-1 variables	GAMS (OA)	Proposed method	GAMS (OA)	Proposed method	(PAI) (%)	(PAI) (%)
4	1917	1351	150	8.55	8.55	224.3	31.1	21.4	5.0
6	2803	1954	225	8.46	8.46	538.0	37.0	22.6	6.1
9	3880	2677	300	25.86	25.86	3936.7	113.8	10.7	3.9
11	4622	3183	350	36.48	36.44	2744.2	227.1	9.1	3.7
14	5699	3906	425	49.40	49.40	3804.8	434.3	8.5	4.0
16	6441	4412	475	54.25	54.55	4944.8	809.0	11.6	7.0
19	7518	5135	550	64.08	64.35	6521.2	887.6	9.8	6.0
21	8260	5641	600	63.86	64.33	11659.3	1163.1	9.7	6.3
23	9002	6147	650	65.34	65.62	9223.3	1211.6	9.8	6.0
25	9744	6653	700	67.90	67.99	19385.9	1423.4	12.6	8.7

Figure 10.4.   Results

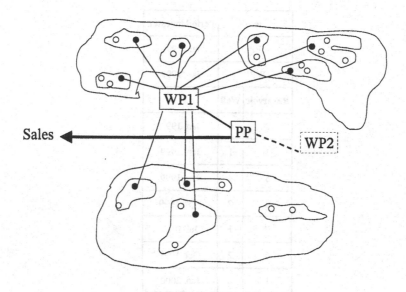

Figure 10.5. The final configuration

be solved in a relatively short time to determine the effect of different reservoir simulations on the outcome. The high initial production is the result of the restriction that the production should be non-increasing, combined with the fact that no significant ramp-up time is imposed by the oil company. Such a ramp-up can be incorporated into the model if required. Finally, it is interesting to compare the optimal solution obtained with a heuristic solution that is based on drilling wells with highest productivity index first to sustain production as long as possible. The net present value of this heuristic is only $ 56.24 million, and the corresponding production profile is shown in Figure 10.7. Thus, it is clear that the benefits of the proposed optimization model are substantial. One may question the use of this heuristic as a benchmark given its weak performance compared to the proposed method, but despite its apparent weakness it is a heuristic often used in practice.

6. Conclusions and Future Work

A multiperiod MINLP model for offshore oilfield infrastructure planning has been presented based on the linear model proposed by Iyer *et al.* [15]. By incorporating nonlinearities directly into the model, the need for using a very large number of binary interpolation variables is eliminated. Non-convexities resulting from the nonlinear equations were dealt with by using convex envelopes.

Item		Period invested
PP		Jan. 1999
WP1		Jan. 1999
Reservoir	Well	
2	4	Jan. 1999
3	1	Jan. 1999
5	3	Jan. 1999
4	2	Apr. 1999
7	1	Jul. 1999
6	2	Oct. 1999
1	2	Jan. 2000
9	2	Jan. 2000
10	1	Jan. 2000

Figure 10.6.   The optimal investement plan

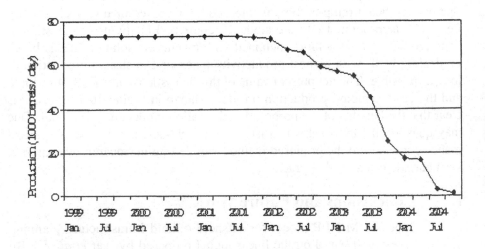

Figure 10.7.   Production profile over six years

Furthermore, an iterative aggregation/disaggregation algorithm, in which logic-based methods, bilevel decomposition and aggregation of time periods are integrated, has been proposed for the solution of this model. A novel dynamic programming subproblem (PAI) has been proposed to update the aggregation scheme after each iteration in order to find a better bound from the aggregate problem.

This algorithm has been applied to an industrial sized example and results show significantly reduced solution times compared to a full space solution strategy, while the quality of the solution is similar to that of the full space solution. Results also show that model (PAI) leads to an aggregation scheme that resembles the disaggregate problem more closely, thereby giving a better indication of the quality of the solution. Benefits compared to a heuristic solution are also shown to be substantial.

More complex nonlinearities need to be incorporated in future work. Another possible extension of the application in future work is to deal with uncertainty in reservoir simulations and future economical factors, as well as the sensitivity of the solution to the future price of crude oil. Work in this direction has been started with a Lagrangean decomposition heuristic for the design and planning of offshore hydrocarbon field infrastructures with more complex economic objectives than had been considered in the past, to capture the effect of taxes, tariffs and royalties on investment decisions [3,4].

7. Acknowledgment

The authors would like to acknowledge financial support from Mobil Technology Company and the Department of Energy under Grant number DE-FG02-98ER14875. Furthermore, the authors would like to thank Greger Ottosson for the useful hint and discussion on dynamic programming, as well as our anonymous reviewer for the insightful comments regarding the oil industry.

8.   Nomenclature

Sets and Indices:

PP	set of production platforms
p	production platform $p \in PP$
$WP(p)$	set of well platforms associated with platform p
π	well platform $\pi \in WP(p)$
F	set of fields
f	field $f \in F$
$R(f)$	set of reservoirs associated with field f
r	reservoir $r \in R(f)$
$W_{WP}(\pi)$	set of wells associated with well platform π
$W_R(r)$	set of wells associated with reservoir r
$W_{WP,R}(r, \pi)$	set of wells associated with reservoir r and well platform π
w	well $w \in W_{(.)}(.)$
t	time periods
τ	aggregated time periods
T	disaggregate time periods
TA	aggregate time periods
t	disaggregate time period $t \in T$
τ	aggregate time period $\tau \in TA$

To denote a specific well w that is associated with a specific well platform π, which is in turn associated with a specific production platform p, we use the index combination (w, π, p). Similarly, the index (π, p) applies to a specific well platform π associated with a specific production platform p. We omit superscripts in the variable definition for the sake of simplicity.

Continuous variables:

x_t	oil flow rate in period t
xc_t	cumulative oil flow up to period t
g_t	gas flow rate (volumetric) in period t
gc_t	cumulative gas flow up to period t
l_t	oil flow (mass) in period t
ϕ_t	gas-to-oil ratio (GOR) in period t
v_t	pressure in period t
δ_t	pressure drop at choke in period t
d_t	design variable in period t
e_t	design expansion variable in period t
Rev_t	sales revenue in period t
CI_t	investment cost in period t(including depreciation)
s_τ	state at the end of aggregate period τ, i.e. number of disaggregate periods available for assignment at the end of aggregate period τ
m_τ	length of aggregate period τ

Boolean variables:

z_t	= true if facility (well, WP or PP) is drilled/installed in period t

Parameters:

ρ	productivity index of well
P_{max}	maximum pressure drop from well bore to well head
GOR_{max}	maximum GOR
m_τ	number of periods in aggregate time period τ
T_a	number of aggregate time periods, $\tau = 1 .. T_a$
M_w	maximum number of wells drilled in a time period
Δt	length of time period t
U	upper bound parameter (defined by the respective constraint)
α	pressure drop coefficient for oil flow rate
β	pressure drop coefficient for GOR
c_{1t}	discounted revenue price coefficient for oil sales
c_{2t}	discounted fixed cost coefficient for capital investment
c_{3t}	discounted variables cost coefficient for capital investment
γ_{p1}	first coefficient for pressure vs. cumulative oil
γ_{p2}	second coefficient for pressure vs. cumulative oil
γ_{g1}	first coefficient for cumulative gas vs. cumulative oil
γ_{g2}	second coefficient for cumulative gas vs. cumulative oil
γ_{g3}	third coefficient for cumulative gas vs. cumulative oil
γ_{gor1}	first coefficient for GOR vs. cumulative oil
γ_{gor2}	second coefficient for GOR vs. cumulative oil
γ_{gor3}	third coefficient for GOR vs. cumulative oil
$f_{inv,t}$	discounting factor for investment in period t

$f_{dpr,t}$	discounting factor for depreciation in period t
$f_{rev,t}$	discounting factor for revenue in period t
I_t	investment costs in period t
R_t	revenue in period t

Superscripts:

(w,π,p)	variables associated with well $w \in W$, with well platform π and production platform p
(π,p)	variables associated with well platform π and production platform p
(p)	variables associated with production platform p
(r)	variables associated with reservoir r

References

(1) Van den Heever, S.A.; Grossmann, I.E., An Iterative Aggregation / Disaggregation Approach for the Solution of a Mixed-Integer Nonlinear Oilfield Infrastructure Planning Model. *Industrial & Engineering Chemistry Research*, **2000**, 39, 1955.

(2) Goel, V. and I.E. Grossmann, ""A Stochastic Programming Approach to Planning of Offshore Gas Field Developments under Uncertainty in Reserves", Computers and Chemical Engineering, 28, 1409-1429 (2004).

(3) Van den Heever, S.A., I.E. Grossmann, S. Vasantharajan and K. Edwards,"Integrating Complex Economic Objectives with the Design an Planning of Offshore Oilfield Facilities," Computers and Chemical Engineering 24, 1049-1056 (2000).

(4) Van den Heever, S.A.; Grossmann, I.E. A Lagrangean Decomposition Heuristic for the Design and Planning of Offshore Hydrocarbon Field Infrastructures with Complex Economic Objectives. *Ind. Ebg. Chem. Res.*, **2001**, 40, 2857.

(5) Garey, M.R.; Johnson, D.S. *Computers and Intractability: A Guide to the Theory of NP-Completeness*; W.H. Freeman and Company: New York, 1978.

(6) Lee, A.S.; Aranofsky, J.S. A Linear Programming Model for Scheduling Crude Oil Production. *Petroleum Transactions*, AIME, **1958**, 213, 389.

(7) Aranofsky, J.S.; Williams, A.C. The Use of Linear Programming and Mathematical Models in Underground Oil Production. *Management Science*, **1962**, 8, 394.

(8) Attra, H.D.; Wise, W.B.; Black, W.M. Application of Optimizing Techniques for Studying Field Production Operations. 34th *Annual Fall Meeting of SPE*, **1961**,82.

(9) Frair, L.C. Economic Optimization of Offshore Oilfield Development. *PhD Dissertation, University of Oklahoma, Tulsa, OK*, **1973**.

(10) Sullivan, J. A Computer Model for Planning the Development of an Offshore Oil Field. *Journal of Petroleum Technology*, **1982**, 34, 1555.

(11) Bohannon, J. A Linear Programming Model for Optimum Development of Multi-Reservoir Pipeline Systems. *Journal of Petroleum Technology*, **1970**, 22, 1429.

(12) Haugland, D.; Hallefjord, Å.; Asheim, H. Models for Petroleum Field Exploitation. *European Journal of Operational Research*, **1988**, 37, 58.

(13) Nygreen, B.; Christiansen, M.; Haugen, K.; Bjørkvoll, T.; Kristiansen, Ø. Modeling Norwegian petroleum production and transportation. *Annals of Operations Research*, **1998**, 82, 251.

(14) Eeg, O.S.; Herring, T. Combining Linear Programming and Reservoir Simulation to Optimize Asset Value. *SPE 37446 presented at the SPE Production Operations Symposium, Oklahoma City, 9-11 March*, **1997**.

(15) Iyer, R.R.; Grossmann, I.E.; Vasantharajan, S.; Cullick, A.S. Optimal Planning and Scheduling of Offshore Oil Field Infrastructure Investment and Operations. *Ind. Eng. Chem. Res.,* **1998**,37, 1380.

(16) Geoffrion, A.M. Generalized Benders Decomposition. *J. Optim. Theory Appl.*, **1972**, 10, 237.

(17) Brooke, A.; Kendrick, D.; Meeraus, A. *GAMS: A User's Guide, Release 2.25;* The Scientific Press: South San Francisco, 1992.

(18) Van den Heever, S.A.; Grossmann, I.E. Disjunctive Multiperiod Optimization Methods for Design and Planning of Chemical Process Systems. *Computers and Chemical Engineering*, **1999**, 23, 1075.

(19) Turkay, M.; Grossmann, I.E.; Logic-Based MINLP Algorithms for the Optimal Synthesis of Process Networks. *Computers and Chemical Engineering*, **1996**, 20, 959.

(20) Duran, M.A.; Grossmann, I.E. An Outer Approximation Algorithm for a Class of Mixed-Integer Nonlinear Programs. *Mathematical Programming*, **1986**, 36, 307.

(21) Balas, E. Disjunctive Programming and a Hierarchy of Relaxations for Discrete Optimization Problems. *SIAM J. alg. Disc. Meth.*, **1985**, 6, 466.

(22) Viswanathan, J.; Grossmann, I.E. A combined penalty function and outer-approximation method for MINLP optimization. *Computers and Chemical Engineering*, **1990**, 14, 769.

(23) Horst, R.; Pardalos, P.M. (eds.) *Handbook of Global Optimization*, Kluwer Academic Publishers, **1995**.

(24) Iyer, R.R.; Grossmann, I.E. A Bilevel Decomposition Algorithm for Long-Range Planning of Process Networks. *Ind. Eng. Chem. Res.* **1998**, 37, 474.

(25) Bellman, R. *Dynamic Programming*, Princeton University Press, **1957**.

(26) Turkay, M.; Grossmann, I.E. Disjunctive Programming Techniques for the Optimization of Process Systems with Discontinuous Investment Costs – Multiple Size Regions. *Ind. Ebg. Chem. Res.*, **1996**, 35, 2611.

Chapter 11

RADIATION TREATMENT PLANNING: MIXED INTEGER PROGRAMMING FORMULATIONS AND APPROACHES*

Michael C. Ferris[1], Robert R. Meyer[1] and Warren D'Souza[2]

[1] *Computer Sciences Department*
University of Wisconsin
1210 West Dayton Street, Madison
Wisconsin 53706
USA

{ ferris,rrm } @cs.wisc.edu

[2] *University of Maryland School of Medicine*
22 South Green Street
Baltimore, MD 21201
USA

wdsou001@umaryland.edu

Abstract   Radiation therapy is extensively used to treat a wide range of cancers. Due to the increasing complexities of delivery mechanisms, and the improved imaging devices that allow more accurate determination of cancer location, determination of high quality treatment plans via trial-and-error methods is impractical and computer optimization approaches to planning are becoming more critical and more difficult.

We outline three examples of the types of treatment planning problem that can arise in practice and strive to understand the commonalities and differences in these problems. We highlight optimization approaches to the problems, and particularly consider approaches based on mixed integer programming. Details of the mathematical formulations and algorithmic approaches are developed and pointers are given to supporting literature that shows the efficacy of the approaches in practical situations.

*This material is based on research partially supported by the National Science Foundation Grants ACI-0113051, DMI-0100220, and CCR-9972372 and the Air Force Office of Scientific Research Grant F49620-01-1-0040.

1.   Introduction

Approximately 1.2 million new cases of cancer are reported each year in the United States, with many times that number occurring worldwide. About 40% of people diagnosed with cancer in the U.S will undergo treatment with radiation therapy. This form of therapy has undergone tremendous improvement from a treatment planning standpoint over the last decade, having benefited significantly from advances in imaging technology for computed tomography (CT), magnetic resonance imaging (MRI) and ultrasound (US). As a result of these advances, there has been an increased trend toward image-based radiation therapy treatment planning.

In treatment planning problems, the objective is to deliver a homogeneous (uniform) dose of radiation to the tumor (typically called the target) area while avoiding unnecessary damage to the surrounding tissue and organs. In many cases, near the target there are several structures (typically called organs at risk (OAR)) for which the dose must be severely constrained due to the probability of damage that will lead to medical complications. Since the target and OAR structures can be more accurately identified, the delivery of radiation that accurately conforms to these tissues becomes an achievable goal.

The planning process is determined, not only by the target and the proximal organs at risk, but also by the physical characteristics of the mechanism that will be used to deliver the dose. Teletherapy [5] is the common collective name for the various kinds of external beam radiation treatment, whereas brachytherapy ("brachy" is a Greek prefix implying a short distance) involves the placement of radioactive source configurations near or within the tumor. The availability of additional data and levels of control of radiation delivery procedures adds great complexity to the problem of determining high quality treatment plans.

Classical radiation therapy treatment planning (sometimes called forward planning) was generally a trial and error process in which improved plans were generated by iteratively experimenting with different incident high-energy beam configurations for teletherapy and with alternative placements of sources for brachytherapy. In recent years, there has been a move toward computer generated plans (sometimes termed inverse planning). These planning procedures increase the allowable complexities when the dose is delivered by radioactive implants (brachytherapy), or when the radiation is fired from a number of angles (external beam therapy). Some further extensions include the use of scans in conjunction with the planning procedure, the use of intensity modulation of portions of the beam (termed pencil beams or beamlets), and the use of stereotactic devices to greatly improve the accuracy of the delivered doses of radiation. There are many techniques available to generate treatment plans for each type of radiation delivery system. However, there are definite commonalities arising in all these problems that we shall endeavor to highlight here.

A unified and automated treatment process has several potential benefits relative to the classical trial-and-error approach. Among these the most important ones are the reduction in planning time and the improvement and uniformity of treatment quality that can be accomplished.

It is usual for the treatment goals to vary from one planner to the next, so a planning tool must be able to accommodate several different goals. Among these goals, the following are typical, although the level of treatment and importance of each may vary.

1 A "homogeneity" goal: An isodose curve is delineated around the volume to ensure delivery of a certain fraction of the maximum delivered dose. A typical homogeneity goal requires that the $x\%$ isodose line encompasses the target volume. Such requirements can be enforced using lower and upper bounds on the dose, or approximated via penalization. (Upper bounds are of lesser importance in brachytherapy since tissues adjacent to the radioactive sources automatically receive high doses.)

2 A "conformity" goal: The overall dosage to the patient is typically limited, and this goal specifies a lower bound on the fraction of that dose to be delivered to the target itself.

3 "Avoidance" goals: These limit the dose delivered to certain sensitive structures (OAR) close to the target.

4 "Simplicity" goals: it is preferable to use as simple a procedure as possible since this reduces "implementation" errors (and frequently reduces treatment time) and allows more patients to be treated with the available resources.

There are often standards established by various professional and advisory groups that specify acceptable homogeneity and conformity requirements.

In all formulations, we need to determine the dose delivered ($Dose$ at a voxel (i, j, k)) from a particular source (e.g. a pencil beam or a radioactive source). A critical feature of all these problem areas is that a functional form for such dose is either highly nonlinear, or is described by a large amount of data that specifies the dose delivered to each voxel of the region of interest. We outline techniques based on both of these approaches.

In a practical setting, many constraints on dose (measured in units called Gray, abbreviated as Gy) are phrased as dose-volume histogram (DVH) constraints of the form:

no more than 30% of volume X should exceed 10 Gy

or

at least 80% of volume Y should exceed 30 Gy

These constraints are hard to enforce due to the fact that the model needs to determine on a per-voxel basis whether the threshold value is exceeded or not. For example, in the first case, the following constraints could be used

$$Dose(i,j,k) \leq 10 + M * Exceed(i,j,k) \quad \forall(i,j,k) \in X$$

$$\sum_{(i,j,k) \in X} Exceed(i,j,k) \leq 0.3 * card(X),$$

where $Exceed(i,j,k)$ is a binary variable. Standard integer programming issues relating to the choice of the constant M ensue. More generally, when the threshold value is $U_{\mathcal{R}}$ and the percentage limit on overdose in region $\mathcal{R}$ is $\beta_{\mathcal{R}}$ we have

$$Dose(i,j,k) \leq U_{\mathcal{R}} + M * Exceed(i,j,k) \quad \forall(i,j,k) \in \mathcal{R}$$

$$\sum_{(i,j,k) \in \mathcal{R}} Exceed(i,j,k) \leq \beta_{\mathcal{R}} * card(\mathcal{R}).$$

(11.1)

These formulations can become quickly impractical due to large numbers of voxels in the regions of interest. In practice, many modelers use approximate techniques to enforce these constraints. Alternatively, subproblems corresponding to subsets of the region of interest may be sequentially solved.

The conformity of the plan presents even more computational difficulties to a modeler since it involves all voxels receiving radiation. The conformity index C is an estimate of the ratio of the dose delivered to the target $\mathcal{T}$, divided by the total dose delivered to the patient. These indices can be used to enforce a conformity requirement using:

$$C \sum_{(i,j,k)} Dose(i,j,k) \leq \sum_{(i,j,k) \in \mathcal{T}} Dose(i,j,k).$$ (11.2)

If conformity is part of the model constraint set, then an appropriate value for C needs to be ascertained beforehand. A reasonable conformity index for a given patient plan is very hard to estimate *a priori* since it depends critically on how complicated the delivery mechanism is allowed to be by the planner and how the volume of the target interacts with the volumes of the allowed delivery.

This paper is not intended to be a complete survey of the use of optimization techniques within treatment planning problems (see [28, 36, 38, 50] for more complete overviews of conformal radiation therapy, for example). In addition to these survey articles, there are a variety of other approaches (for which we cite representative papers) including those based on optimal control primitives [1], or simulated annealing [26, 32, 43, 44], iterative (relaxation) techniques [10], approaches using biological objectives [7, 23, 33] techniques of multi-objective [21] and neuro-dynamic programming [19]. In this paper we specifically outline three particular problem areas that arise in treatment planning and highlight the discrete nature of some of the decisions that need to be made. Some detail of the underlying applications are given, along with an overview of several solution approaches we feel are promising. This survey is organized as follows: the following two sections describe successful optimization applications to Gamma Knife teletherapy and to brachytherapy for prostate cancer. We then consider extensions of these and related approaches to other teletherapy mechanisms such as IMRT (intensity modulated radiation therapy), and conclude with an assessment of future research directions in radiation treatment planning.

2. Gamma Knife Radiosurgery

The Gamma Knife is a highly specialized treatment unit that provides an advanced stereotactic approach to the treatment of tumor and vascular malformations within the head [20]. The Gamma Knife (see Figure 11.1(a)) delivers a single, high dose of radiation emanating from 201 Cobalt-60 unit sources. All 201 beams simultaneously intersect at the same location in space to form an approximately spherical dose region that is typically termed a shot of radiation. A typical treatment consists of a number of shots, of possibly different sizes and different intensities, centered at different locations in the tumor, whose cumulative effect is to deliver a certain dose to the treatment volume while minimizing the effect on surrounding tissue.

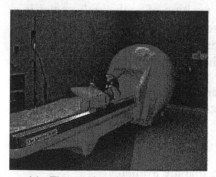

(a) The couch and treatment area   (b) The head-frame and helmet

Figure 11.1. The Gamma Knife Treatment Unit. A focusing helmet is attached to the frame on the patient's head. The patient lies on the couch and is moved back into the shielded treatment area

Gamma Knife radiosurgery begins (after administering local anesthesia) by fixing a stereotactic coordinate head frame to the patient's head using adjustable posts and fixation screws. This frame establishes a coordinate system within which the target location is known precisely and also serves to immobilize the patient's head within an attached focusing helmet during the treatment (see Figure 11.1(b)). An MRI or CT scan is used to determine the position of the treatment volume in relation to the coordinates determined by the head frame. Once the location and the volume of the tumor are identified, a neurosurgeon, a radiation oncologist, and a physicist work together in order to develop the patient's treatment plan. Typically, the plan should be completed in no more than 30 minutes, primarily due to patient comfort considerations.

The determination of plans varies substantially in difficulty. For example, some tumors are small enough to require only one shot of radiation. On the other hand, when the shape of the tumor is large or has an irregular shape or is close to a sensitive structure, many shots of different sizes could be needed to achieve a high dose of radiation to the intracranial target volume while sparing the surrounding tissue. Further description of the treatment process, along with some more explanatory figures can be found in [18].

A number of researchers have studied techniques for automating the Gamma Knife treatment planning process. One approach incorporates the assumption that each shot of radiation can be modeled as a sphere. The problem is then reduced to one of geometric coverage, and a ball packing approach [42, 41, 49, 48, 47] can be used to determine the shot locations and sizes. The use of a modified Powell's method in conjunction with simulated annealing has also been proposed [30, 52]. A mixed integer programming (MIP) and a nonlinear programming approach for the problem are presented in [16, 17, 18, 37, 39].

In the model we propose, there are three types of decision variables:

1 *A set of coordinates* (x_s, y_s, z_s): the position of each shot's center is a continuous variable to be chosen. We assume that $S = \{1, 2, \cdots, m\}$ denotes the set of m shots to be considered in the optimization. Normally, we have to choose $n < m$ of these shots to be used.

2 *A discrete set of collimator sizes*: There are four focusing helmets available that generate different width shots.

$$W = \{4mm, 8mm, 14mm, 18mm\}$$

denotes the choices of discrete widths that are available.

3 *Radiation exposure time*: $t_{s,w}$ is the time each shot (s, w) is exposed. It is known that the total dose delivered is a linear function of the exposure time.

The dose delivered at a voxel (i, j, k) in unit time from a shot centered at (x_s, y_s, z_s) can be modeled by a nonlinear function $D_w(x_s, y_s, z_s, i, j, k)$ [11, 24, 43]. The total dose delivered to a voxel (i, j, k) from a given set of shots can then be calculated as

$$Dose(i, j, k) = \sum_{(s,w) \in \mathcal{S} \times \mathcal{W}} t_{s,w} D_w(x_s, y_s, z_s, i, j, k). \qquad (11.3)$$

We use the following functional form

$$D_w(x_s, y_s, z_s, i, j, k) =$$

$$\sum_{p=1}^{2} \lambda_p \left(1 - \mathrm{erf} \left(\frac{\sqrt{(i - x_s)^2 + \mu_p^y(j - y_s)^2 + \mu_p^z(k - z_s)^2} - r_p}{\sigma_p} \right) \right)$$

and for each value of $w \in \mathcal{W}$ we fit ten parameters λ_p, μ_p^y, μ_p^z, r_p and σ_p (for $p = 1, 2$) to observed data via least-squares. The notation erf (x) represents the integral of the standard normal distribution from $-\infty$ to x; the particular functional form is suggested in the literature.

In many cases, the planner wishes to limit the use of certain resources. In the Gamma Knife case, there is a bound of n shots, and each shot must have a specified width. If we introduce a binary variable $\psi_{s,w}$ that indicates whether shot s uses width w or not, the following constraints implement the above requirement:

$$0 \leq t_{s,w} \leq \psi_{s,w} \bar{t}$$
$$\sum_{(s,w) \in \mathcal{S} \times \mathcal{W}} \psi_{s,w} \leq n. \qquad (11.4)$$

Note that the planner needs to specify an upper bound $\bar{t}$ on the exposure time of any shot. The typical range of values for n is $1 - 15$. It would be possible to impose additional constraints of the form

$$\sum_{w \in \mathcal{W}} \psi_{s,w} \leq 1, \quad \forall s \in \mathcal{S},$$

but we believe it is better to enforce the above aggregation instead.

It is easy to specify homogeneity in models simply by imposing lower and upper bounds on the dose delivered to voxels in the target $\mathcal{T}$. Similar bounding techniques can be used for avoidance requirements. The imposition of rigid bounds typically leads to plans that are too homogeneous and not conformal enough, that is, they provide too much dose outside the target. To overcome this problem, the notion of "underdose" was suggested in [17]:

$$Underdose(i, j, k) := \max\{0, \theta - Dose(i, j, k)\}.$$

Informally, underdose measures how much the delivered dose is below the prescribed dose, θ, on the target voxels. In practice, θ is typically given as a scalar, but could be specified on a per-voxel basis if desired. Provided we minimize

Underdose we can implement this construct using linear constraints:

$$\theta \leq Underdose(i, j, k) + Dose(i, j, k)$$
$$0 \leq Underdose(i, j, k). \tag{11.5}$$

The basic model attempts to minimize the underdose subject to the afore-mentioned constraints on conformity, homogeneity and avoidance.

$$
\begin{aligned}
\min \quad & \sum_{(i,j,k) \in \mathcal{T}} Underdose(i, j, k) \\
\text{subject to} \quad & (11.3) \text{ dose definition} \\
& (11.5) \text{ underdose enforcement constraints} \\
& (11.2) \text{ conformity constraints} \\
& (11.1) \text{ dose volume histogram constraints} \\
& (11.4) \text{ resource use constraints} \\
& x_s, y_s, z_s, t_{s,w} \in \mathbf{R} \\
& 0 \leq Dose(i, j, k) \leq U, \quad \forall (i, j, k) \in \mathcal{T} \\
& \psi_{s,w}, \; Exceed(i, j, k) \in \{0, 1\}
\end{aligned}
\tag{11.6}
$$

This model is a nonlinear, mixed integer programming problem. If we choose a fixed set of shot locations (x_s, y_s, z_s) then the model becomes a linear mixed integer programming problem with a collection $(4m)$ of large data matrices $D_w^s(i, j, k)$ replacing the nonlinear functions $D_w(x_s, y_s, z_s, i, j, k)$ in (11.3). Our investigations found such approaches to be impractical and not as accurate as the scheme outlined below. For realistic instances, the data and number of binary variables becomes very large and the models are unsolvable within the available time limit. Furthermore, the solution quality requires the shot centers to be determined accurately. We therefore resort to methods that include the shot center locations as variables. Similar issues arise in the brachytherapy application we describe later, and we outline a different approach for the solution of that problem.

To avoid the combinatorial issues associated with $\psi_{s,w}$ we use a smooth nonlinear approximation H_α to the Heaviside step function:

$$H_\alpha(t) := \frac{2 \arctan(\alpha t)}{\pi}.$$

For increasing values of α, H_α becomes a closer approximation to the step function for $t \geq 0$. We replace (11.4) with

$$n = \sum_{(s,w) \in \mathcal{S} \times \mathcal{W}} H_\alpha(t_{s,w}). \tag{11.7}$$

To reduce the number of voxels considered we use a coarse grid $\mathcal{G}$ of voxels in the critical regions and refine the grid appropriately as the calculations progress. To avoid the problems associated with calculating the dose at every voxel in the volume, we approximate the total dose to the volume using

$$\sum_{(s,w)\in\mathcal{S}\times W} \bar{D}_w t_{s,w} \;,$$

where $\bar{D}_w$ is the (measured) dose delivered by a shot of size w to a "phantom". The quantities can be determined once, and used to generate a very good estimate of the total dose delivered to the volume without performing any dose calculations outside the target and the critical organ regions. This leads to the following approximation to the conformity constraint (11.2):

$$C \sum_{(s,w)\in\mathcal{S}\times W} \bar{D}_w t_{s,w} \le \frac{\mathcal{N}_{\mathcal{T}}}{\mathcal{N}_{\mathcal{G}\cap\mathcal{T}}} \sum_{(i,j,k)\in\mathcal{G}\cap\mathcal{T}} Dose(i,j,k), \qquad (11.8)$$

where $\mathcal{N}_X$ represents the number of voxels in the volume X.

Thus, the nonlinear programming model that we use to approximate the solution of (11.6) has the following form:

$$\min \sum_{(i,j,k)\in\mathcal{G}\cap\mathcal{T}} Underdose(i,j,k)$$

subject to   (11.3) dose definition
(11.5) underdose enforcement constraints
(11.8) approximated conformity constraints   (11.9)
(11.7) approximated resource use constraints
$x_s, y_s, z_s, t_{s,w} \in \mathbf{R}$
$0 \le Dose(i,j,k) \le U, \quad \forall (i,j,k) \in \mathcal{G}\cap\mathcal{T}$
$0 \le t_{s,w} \le \bar{t}$

A series of five optimization problems are solved to determine the treatment plan. The model is solved iteratively (steps 2, 3, and 4 below) to reduce the total time to find the solution. Our experience shows that combining those three steps into one takes at least three times longer to converge (and can take up to 90 minutes for complex target shapes and large numbers of shots), which is often not clinically acceptable.

1 Conformity estimation. In order to avoid calculating the dose outside of the target, we first solve an optimization problem on the target to estimate an "ideal" conformity for the particular patient for a given number of shots. The conformity estimate C is passed to the basic model as an input parameter. Details can be found in [16].

2 Coarse grid estimate. Given the estimate of conformity C, we then specify a series of optimization problems whose purpose is to minimize the

total underdose on the target for the given conformity. In order to reduce the computational time required to determine the plan, we first solve (11.9) on a coarse grid subset of the target voxels. We have found it beneficial to use one or two more shot locations in the model than the number requested by the user, that is $S := \{1, ..., n + 2\}$, to allow the optimization to choose not only useful sizes but also to discard the extraneous shot locations.

3 Refined grid estimate. To keep the number of voxels in the optimization as small as possible, we only add to the coarse grid those voxels on a finer grid for which the homogeneity (bound) constraints are violated. This procedure improves the quality of the plan without greatly increasing the execution time.

4 Shot reduction problem. In the solution steps given above, we use a small value of α, typically 6 to impose the constraint (11.7) in an approximate manner. In the fourth solve, we increase the value of α to 100 in an attempt to force the planning system to choose which size/location pairs to use. At the end of this solve, there may still exist some size/location pairs that have very small exposure times t. Also note that our solution technique does not guarantee that the shots are centered at locations within the target.

5 Fixed location model. The computed solution may have more shots used than the user requested and furthermore may not be implementable on the Gamma Knife since the coordinate locations cannot be keyed into the machine. Our approach to adjust the optimization solution to generate implementable coordinates for the shot locations is to round the shot location values and then fix them. Once these locations are fixed, the problem becomes linear in the intensity values t. We reoptimize using (11.6) and force the user requested number of size/location pairs precisely using a mixed integer program solver.

Note that the starting point for each of the models is the solution point of the previous model. Details on how to generate an effective starting point for the first model are given in [16]. All the optimization models are written using the General Algebraic Modeling System (GAMS) [9] and solved using CONOPT [12] or CPLEX [22]. Typical solution times range from seconds for simple plans up to 20 minutes for very complex treatment plans.

A subset of the large number of holes in the focusing helmet can be blocked in order to (locally) reduce the amount of radiation delivered or to change the shape of the shot. By determining the amount of radiation that each blocked hole removes and updating D_w appropriately, an extension of the mixed integer model above can be used to further enhance the treatment plan, and spare sensitive structures even more. The "block or not" decision can be modeled using the approach outlined in (11.4), but this has yet to be implemented in practice due to concerns from clinicians regarding the chances of errors in the physical delivery process.

3.   Brachytherapy Treatment Planning

Brachytherapy involves the use of radioactive sources such as catheters or pellets (the latter are referred to as "seeds" below) that are placed within or close to the tumor. A disease site that has been receiving a great deal of attention for conformal treatment planning in brachytherapy is the prostate. The number of diagnosed prostate cancer cases has increased due to the widespread use of the PSA (prostate specific antigen) test. An option for treating prostate cancer is permanent radioactive implant brachytherapy under ultrasound guidance (that is, using needles for injection, the radioactive seeds are permanently implanted in the prostate). While image-guided 3-D conformal treatment planning in brachytherapy is still in its infancy, ultrasound-guided implantation of the prostate is one of the fastest growing medical procedures in the country. The number of such implants is projected to increase to over 100,000 by the year 2005 [2].

In contrast to the Gamma Knife model of the preceding section, the radiation delivery variables for the brachytherapy model consist of only binary variables $Seed(r, s, t)$ that take the value 1 if a seed is placed in voxel (r, s, t) and 0 otherwise (note that seeds may only be placed within the target T). For each possible seed position (r, s, t), a nonlinear dose function (essentially modeling an inverse square law) may then be used to compute a matrix $D_{r,s,t}$ of corresponding radiation doses for all voxels in the region of interest. (Note that the entries of this matrix need only be computed once from the nonlinear radiation function, since translations of this matrix will yield dose matrices for other seed positions.) The total dose at voxel (i, j, k) is then given as the weighted sum of dose contributions to (i, j, k) from all seeds:

$$Dose(i, j, k) = \sum_{(r,s,t)\in T} D_{r,s,t}(i, j, k) * Seed(r, s, t). \qquad (11.10)$$

The brachytherapy model also includes non-negative continuous underdose variables as defined in (11.5) for the target, as well as bounded non-negative overdose variables for each voxel in each organ at risk (OAR):

$$Dose(i, j, k) - Overdose(i, j, k) \leq U_{OAR}$$
$$0 \leq Overdose(i, j, k) \leq M - U_{OAR} \qquad (11.11)$$

These relations place both soft and hard constraints on the doses to the urethra and rectum. For I-125 implants the urethral and rectal threshold doses U_{OAR}

are set to 217.5 Gy and 101.5 Gy respectively. For I-125 the upper dose limits M are set at 275 Gy and 145 Gy for the urethra and rectum respectively by imposing the appropriate upper bounds on the overdose variables.

Finally, the model contains binary variables $Needle(i, j)$ that are forced to have value 1 if a seed is placed at position (i, j) in any plane k:

$$Seed(i, j, k) \leq Needle(i, j). \tag{11.12}$$

These needle constraints model the mechanics of the implantation process in which a template is used to position the seed-carrying needles and a needle in position (i, j) in the template can then implant seeds in position (i, j) in several different planes k. Since it is undesirable to use a large number of needles, a resource use term representing a weighted sum of $Needle$ variables is used in the objective.

The overall model that we would like to solve then becomes:

$$\min \quad \alpha * \sum_{(i,j,k) \in T} Underdose(i, j, k) + \beta *$$

$$\sum_{(i,j,k) \in OAR} Overdose(i, j, k) + \gamma * \sum_{(i,j) \in template} Needle(i, j)$$

subject to   (11.10) dose definition   (11.13)

(11.5) underdose constraints,

(11.11) overdose constraints,

(11.12) needle use constraints

$$Seed(i, j, k), \ Needle(i, j) \in \{0, 1\}$$

where α, β, γ are appropriately chosen weights.

Since this model involves so many variables, it is impractical to solve it as a single MIP. Thus, we consider a collection of problems instead, each of which focuses on a subset of variables and constraints that essentially reflects seed implant decisions for a single plane k, assuming that some radiation is already delivered to the voxels in plane k from seeds in the other planes. We use the term "sequential optimization" for this process of cycling through the planes one at a time, adjusting seed placement to optimize incremental doses only in the plane currently under consideration (see [15] for details, and for an alternative MIP based on coarse grids see [29]). For a fixed value of the plane index k we thus need only the pair (i, j) to represent a voxel, and incremental dose within a plane is modeled by replacing the dose equation (11.10) by

$$Dose(i, j) = InterplaneDose(i, j) + \sum_{(r,s) \in \mathcal{T}_k} D_{r,s,k}(i, j, k) * Seed(r, s, k),$$

(11.14)

where $InterplaneDose(i, j)$ represents the dose contributed to voxel (i, j, k) from seeds in the other planes (as seed positions are changed, these dose contributions are updated) and $\mathcal{T}_k$ is the subset of the target in plane k. Another change needed in the overall model during application of the sequential approach is to update the needle constraints to account for needle positions already in use in the other planes (so that no constraint or corresponding objective function penalty is imposed for those template positions already used in other planes). This simply requires a change in index sets. Since the optimization is performed in sequential plane-by-plane manner, it cannot accurately account for inter-plane variations in the location of the urethra. The urethra is 3-dimensional and shows significant curvature. Poor placement of the seeds in one plane may adversely affect the urethra in other planes. To circumvent this problem, it is possible to add a small number of additional constraints to a 2-D problem to reflect critical structures in nearby planes. Details of this process may be found in [13]. These constraints are termed positioning constraints. The net effect of these constraints is that seeds are less likely to be placed in close proximity to the urethra thereby also reducing the risk of significant impact on the dose distribution to the critical structures in the event that the seeds are misplaced.

Sequential optimization for brachytherapy at the initial pre-plan stage using mixed-integer programming and branch-and-bound has been previously discussed [15]. In that research, the initial values of $InterplaneDose(i, j)$ were based on estimated contributions from other planes, whereas here we describe the method as modified for operating room (OR) based treatment planning, in which the initial values of $InterplaneDose(i, j)$ are based on seed placements from the pre-plan. Our approach is also based on the premise that the volume of the prostate (target) does not change significantly (more than 5%) between the initial volume study performed weeks before and the actual implant, and this is generally the case for early stage (T1-T2) patients. This allows us to "hot start" the solution process not only by starting with seed placements from the pre-plan, but also provides for faster solutions by limiting seed placement deviations relative to that pre-plan.

Prior to the start of the optimization process, the locations of seeds obtained from the pre-treatment plan are reproduced in the new ultrasound image data set. During optimization, we consider the dose contribution from seeds in other planes up to a distance of 4 cm from the current plane under consideration. Contributions from seeds at a distance greater than 4 cm is negligible [15]. Optimization starts with the outermost planes on each side of the central plane, and proceeding toward the center planes from alternating directions. We found this ordering to be desirable because the small size of the outermost planes implies that there are few possible seed positions, and in some cases it was difficult to deal with these planes once the seed positions in the other slices had been fixed. We also inhibit the introduction of extra needles by increasing the weight γ on the number of needles in the objective function as additional planes are optimized.

The optimization process itself is carried out using GAMS [9] and the commercial mixed-integer programming branch-and-bound solver CPLEX [22]. Using this software, solutions with a relative gap (the gap between the best feasible solution generated and the lower bound of the relaxed mixed-integer programming problem) of $< 1\%$ were obtained in all of the 2-D problems.

In this research, data from 10 patients was used to test this re-optimization framework. Below we consider only those 7 patients receiving radioactive iodine implants (see [13] for data on the additional patients, who received palladium implants, for whom a slightly different analysis is needed). The dose distributions from the pre-treatment (both optimized and manual) plans for the original contours were applied to the new set of contours. These dose distributions were then compared with the dose distribution obtained from the re-optimized plans.

The sequential re-optimization process in the OR can be executed in a single iteration (sweep through all 2-D planar sections) with very good results. Since the pre-treatment optimized plans generally result in loss of target coverage when applied to the OR contours, one of the goals of re-optimizing the pre-treatment optimized plan is increasing (recovering) target coverage. At the same time, it is imperative that while achieving this end, no unnecessary deterioration in the dose distribution to the critical structures should occur. Although no undesirable dose increases were observed for the urethra and rectum from the application of the pre-treatment plan to the simulated OR contours, a significant improvement in the dose distribution to the critical structures is achieved by re-optimizing. When compared with the pre-treatment optimized plan, the volume of the rectum exceeding the threshold of 101.5 Gy is reduced in most patients. In the other patients, the increase in this volume is $< 4\%$ and the overdose fraction remains under 20%. In cases in which the dose to the critical structures increases by re-optimizing the pre-treatment optimized plan, this increase is counter-balanced by the increase in target coverage.

Table 11.1 provides a partial comparison between pre-plan and re-optimization results. Note that the OR plan always achieves at least 93% target coverage, and in those cases in which this does not represent an improvement relative to both pre-plan results, the re-optimization provides a significant improvement to the OAR dose (see [13] for additional tables of results and DVH plots that give data for the urethra and rectum). For example, for patient 1,

Patient #	Pre-man. (%)	Pre-opt. (%)	OR re-opt. (%)	seeds (av.)	needles (av)
1	97.4	94.5	95.3	101	24
2	92.5	95.5	97.2	104	24
3	90.4	93.3	94.8	102	24
4	95.2	93.6	97.0	100	22
5	96.1	93.8	94.7	79	18
6	91.7	92.7	93.3	102	27
7	93.7	95.2	95.2	102	26

Table 11.1. Target coverage of manual pre-plan vs optimized pre-plan vs OR re-optimized plan

the urethra overdose volume fraction is 15% for the both pre-plans and 0% for the re-optimized plan. Note also the average (over all three plans) counts for seeds and needles for each patient, which provide insights into the complexity of treatment planning (seed and needle counts do not vary significantly in the three plans).

Figure 11.1 shows the difference between the target underdoses for the pre-treatment optimized and re-optimized OR plans for the base and apex planes in the prostate for a representative patient (patient 3). The difference between the plans is most pronounced in these planes. The underdoses (cold spots) would be significant if the pre-treatment plan was used on the day of the procedure. It can be seen from the figure that the re-optimized plan is highly conformal and that significant cold spots are eliminated.

In summary, treatment planning time in the OR is of great importance. The mixed-integer sequential optimization framework allows the pre-plan to be used to provide initial estimates of seed positions and hence allows the OR-based plan to be performed in about 1/3 of the pre-plan time (i.e., approximately 4 to 8 minutes using a 440 MHz processor).

For our collection of patient data, re-optimization of the treatment plan using simulated OR contours resulted in an increase in target coverage relative to the optimized pre-plan in all cases (maximum increase was 3.4%). Critical structure dose distribution was also improved appreciably. We also found that the addition of positioning constraints to the basic optimization model produced treatment plans that are more robust with respect to possible seed misplacement (simulated by small random displacements) in the OR.

4. IMRT

Intensity modulated radiation therapy (IMRT) represents a rather sophisticated approach in which each radiation treatment (out of a total of perhaps 10-45 such treatments for a patient) involves the application of intensity-modulated

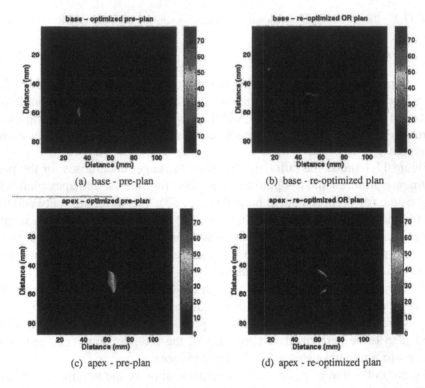

(a) base - pre-plan

(b) base - re-optimized plan

(c) apex - pre-plan

(d) apex - re-optimized plan

Figure 11.2.   Underdose of target regions for (a), (c) the pre-treatment plan and (b), (d) the re-optimized plan. (a) and (b) show the base plane, while (c) and (d) show the apex plane

beams of radiation from 5-9 different angles (relative to the patient) [4, 6, 8, 35, 44, 45, 46, 51]. The 10X10 centimeter beam cross-section consists of a grid of 100 or more small beamlets of radiation. Intensity modulation is usually accomplished via repeated adjustments of a multi-leaf collimator (a beam blocking mechanism comprising 40 opposing pairs of tungsten strips or leaves) in which the positions of beam-blocking tungsten leaves are set to allow the passage of radiation for a specified amount of time only in those positions corresponding to the desired beamlets. From an algebraic viewpoint, these binary radiation patterns are weighted by radiation intensities (determined by the length of time radiation is emitted) and the resulting weighted patterns are added together to produce the desired intensity matrix for each angle. This fine level of control of the radiation yields optimization models involving the choice of beam angles and beamlet intensities.

At least two types of optimization problems arise in IMRT treatment planning. The first is the intensity matching problem, and the second is the overall treatment planning problem described above. In the intensity matching problem, a desired integer matrix of beamlet intensities is specified for a given beam angle, and this matrix must be optimally decomposed into an intensity-weighted sum of binary shape matrices representing beamlet patterns that are realizable via potential leaf positions in the collimator. (There are a number of types of IMRT devices currently in clinical use, each of which has slightly different physical constraints that determine the leaf positions that are possible for that device.)

A simple model for an intensity matching problem is

$$\min \quad f(t, c)$$
$$\text{subject to} \quad \sum_k t_k * S_k = I, \quad 0 \le t \le Mc \qquad (11.15)$$
$$c_k \in \{0, 1\},$$

where I is the given integer intensity map matrix, the S_k are realizable binary shape matrices, t_k is a non-negative variable corresponding to the radiation intensity (determined by beam-on time and bounded above by M) associated with shape S_k, and c_k is a binary "shape counter" variable that is forced to 1 if shape S_k is used (which is equivalent to $t_k > 0$). A variety of objective functions f (all of which deal with treatment time and complexity) have been considered for the intensity matching problem, and we describe some of these alternatives after providing some additional background about this problem. Although this formulation has the advantage of simplicity, it is impractical because of the enormous number of shapes S_k.

The intensity matching problem is of critical importance because existing treatment planning software often produces plans that are clinically unacceptable because they require too many changes in the leaf positions at a beam angle. For example, on some equipment the set-up time required to recycle accelerator power (which must be done at each change in leaf positions) is about 7 seconds, so a plan involving 50 leaf positions at each of 9 beam angles (a not uncommon occurrence) would translate into a total treatment time of about one hour, which is clinically undesirable because of patient motion/discomfort problems and because of the need to treat large numbers of patients on a daily basis. (There are other accelerators with much smaller recycle times for which such a complex treatment plan would result in a more acceptable 30 minute or less treatment time, but would still be undesirable since the plan would produce unnecessary wear on the equipment (relative to plans with fewer leaf adjustments) and less accuracy in the delivery of radiation therapy). Langer [27] develops a different formulation based on leaf position variables and considers two objectives: beam-on-time ($f(t, c) = \sum_k t_k$) and cardinality ($f(t, c) = \sum_k c_k$). He circumvents the difficulty of enumerating shape matrices by using binary variables $l(i, j, k)$ and $r(i, j, k)$ corresponding to coverage of bixel (for beam pixel, or position in the shape matrix) (i, j) in shape k by portions of left and right leaves respectively. The continuity of the left leaf in row i in shape k is then enforced by constraints of the form $l(i, j, k) \geq l(i, j + 1, k)$ and analogous constraints apply to the right leaves. (These binary variables may also be used to enforce additional constraints known as "tongue and groove" constraints that relate to adjacent bixels and, for certain types of IMRT machines, "leaf collision" constraints that disallow overlap of left and right leaves in adjacent rows.) Finally, the following constraint guarantees that each bixel is either open to deliver radiation or covered by a leaf: $l(i, j, k) + r(i, j, k) + b(i, j, k) = 1$, where $b(i, j, k)$ is a binary variable that assumes value 1 if radiation is delivered through position (i, j) for a unit time interval. Intensity match corresponds to the constraints

$$\sum_k b(i, j, k) = I(i, j) \quad \forall (i, j). \tag{11.16}$$

To deal with the minimum cardinality problem in which the number of shapes is minimized ($f(t, c) = \sum_k c_k$), the binary shape change variable c_k is subjected to the constraints $-c_k \leq b(i, j, k) - b(i, j, k + 1) \leq c_k$, forcing it to 1 if any bixel changes from open to covered or vice-versa in the transition from time period k to time period $k + 1$. Since large system of constraints must be constructed for every time interval (that is, the range of k is at least as long as the total treatment time), this approach gives rise to very large constraint sets, but has been effective for problems in which the intensity maps are not very complex.

An interesting shortest path approach to the intensity matching problem has been developed recently by Boland, et al. [3]. They consider the easier (from an optimization viewpoint) objective of minimizing total beam-on time ($f(t, c) = \sum_k t_k$) and show that optimal solutions can be obtained very quickly by solving a problem in a graph with certain side constraints. Boland, et al. avoid the replication of constraint systems by constructing a layered graph whose nodes represents possible pairs of left/right leaf positions for each successive row, and whose arcs represent allowable transitions to the leaf pairs in the next row (leaf collision constraints are enforced by excluding certain arcs). They then observe that a path from the top row to the bottom row in this graph corresponds to an allowable shape. The variables are the flows on these paths, which correspond to intensity variables for the left/right row segments. Side constraints are used to ensure intensity match. By considering conformal decomposition of the flows into paths, it is easy to see that minimizing total flow from sources to sinks is equivalent to minimizing beam-on time. The resulting problem is a linear program that can usually be solved in under a minute with good LP software. However, this approach does not extend readily to the minimum cardinality case, because the graph model is based on flows and the cardinality problem would require a count of the number of paths used. Thus, there are clearly opportunities for further research in this area, since Langer observes from his numerical results that minimal beam-on time solutions may have relatively large cardinality. Column generation approaches that approximate the problem [34] may be promising in this regard.

The IMRT treatment planning problem at the bixel level may be defined in terms of intensity variables $I(r, s, a)$ where (r, s) are bixel indices and a is a beam angle index and $D_{r,s,a}(i, j, k)$ is the dose at (i, j, k) resulting from a unit value of the integer variable $I(r, s, a)$. The resulting model is similar to the brachytherapy model in that the dose definition is given by:

$$Dose(i, j, k) = \sum_{(r,s,a)} D_{r,s,a}(i, j, k) * I(r, s, a). \qquad (11.17)$$

The optimization model may now be stated as:

$$\min \alpha * \sum_{(i,j,k)\in\mathcal{T}} Underdose(i, j, k) + \beta * \sum_{(i,j,k)\in OAR} Overdose(i, j, k)$$

$$
\begin{aligned}
\text{subject to} \quad &(11.17) \text{ dose definition} \\
&(11.5) \text{ underdose constraints} \qquad (11.18) \\
&(11.11) \text{ overdose constraints} \\
&I(r, s, a) \text{ integer.}
\end{aligned}
$$

Key difficulties with this model in addition to its overall size and number of integer variables include choosing promising beam angles. Note that a solution of this model will result in bixel intensity maps (for each shot angle), and these must then be decomposed via solutions of intensity matching problems in order to obtain the final treatment plan. New techniques for the full IMRT treatment planning problem have been proposed in [40] (using simulated annealing) and [34] (using column generation). Possibilities for future research include slicing approaches analogous to those that we have previously successfully employed for brachytherapy [14, 31]. In the slicing approach, radiation delivery would be optimized for a selected beam angle, assuming a certain total amount of radiation to tissues from the remaining beam angles, and then this optimization procedure would be repeated one beam angle at a time for a promising list of beam angles. The slicing approach has the advantage of significantly reducing the size of the optimization problems considered, but requires the construction of a good set of initial conditions, including a good priority list for beam angles. Ongoing work is focusing on methodology to rank beam angle suitability for a given treatment. Nested partitions [25] may also be useful in this context in terms of providing a structured approach for stochastically sampling the space of beam angles and shape matrices.

5.   Conclusions and Directions for Future Research

The large-scale combinatorial problems arising from radiation treatment planning offer significant challenges to the optimization community because they contain large numbers of variables and constraints as well as large amounts of data. We have shown here that carefully tailored optimization approaches that approach the "ideal" version of treatment planning problems through a series of simpler problems and "hot starts" can yield high-quality treatment plans in areas of both teletherapy and brachytherapy. However, there remain many open problems because of the diversity and ever increasing complexity of radiation delivery mechanisms. In particular, treatment planning problems arising in intensity modulated radiation therapy represent an extremely complex class of optimization problems for which fast techniques that provide results of guaranteed quality are currently needed. Given the critical importance of utilizing radiation technology in a manner that maximizes tumor control and minimizes harmful radiation to the patient, and the impracticality of dealing with this complex technology by trial-and-error, we are confident that researchers in optimization will be able to develop the effective new software tools that are needed.

References

[1] G. Arcangeli, M. Benassi, L. Nieddu, C. Passi, G. Patrizi, and M.T. Russo. Optimal adaptive control of treatment planning in radiation therapy. *European Journal of Operational Research*, 140:399–412, 2002.

[2] BBIN. *Biomedical Business International Newsletter*, pages 72–75, 1996.

[3] N. Boland, H.W. Hamacher, and F. Lenzen. Minimizing beam-on time in cancer radiation treatment using multileaf collimators. *Networks, forthcoming*, 2004.

[4] T. R. Bortfeld, D. L. Kahler, T. J. Waldron, and A. L. Boyer. X-ray field compensation with multileaf collimators. *International Journal of Radiation Oncology, Biology and Physics*, 28(3):723–730, 1994.

[5] A. Brahme. Biological and physical dose optimization in radiation therapy. In J.G. Fortner and J.E. Rhoads, editors, *Accomplishments in Cancer Research*, pages 265–298. General Motors Cancer Research Foundation, 1991.

[6] A. Brahme. Optimization of radiation therapy and the development of multileaf collimation. *International Journal of Radiation Oncology, Biology and Physics*, 25:373–375, 1993.

[7] A. Brahme. Treatment optimization: Using physical and radiobiological objective functions. In A. R. Smith, editor, *Radiation Therapy Physics*, pages 209–246. Springer-Verlag, Berlin, 1995.

[8] L. Brewster, R. Mohan, G. Mageras, C. Burman, S. Leibel, and Z. Fuks. Three dimensional conformal treatment planning with multileaf collimators. *International Journal of Radiation Oncology, Biology and Physics*, 33(5):1081–1089, 1995.

[9] A. Brooke, D. Kendrick, and A. Meeraus. *GAMS: A User's Guide*. The Scientific Press, South San Francisco, California, 1988.

[10] Y. Censor. Parallel application of block-iterative methods in medical imaging and radiation therapy. *Mathematical Programming*, 42:307–325, 1988.

[11] P. S. Cho, H. G. Kuterdem, and R. J. Marks. A spherical dose model for radiosurgery treatment planning. *Physics in Medicine and Biology*, 43:3145–3148, 1998.

[12] A. Drud. CONOPT: A GRG code for large sparse dynamic nonlinear optimization problems. *Mathematical Programming*, 31:153–191, 1985.

[13] W. D. D'Souza and R. R. Meyer. An intraoperative reoptimization framework for prostate implant treatment plans. Technical report, Computer Sciences Department, University of Wisconsin-Madison, in preparation 2003.

[14] W. D. D'Souza, R. R. Meyer, M. C. Ferris, and B. R. Thomadsen. Mixed integer programming models for prostate brachytherapy treatment optimization. *Medical Physics*, 26(6):1099, 1999.

[15] W. D. D'Souza, R. R. Meyer, B. R. Thomadsen, and M. C. Ferris. An iterative sequential mixed-integer approach to automated prostate brachytherapy treatment optimization. *Physics in Medicine and Biology*, 46:297–322, 2001.

[16] M. C. Ferris, J.-H. Lim, and D. M. Shepard. Optimization approaches for treatment planning on a Gamma Knife. *SIAM Journal on Optimization*, 13:921–937, 2003.

[17] M. C. Ferris, J.-H. Lim, and D. M. Shepard. Radiosurgery treatment planning via nonlinear programming. *Annals of Operations Research*, 119:247–260, 2003.

[18] M. C. Ferris and D. M. Shepard. Optimization of Gamma Knife radiosurgery. In D.-Z. Du, P. Pardalos, and J. Wang, editors, *Discrete Mathematical Problems with Medical Applications*, volume 55 of *DIMACS Series in Discrete Mathematics and Theoretical Computer Science*, pages 27–44. American Mathematical Society, 2000.

[19] M. C. Ferris and M. M. Voelker. Neuro-dynamic programming for radiation treatment planning. Numerical Analysis Group Research Report NA-02/06, Oxford University Computing Laboratory, Oxford University, 2002.

[20] J. C. Ganz. *Gamma Knife Surgery*. Springer-Verlag Wien, Austria, 1997.

[21] H. W. Hamacher and K.-H. Küfer. Inverse radiation therapy planning — a multiple objective optimization approach. *Discrete Applied Mathematics*, 118:145–161, 2002.

[22] ILOG CPLEX Division, 889 Alder Avenue, Incline Village, Nevada. *CPLEX Optimizer*. http://www.cplex.com/.

[23] L.C. Jones and P.W. Hoban. Treatment plan comparison using equivalent uniform biologically effective dose (EUBED). *Physics in Medicine and Biology*, pages 159–170, 2000.

[24] H. M. Kooy, L. A. Nedzi, J. S. Loeffler, E. Alexander, C. Cheng, E. Mannarino, E. Holupka, and R. Siddon. Treatment planning for streotactic radiosurgery of intra-cranial lesions. *International Journal of Radiation Oncology, Biology and Physics*, 21:683–693, 1991.

[25] L. Shi L. and S. Olafsson. Nested partitions method for global optimization. *Operations Research*, 48:390–407, 2000.

[26] M. Langer, S. Morrill, R. Brown, O. Lee, and R. Lane. A comparison of mixed integer programming and fast simulated annealing for optimized beam weights in radiation therapy. *Medical Physics*, 23:957–964, 1996.

[27] M. Langer, V. Thai, and L. Papiez. Improved leaf sequencing reduces segments or monitor units needed to deliver IMRT using multileaf collimators. *Medical Physics*, 28(12):2450–2458, 2001.

[28] E. K. Lee, T. Fox, and I. Crocker. Optimization of radiosurgery treatment planning via mixed integer programming. *Medical Physics*, 27:995–1004, 2000.

[29] E. K. Lee, R. J. Gallagher, D. Silvern, C. S. Wuu, and M. Zaider. Treatment planning for brachytherapy: an integer programming model, two computational approaches and experiments with permanent prostate implant planning. *Physics in Medicine and Biology*, 44:145–165, 1999.

[30] L. Luo, H. Shu, W. Yu, Y. Yan, X. Bao, and Y. Fu. Optimizing computerized treatment planning for the Gamma Knife by source culling. *International Journal of Radiation Oncology, Biology and Physics*, 45(5):1339–1346, 1999.

[31] R. R. Meyer, W. D. D'Souza, M. C. Ferris, and B. R. Thomadsen. MIP models and BB strategies in brachytherapy treatment optimization. *Journal of Global Optimization*, 25:23–42, 2003.

[32] S. M. Morrill, K. S. Lam, R. G. Lane, M. Langer, and I. I. Rosen. Very fast simulated annealing in radiation therapy treatment plan optimization. *International Journal of Radiation Oncology, Biology and Physics*, 31:179–188, 1995.

[33] A. Niemierko. Radiobiological models of tissue response to radiation in treatment planning systems. *Tumori*, 84:140–143, 1998.

[34] F. Preciado-Walters, R. Rardin, M. Langer, and V. Thai. A coupled column generation, mixed-integer approach to optimal planning of intensity modulated radiation therapy for cancer. Technical report, Industrial Engineering, Purdue University, 2002.

[35] W. Que. Comparison of algorithms for multileaf collimator field segmentation. *Medical Physics*, 26:2390–2396, 1999.

[36] W. Schlegel and A. Mahr, editors. *3D Conformal Radiation Therapy - A Multimedia Introduction to Methods and Techniques*. Springer-Verlag, Berlin, 2001.

[37] D. M. Shepard, L. S. Chin, S. J. DiBiase, S. A. Naqvi, J. Lim, and M. C. Ferris. Clinical implementation of an automated planning system for Gamma Knife radiosurgery. *International Journal of Radiation Oncology, Biology, Physics*, 56:1488–1494, 2003.

[38] D. M. Shepard, M. C. Ferris, G. Olivera, and T. R. Mackie. Optimizing the delivery of radiation to cancer patients. *SIAM Review*, 41:721–744, 1999.

[39] D. M. Shepard, M. C. Ferris, R. Ove, and L. Ma. Inverse treatment planning for Gamma Knife radiosurgery. *Medical Physics*, 27:2748–2756, 2000.

[40] D.M. Shepard, M.A. Earl, X.A. Li, and C. Yu. Direct aperture optimization: A turnkey solution for step-and-shoot IMRT. *Medical Physics*, 29:1007–1018, 2002.

[41] R. A. Stone, V. Smith, and L. Verhey. Inverse planning for the Gamma Knife. *Medical Physics*, 20:865, 1993.

[42] J. Wang. Packing of unequal spheres and automated radiosurgical treatment planning. *Journal of Combinatorial Optimization*, 3:453–463, 1999.

[43] S Webb. Optimisation of conformal radiotherapy dose distributions by simulated annealing. *Physics in Medicine and Biology*, 34(10):1349–1370, 1989.

[44] S. Webb. Inverse planning for imrt: the role of simulated annealing. In E. Sternick, editor, *The Theory and Practice of Intensity Modulated Radiation Therapy*. Advanced Medical Publishing, 1997.

[45] S. Webb. Configuration options for intensity-modulated radiation therapy usi ng multiple static fields shaped by a multileaf collimator. *Physics in Medicine and Biology*, 43:241–260, 1998.

[46] S. Webb. Configuration options for intensity-modulated radiation therapy using multiple static fields shaped by a multileaf collimator. II: Constraints and limitations on 2D modulation. *Physics in Medicine and Biology*, 43:1481–1495, 1998.

[47] A. Wu, G. Lindner, and A. H. Maitz et al. Physics of gamma knife approach on convergent beams in stereotactic radiosurgery. *International Journal of Radiation Oncology, Biology and Physics*, 18(4):941–949, 1990.

[48] Q. J. Wu. Sphere packing using morphological analysis. In D.-Z. Du, P. Pardalos, and J. Wang, editors, *Discrete Mathematical Problems with Medical Applications*, volume 55 of *DIMACS Series in Discrete Mathematics and Theoretical Computer Science*, pages 45–54. American Mathematical Society, 2000.

[49] Q. J. Wu and J. D. Bourland. Morphology-guided radiosurgery treatment planning and optimization for multiple isocenters. *Medical Physics*, 26(10):2151–2160, 1999.

[50] Q. J. Wu, J. Wang, and C. H. Sibata. Optimization problems in 3D conformal radiation therapy. In D.-Z. Du, P. Pardalos, and J. Wang, editors, *Discrete Mathematical Problems with Medical Applications*, volume 55 of *DIMACS Series in Discrete Mathematics and Theoretical Computer Science*, pages 183–194. American Mathematical Society, 2000.

[51] P. Xia and L.J. Verhey. Multileaf collimator leaf sequencing algorithm for intensity modula ted beams with multiple static segments. *Medical Physics*, 25(8):1424–1434, 1998.

[52] Y. Yan, H. Shu, and X. Bao. Clinical treatment planning optimization by Powell's method for Gamma unit treatment system. *International Journal of Radiation Oncology, Biology and Physics*, 39:247–254, 1997.

Chapter 12

MULTIPLE HYPOTHESIS CORRELATION IN TRACK-TO-TRACK FUSION MANAGEMENT

Aubrey B Poore[1,2], Sabino M Gadaleta[2] and Benjamin J Slocumb[2]

[1] *Department of Mathematics*
Colorado State University
Fort Collins, 80523
USA

aubrey.poore@colostate.edu

[2] *Numerica*
PO Box 271246
Fort Collins, CO 80527-1246
USA

{ abpoore,smgadaleta,bjslocumb } @numerica.us

Abstract   Track to track fusion systems require a capability to perform track matching across the reporting sensors. In conditions where significant ambiguity exists, for example due to closely spaced objects, a simple single frame assignment algorithm can produce poor results. For measurement-to-track fusion this has long been recognized and sophisticated multiple hypothesis, multiple frame, data association methods considerably improve tracking performance in these challenging scenarios. The most successful of the multiple frame methods are multiple hypothesis tracking (MHT) and multiple frame assignments (MFA), which is formulated as a multidimensional assignment problem. The performance advantage of the multiple frame methods over the single frame methods follows from the ability to hold difficult decisions in abeyance until more information is available and the opportunity to change past decisions to improve current decisions. In this chapter, the multiple source track correlation and fusion problem is formulated as a multidimensional assignment problem. The computation of cost coefficients for the multiple frame correlation assignments is based on a novel batch MAP estimation approach. Based on the multidimensional assignments we introduce a novel multiple hypothesis track correlation approach that allows one to make robust track management decisions over multiple frames of data. The use of the proposed multiple hypothesis, multiple frame correlation system, is expected to improve the fusion system performance in scenarios where significant track assignment ambiguity exists. In the same way that multiple frame processing has shown improvements in the tracking performance in measurement-to-track fusion applications, we expect to achieve improvements in the track-to-track fusion problem.

Keywords: Track fusion, multiple hypothesis track correlation, multidimensional assignment

Introduction

Multiple target tracking is a subject devoted to the production of track states (e.g., position and velocity) of objects that generally do not identify themselves with, for example, a beacon code. This includes airplanes, ground vehicles, surface ships, and missiles. The central problem is that of data association in which to partition the data (measurements or track reports) into tracks and false alarms. Data association methods divide into two broad classes, namely single frame and multiple frame methods. The single frame methods include nearest neighbor, global nearest neighbor and JPDA (joint probabilistic data association). The most successful of the multiple frame methods are multiple hypothesis tracking (MHT) [Blackman and Popoli, 1999] and multiple frame assignments (MFA) [Poore, 1994, Poore et al., 2001b], which is formulated as a multidimensional assignment problem. The performance advantage of the multiple frame methods over the single frame methods follows from the ability to hold difficult decisions in abeyance until more information is available and the opportunity to change past decisions to improve current decisions. In dense tracking environments the performance improvements of multiple frame methods over single frame methods are very significant, making it the preferred solution for such tracking problems.

The application of multiple frame tracking methods must consider an architecture in which the sensors are distributed across multiple platforms. Such geometric and sensor diversity has the potential to significantly enhance tracking. A centralized architecture in which all reports are sent to one location and processed with composite tracks being transmitted back to the different platforms is a simple one that is probably optimal in that it is capable of producing the best track quality (e.g., purity and accuracy) and a consistent air picture. The centralized tracker is, however, unacceptable for several reasons, notably the communication overloads and single-point-failure. Thus, one must turn to a distributed architecture for both estimation/fusion [Chong et al., 1990] and data association. In these architectures, measurements are generally set over a communications network so that both local and remote measurements are used in the fusion and association process.

Many papers have been written on this approach to tracking. Good references include the books by Blackman and Popoli [Blackman and Popoli, 1999]

and the articles in the books by Bar-Shalom [Bar-Shalom, 1990, Bar-Shalom, 1992, Bar-Shalom and Blair, 2000] and references therein. Distributed tracking architectures with single frame processing in mind are covered in these books. For multiple frame processing, the paper [Poore et al., 2001a] introduces several distributed architectures and gives a brief comparison on air defense scenarios.

Most current systems, however, communicate track states (and sometimes partial or full covariances) over a network rather than measurements. The problem of associating and fusing track states from multiple sources is probably an even more important one than measurement association and fusion. In addition, the needs for track-to-track association and fusion are very different from that of measurement-to-measurement or measurement-to-track. For example, a particular sensor or source may already have excellent tracks and thus the problem is one more of determining which tracks go together (i.e., associate, match, or correlate). Instead of fusing tracks, one could then select a representative track out of this set of associated track groups.

With respect to architectures for track-to-track correlation and fusion, one can distinguish between two fundamentally different architectures for track fusion: (1) *sensor-to-sensor track fusion* and (2) *sensor-to-system track fusion*. This chapter focuses on developing a track fusion within the sensor-to-sensor processing architecture. This architecture has the fundamental advantage that only the cross-correlations among the sensor-level tracks need to be addressed. However, since system level tracks are not directly maintained, the architecture requires a mechanism for matching and maintaining system-level tracks over time. In principle, the sensor-to-sensor processing architecture has the potential to replace incorrect fusion decisions made in the past with corrected, current, information. This however implies also that the architecture is potentially susceptible to incorrect current data. Thus, a mechanism is needed that allows to make system-level track management decisions robustly over multiple frames, incorporating both previous and current information into the frame-to-frame matching decision process. The purpose of this chapter is to document a *multiple hypothesis, multiple frame, correlation approach* for system-level track management in sensor-to-sensor track fusion architectures that provides such a robust frame-to-frame matching mechanism. It is also based on the multidimensional assignment problem.

The chapter is structured as follows: Section 12.1 briefly describes two fusion processing architectures. Section 12.2 describes the problem of frame-to-frame matching in sensor-to-sensor track fusion. Section 12.3.3 introduces the proposed multiple hypothesis, multiple frame, correlation approach for system-level track management in sensor-to-sensor track fusion architectures. Section 12.5 summarizes the chapter.

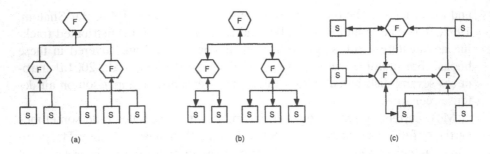

Figure 12.1. Diagrams of the (a) hierarchical architecture without feedback, (b) hierarchical architecture with feedback, and (c) fully distributed architecture. S-nodes are sensor/tracker nodes, while F-nodes are system/fusion nodes

1. Track Fusion Architectures

To set the context for this chapter, we briefly review here various concepts for constructing a track fusion system architecture. We delineate both the *network* and the *processing* architectures that may be employed.

1.1 Network Architectures

There are three fundamental fusion network architectures that describe the flow of data among nodes in the network [Blackman and Popoli, 1999, Chong et al., 1990, Chong et al., 2000, Liggens et al., 1997]. The three architectures are described in Figure 12.1 and the following list summarizes the key architecture concepts.

- *Hierarchical Architecture without Feedback.* In this architecture type, fusion nodes are arranged in a hierarchy with higher-level nodes processing results from lower-level nodes. Data (information) flows from the lower-level nodes to the upper-level nodes without estimation feedback from the upper nodes to lower nodes. Considerable savings in computation can be achieved if the communication rate to the higher-level nodes is slower than the sensor observation rate.

- *Hierarchical Architecture with Feedback.* This architecture type is the same as the previous, however, feedback of fused state data from upper nodes to lower nodes in the hierarchy is allowed. Since new track reports received by upper nodes from lower nodes contains feedback data, the fusion algorithm must remove the feedback data during fusion, otherwise double counting of information will lead to overly optimistic fusion results.

- *Fully Distributed Architecture.* In this architecture type, there is no fixed superior/subordinate relationship among the fusion nodes. Information may flow from each node to all others, or to some select subset of the other nodes. Feedback may or may not be used.

When one develops a fusion algorithm for each of these network architectures, the flow of information should be taken into account. To do so, an *information graph* [Chong et al., 1990, Liggens et al., 1997] can be used to represent information events in the fusion system and their interaction. Common information shared by nodes can be identified so that redundancies can be removed. The information graph contains four types of nodes: observation nodes, data reception nodes, communication transmission nodes, and communication reception nodes (fusion nodes).

For this chapter, we are only concerned with the Hierarchical Architecture with Feedback. Hence, information is not reused and removal of common information is not of concern in our development.

1.2   Processing Architectures

At each track fusion node in any of the above network architectures, two types of track fusion processing architectures are possible [Chong et al., 2000]: *sensor-to-sensor* track fusion, and *sensor-to-system* track fusion. The focus for this chapter is on the former architecture, but we provide here a description of both to differentiate the two approaches.

1.2.1   Sensor-to-Sensor Track Fusion Architecture.   This processing architecture is shown in Figure 12.2. A two-sensor fusion system is shown, but this easily generalizes to more sensors. At each fusion time, sensor-level tracks are communicated to the fusion node where they are combined (matched and fused) into system-level tracks. To perform matching and fusion all sensor-level tracks must be transformed to a common coordinate frame. As shown in Figure 12.2, the system-level updates can occur each time one of the sensor-level trackers updates its tracks (this defines the maximum communication rate), or at some longer time interval. In the example in Figure 12.2, updates alternate between Sensor-1 and Sensor-2. Previously formed system-level tracks are *replaced* by the newly fused sensor-level tracks.

Sensor-to-sensor fusion has the advantage that the algorithm only needs to address the cross-correlations among the sensor-level tracks. Furthermore, this processing architecture will extinguish incorrect fusion decisions made in the past, i.e., fusion mistakes in the past will not degrade current fusion results. The drawback in this architecture is that there is no direct mechanism for managing system-level tracks from one update to the next. If the fusion of sensor-level tracks changes from one update to the next (e.g., due to mis-

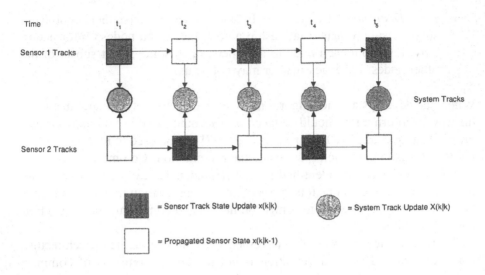

Figure 12.2.   Diagram showing the sensor-to-sensor fusion process

assignments), then any identifying properties such as track numbers, features, and attributes will *swap* and therefore degrade the track picture. Hence, in the sensor-to-sensor fusion architecture one requires a mechanism for *matching and maintaining* system-level tracks across the update times. Development of a multi-hypothesis approach to this problem is the focus of this chapter.

1.2.2   Sensor-to-System Track Fusion Architecture.   This processing architecture is shown in Figure 12.3. Here, sensor-level tracks directly update the continuously maintained system-level tracks. Whenever an updated sensor-level track is received at the fusion node, the system-level tracks are propagated to the time of this update and the sensor-level tracks are associated and fused to the system-level tracks. In this architecture, standard track association (assignment) algorithms may be used. However, the fusion processing algorithm must address cross-correlation between the sensor-level and system-level tracks. Also, since system-level tracks are carried forward in time, any past processing errors will impact future fusion performance. Hence, this processing architecture has the advantage that track identity management is more straightforward, but the disadvantage that sensor-level tracking errors will continuously degrade the fusion result (in other words, recovery from bad data is more difficult).

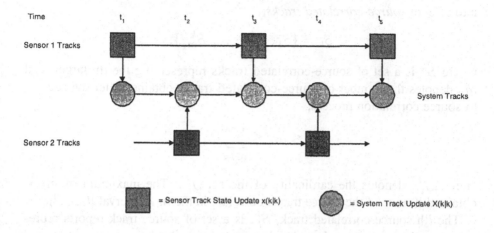

Figure 12.3.   Diagram showing the sensor-to-system fusion process

2.   The Frame-to-Frame Matching Problem

As discussed, our objective is to develop a multi-hypothesis frame-to-frame matching mechanism in the sensor-to-sensor track fusion architecture. To set up our solution, we introduce some notation in this section. Let $\mathcal{I}_S$ denote the set of indices of the contributing sensors (or sources) in the track fusion network. Each sensor is assumed to provide source track reports over a time window to the fusion node. During a given time interval, the ith source may report multiple updates on the jth source track when the track update interval is shorter than the time window. Let $\mathbf{z}_u^{ij,k}$ denote the uth update of the jth track from the ith source in time window $[t_{k-1}, t_k)$. Let $\mathcal{U}_{ijk}$ denote the enumeration of the multiple source track updates in the time interval, and let $\mathcal{Z}_j^{ik}$ denote the set of all reports for track j from sensor i received in the time interval t_{k-1} through t_k. Therefore, we have

$$\mathcal{Z}_j^{ik} = \{\mathbf{z}_u^{ij,k}\}_{u \in \mathcal{U}_{ijk}}.$$

Now let $\mathcal{J}_i^k$ denote the enumeration of all source tracks of the ith sensor reported in the time interval $[t_{k-1}, t_k)$. We define a *frame of track reports*, $\mathcal{F}_k^i$, as the set of source track updates received at the fusion node from the ith sensor in the time interval $[t_{k-1}, t_k)$,

$$\mathcal{F}_k^i = \{\mathcal{Z}_j^{ik}\}_{j \in \mathcal{J}_i^k}.$$

The task of source-to-source correlation within the sensor-to-sensor track fusion architecture is to correlate the frames of tracks from all sources, $\{\mathcal{F}_k^i\}_{i \in \mathcal{I}_S}$, into a set of *source-correlated tracks*,

$$S_k = \{S_1^k, S_2^k, \ldots, S_{m_k}^k\},$$

where S_l^k is a set of source-correlated tracks representing the lth target, and m_k denotes the number of source-correlated tracks obtained after the source-to-source correlation process,

$$0 \leq m_k \leq \sum_{i \in I_S} |\mathcal{J}_i^k|.$$

Here, $|\mathcal{J}_i^k|$ denotes the cardinality of the set $\mathcal{J}_i^k$. The maximum of m_k is obtained if none of the source tracks correlate in the time interval $[t_{k-1}, t_k)$.

The lth source-correlated track, S_l^k, is a set of source track reports representing a single object and can be represented as follows,

$$S_l^k = \{\mathcal{Z}_j^{ik}\}_{i \in I_l^k}^{j \in J_l^k}, \quad \text{for } I_l^k \subseteq \mathcal{I}_S, \quad J_l^k \subseteq (\bigoplus_{i \in \mathcal{I}_S} J_i^k),$$

where $\bigoplus J_i^k$ denotes the disjoint union of the source track enumerations. A valid S_l^k satisfies the following two constraints:

- Each source-correlated track contains at most a one track report from each source.

- Each source track is required to be part of exactly one source-correlated track. This assumes that a singleton source track will form "its own" source-correlated track.

The source-to-source assignment problem is solved in a manner that satisfies the two constraints.

Figure 12.4 illustrates the source-to-source correlation process through a simple example. We assume two sources, source A and source B. Source A reports track updates on three source tracks and source B reports on two source tracks. The local tracker data association algorithms provide for a level of consistency between source track report IDs in different time frames. However, in certain cases the source track data association is wrong. Thus, a fusion algorithm should not extensively rely on the source track ID numbers. The task of sensor-to-sensor processing is to correlate the source tracks from the sensors within corresponding time frames. In the example, on time frame k, the source-to-source correlation produces three source-correlated tracks. The correlation between source A, track 1, and source B, track 1, (denoted in the figure with 1-1) is arbitrarily called track a. Similarly, the source correlation between source A, track 2, and source B, track 2, denoted (2-2), is called track b. Source A, track 3 does not correlate to any source B track. This

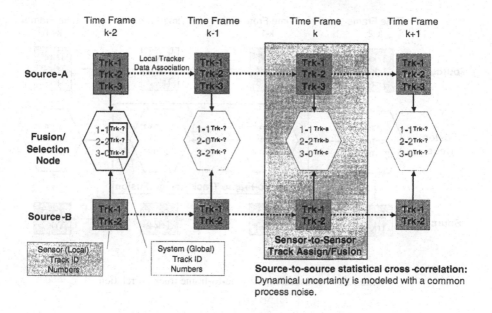

Figure 12.4.   Illustration of source-to-source track correlation

singleton correlated track is called track c. How source-to-source correlation is performed is an important problem, however, our focus is on a methodology that carries the track labels, $\{a, b, c\}$, in a consistent manner across frames.

Figure 12.5 illustrates the frame-to-frame matching problem where correlated source tracks on frame $k - 1$ are connected to correlated source tracks on frame k. We assume that at time frame $k - 1$ that source-correlated tracks a, b, and c are established; we refer to these tracks as *system tracks* or *global tracks*. We also assume that three source-correlated tracks are given on both frames k and $k + 1$. The essence of the frame-to-frame matching problem is to identify which source-correlated track, if any, on frame k should be called track a, i.e., correlated to global track a on frame $k - 1$. Similarly, we seek to determine which tracks on frame k should be called b and c. The frame-to-frame matching problem extends similarly to frame $k + 1$. Further, the methodology should correlate source tracks on consecutive frames such that the matching *most likely* corresponds to the same truth target and provides the system operator with information in a robust manner.

Two subtle but important issues arise in the frame-to-frame processing. First, the methodology needs to support new global track initiation when no established global track on frame $k - 1$ correlates to a new source-correlated track on frame k. Second, global tracks need to be able to *coast* when estab-

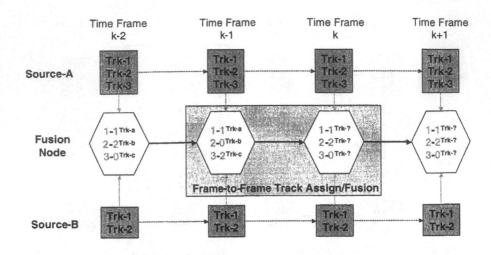

Figure 12.5.   Illustration of frame-to-frame track correlation

lished global tracks do not correlate to new source-correlated tracks. Coasting is necessary if the current time frame updates did not report on targets that are represented by established global tracks.

Nomenclature for the generated *global* or *system* track is as follows. Define global track i at time k as a set G_i^k that consists of an ID (global track ID), the source-correlated track $S_{l_i}^{k_i}$ that correlated to global track i at the time k_i of its last update, and a *past* association history $\mathcal{A}_i^k$ that contains all source-correlated tracks that updated the global track i prior to its last update,

$$G_i^k := \left\{ \mathrm{ID}_i,\; S_{l_i}^{k_i},\; \mathcal{A}_i^k \right\}.$$

We note that k_i may be different from k since no update by source-correlate track may have occurred on frame k, i.e., the global track was coasted. This notation is convenient since we typically will only use the most recent source-correlated track of a global track in the frame-to-frame matching process. Further, let $\mathcal{G}_k = \{G_1^k,\; G_2^k, \ldots, G_m^k\}$ denote the set of m established global tracks at time frame k.

3.   Assignment Problems for Frame-to-Frame Matching

We now turn our focus to the track-to-track assignment problems in frame-to-frame matching. In Section 12.3.1, we begin with the two-dimensional assignment formulation where a set of established system tracks is correlated to a new set of source-correlated tracks. In Section 12.3.2 we present the general

N-dimensional assignment problem formulation for this problem. This allows us to introduce the multiple hypothesis, multiple frame, approach to frame-to-frame matching in Section 12.3.3.

3.1   The Two-Dimensional Assignment Problem

The two-dimensional assignment problem is formulated between the set of global tracks established at the previous frame (denoted with k) and the set of new source-correlated tracks for the current frame (denoted with $k+1$). Let us assume that the set $\mathcal{G}_k$ contains m global tracks and that the set $\mathcal{S}_{k+1}$ contains n source-correlated tracks. The two-dimensional one-to-one assignment problem for frame-to-frame matching then is

$$\text{Minimize} \sum_{i=0}^{m} \sum_{j=0}^{n} c_{ij} x_{ij},$$

$$\text{Subject To:} \sum_{j=0}^{n} x_{ij} = 1 \ (i = 1, ..., m),$$

$$\sum_{i=0}^{m} x_{ij} = 1 \ (j = 1, ..., n), \tag{12.1}$$

$$x_{ij} \in \{0, 1\}.$$

The c_{ij} denotes the cost coefficient for correlating global track i with source-correlated track j. Section 12.4 describes the computation of cost coefficients through a batch scoring approach that uses the actual source track data represented by the source-correlated track j, i.e., the source track updates in the set $\mathcal{S}_{k+1}$, and the source track data from the association history of the global track i, e.g., the source track updates from the most recent source-correlated track $S_{l_i}^{k_i}$ of global track i (see Section 12.2). One also has the additional cost c_{i0} of not assigning global track i and c_{0j} for not assigning source-correlated track j.

The *assignment variable*, $x_{ij} \in [0, 1]$, is equal to one if global track i correlates to source-correlated track j, and is zero otherwise. Given the solution of the assignment problem, *global track data management* can be performed. If $x_{ij} = 1$, this implies that we can *extend* the global track G_i^k with the source-correlated track S_j^{k+1}:

$$G_i^k = \left\{ \text{ID}_i, \ S_{l_i}^{k_i}, \ \mathcal{A}_i^k \right\} \longrightarrow G_i^{k+1} = \left\{ \text{ID}_i, \ S_j^{k+1}, \ S_{l_i}^{k_i}, \ \mathcal{A}_i^k \right\}$$

$$= \left\{ \text{ID}_i, \ S_j^{k+1}, \ \mathcal{A}_i^{k+1} \right\}.$$

In the one-to-one assignment, each global track can be assigned to at most one source-correlated track and vice-versa. The zero variables are present because a global track may or may not be assigned to a source-correlated track (e.g., a target may not have been reported or detected), or a source-correlated track may or may not be assigned to an established global track (e.g., it may be a false track or a new target). In other words, if a global track is not assigned to a source-correlated track ($x_{i0} = 1$), then we can *coast* the global track:

$$G_i^k \longrightarrow G_i^{k+1} = G_i^k.$$

If a source-correlated track is not assigned to a global track ($x_{0j} = 1$), then we can *initiate* a new global track:

$$\emptyset \longrightarrow G_{\text{new}}^{k+1} = \left\{ \text{ID}_{\text{new}}, \; S_j^{k+1}, \; A_{\text{new}}^k \right\}, \text{ with } A_{\text{new}}^k = \emptyset.$$

Figure 12.6 illustrates this two-dimensional assignment problem for three global tracks and three source-correlated tracks.

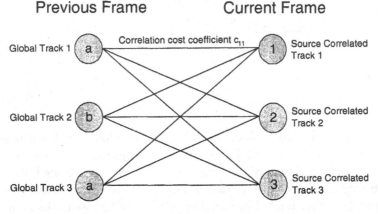

Figure 12.6.   Illustration of two-dimensional assignment problem for frame-to-frame matching

The above formulation Eqn. (12.1) is commonly called the "dense" form since all arcs (i, j) are included. In tracking as well as in frame-to-frame matching, gating functions eliminate most of theses arcs, leaving a "sparse" problem. To denote these dynamically infeasible arcs one frequently sets costs $c_{ij} = \infty$ or some very large number. However, instead of using the dense formulation, the sparse formulation is more appropriate if an efficient gating algorithm is available.

To explain the sparse formulation[1], let $\mathcal{I} = \{0, 1, \ldots, m\}$ and $\mathcal{J} = \{0, 1, \ldots, n\}$ and $\mathcal{P} \subset \{(i, j) \, | \, (i, j) \in \mathcal{I} \times \mathcal{J}\}$ denote the collection of arcs, each with a cost c_{ij}. Also, define $A(i) = \{j \, | \, (i, j) \in \mathcal{P}\}$ and

---

[1]To ease notation, we reuse some previously introduced variables.

$B(j) = \{i \mid (i,j) \in \mathcal{P}\}$ and require $0 \in A(i)$ for all $i \in \mathcal{I}$, $0 \in B(j)$ for all $j \in \mathcal{J}$, i.e., $A(0) = \mathcal{J}$, and $B(0) = \mathcal{I}$. The resulting two-dimensional, one-to-one, assignment problem in the sparse formulation then is

$$
\text{Minimize} \quad \sum_{(i,j) \in \mathcal{P}} c_{ij} x_{ij},
$$

$$
\text{Subject To:} \quad \sum_{j \in A(i)} x_{ij} = 1 \ (i \in \mathcal{I}),
$$

$$
\sum_{i \in B(j)} x_{ij} = 1 \ (j \in \mathcal{J}), \tag{12.2}
$$

$$
x_{ij} \in \{0, 1\}.
$$

3.2   Three- and N-Dimensional Assignment Problem

The two-frame frame-to-frame matching discussed in Section 12.3.1 can be extended toward general frame-to-frame matching over N time frames. In this formulation we assume that established global tracks are given $(N - 1)$ time frames back. The frame-to-frame matching is formulated as an N-dimensional assignment problem over the set of established tracks on frame k and the $(N - 1)$ recent frames, $\{k + 1, k + 2, \ldots, k + N - 1\}$. For a detailed derivation of the N-dimensional assignment problem that will be used in this section see [Poore et al., 2001b].

Figure 12.7 illustrates this assignment problem for $N = 3$, i.e., the set of global tracks at frame k and source-correlated tracks at frame $k + 1$ and frame $k + 2$. The cost coefficients $c_{i\,j_1 j_2}$ are three dimensional in the sense that, in general, $c_{i\,j_1 j_2} \neq c_{i\,j_1} + c_{j_1 j_2}$. Otherwise the problem is said to be decomposable.

To give a mathematical formulation of the three dimensional assignment problem we start with three sets of objects to be correlated. Denote with $\mathcal{I} = \{0, 1, \ldots, m\}$ an enumeration of the set of established global tracks at frame k. Denote with $\mathcal{J}_1 = \{0, 1, \ldots, n_1\}$, and $\mathcal{J}_2 = \{0, 1, \ldots, n_2\}$ and enumeration of the set of source-correlated tracks at frame $k + 1$ and frame $k + 2$, respectively. Then, the feasible arcs (connections between the objects) is defined as $\mathcal{P} = \{(i, j_1, j_1) \in \mathcal{I} \times \mathcal{J}_1 \times \mathcal{J}_2\}$. In addition, one needs the following notation

$$
P(\cdot, j_1, j_2) = \{i \mid (i, j_1, j_1) \in \mathcal{P}\}
$$
$$
P(\cdot, \cdot, j_2) = \{(i, j_1) \mid (i, j_1, j_2) \in \mathcal{P}\}.
$$

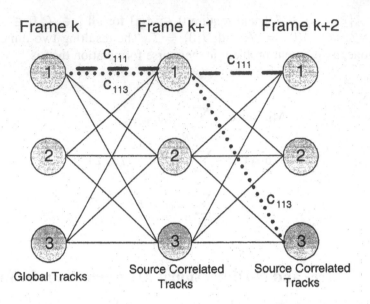

Figure 12.7.   Illustration of three-dimensional assignment problem for frame-to-frame matching

Permutations of these sets are also used in the subsequent development. Then, the three dimensional assignment problem can be posed as

$$\text{Min} \sum_{(i,j_1,j_1) \in \mathcal{P}} c_{i\,j_1j_2} x_{i\,j_1j_2}$$

Subj. To:
$$\sum_{(j_1,j_2) \in P(i,\cdot,\cdot)} x_{i\,j_1j_2} = 1 \ (i = 1, ..., m), \qquad \sum_{(j_1,j_2) \in P(0,\cdot,\cdot)} x_{0\,j_1j_2} = n_1 + n_2,$$

$$\sum_{(i,j_2) \in P(\cdot,j_1,\cdot)} x_{i\,j_1j_2} = 1 \ (j_1 = 1, ..., n_1), \qquad \sum_{(i,j_2) \in P(\cdot,0,\cdot)} x_{i\,0j_2} = m + n_2,$$

$$\sum_{(i,j_1) \in P(\cdot,\cdot,j_2)} x_{i\,j_1j_2} = 1 \ (j_2 = 1, ..., n_2), \qquad \sum_{(i,j_1) \in P(\cdot,\cdot,0)} x_{i\,j_10} = m + n_1,$$

$$x_{i\,j_1j_2} \in \begin{cases} \{0, 1\}, \text{if at least one of } i, j_1, j_2 \text{ is nonzero} \\ \{0, ..., \min\{m + n_1, n_1 + n_2, m + n_2\}\}, \text{for } (i, j_1, j_2) = (0, 0, 0). \end{cases}$$

$$(12.3)$$

The zero index in this problem has more than one meaning. First, a variable with two nonzero and one zero index such as $x_{i\,0j_2}$ stands for missing data in the second collection of objects. The second case corresponds to slack variables of the form $x_{i\,00}, x_{0\,j_1 0}, x_{0\,0j_2}$ just as x_{i0} and x_{0j} are in the two dimensional problem.

The extension to the general N-dimensional assignment problem for the frame-to-frame matching of $N - 1$ source-correlated frames of data at times $k + 1, \ldots, k + N - 1$ with a set of global tracks established at time frame k is straightforward. We give only the dense formulation and assume that there are m established tracks at time frame k. Furthermore we assume that there are $n_1, \ldots, n_{N-1}$ source-correlated tracks given at time frames $k + 1, \ldots, k + N - 1$, respectively. The multi-dimensional one-to-one assignment problem for frame-to-frame matching then becomes:

$$\text{Minimize} \quad \sum_{i=0}^{m} \sum_{j_1=0}^{n_1} \cdots \sum_{j_{N-1}=0}^{n_{N-1}} c_{i\,j_1 \cdots j_{N-1}} x_{i\,j_1 \cdots j_{N-1}}$$

$$\text{Subj. To} \quad \sum_{j_1=0}^{n_1} \cdots \sum_{j_{N-1}=0}^{n_{N-1}} x_{i\,j_1 \cdots j_{N-1}} = 1, \quad i = 1, \ldots, m,$$

$$\sum_{i=0}^{m} \sum_{j_2=0}^{n_2} \cdots \sum_{j_{N-1}=0}^{n_{N-1}} x_{i\,j_1 \cdots j_{N-1}} = 1, \quad j_1 = 1, \ldots, n_1,$$

$$\sum_{i=0}^{m} \sum_{j_1=0}^{n_1} \cdots \sum_{j_{l-1}=0}^{n_{l-1}} \sum_{j_{l+1}=0}^{n_{l+1}} \cdots \sum_{j_{N-1}=0}^{n_{N-1}} x_{i\,j_1 \cdots j_{N-1}} = 1,$$

$$\text{for} \quad j_l = 1, \ldots, n_l \text{ and } l = 2, \ldots, N - 2,$$

$$\sum_{i=0}^{m} \sum_{j_1=0}^{n_1} \cdots \sum_{j_{N-2}=0}^{n_{N-2}} x_{i\,j_1 \cdots j_{N-1}} = 1, \quad j_{N-1} = 1, \ldots, n_{N-1},$$

$$x_{ij_1 \cdots j_{N-1}} \in \{0, 1\} \text{ for all } i, j_1, \ldots, j_{N-1}.$$

Note that the association problem involving N source-correlated frames of observations is an $(N + 1)$-dimensional assignment problem for frame-to-frame matching (i.e., global track maintenance).

3.3 Multiple Hypothesis, Multiple Frame, Frame-to-Frame Matching

The above development introduced the two-, three-, and N-dimensional assignment formulation for frame-to-frame matching of established global tracks with source-correlated tracks. This formulation assumed a single set of source-correlated tracks per time frame. That is, we considered only source-correlated tracks obtained from the best solution of the source-to-source assignment problem. If ambiguity in the source data association is present, one will frequently observe that other solutions to the source-to-source correlation for a given time frame are *almost as likely* as the best source-to-source correlation for the respective time frame. This section develops a multiple hypothesis, multiple frame, frame-to-frame matching methodology that attempts to make better frame-to-frame matching decisions by considering several source-correlation solution *hypotheses* per time frame. A solution is selected based on the information from multiple frames of data. This approach is similar to an approach that was originally developed for group-cluster tracking [Gadaleta et al., 2002, Gadaleta et al., 2004].

Assume a given set of established global tracks at time frame k, $\mathcal{G}_k$, and let $\mathcal{H}_{k+l} = \{\mathcal{S}_1^{k+l}, \mathcal{S}_2^{k+l}, \ldots, \mathcal{S}_{n_l}^{k+l}\}$ denote a set of n_l distinct source-to-source correlation hypotheses for correlation of all source tracks within the time frame $k + l$. The multiple source-to-source correlation hypotheses can be generated using Murty's algorithm [Murty, 1968] to find ranked assignment solutions of the source-to-source correlation assignment problem, or more generally, by solving a set of sequential two-dimensional assignment problems similar to the hypotheses generation in a multiple hypothesis tracking (MHT) system [Blackman and Popoli, 1999]. We do not discuss such an approach further here.

The objective of multiple hypothesis, multiple frame, frame-to-frame matching is to find an extension of global tracks $\mathcal{G}_k$ that selects one $\mathcal{S}_j^{k+l}$ per hypothesis $\mathcal{H}_{k+l}$. Figure 12.8 illustrates this approach to multiple hypothesis frame-to-frame matching. The example assumes three reporting sensors and shows three consecutive time frames for each sensor. For each time frame we assume that three ranked source-to-source correlation hypotheses are formed. The lower part of the figure illustrates the resulting frame-to-frame matching correlation problem. The desired solution is the *shortest path*, i.e., lowest cost path, through the shown *trellis*.

Figure 12.9 provides a second, more detailed illustration of the multiple hypothesis frame-to-frame matching approach. We assume again that three sources report to the fusion node. For the two time frames $k+1$ and $k+2$, the illustration shows three ranked source-to-source correlation hypotheses, respectively. On time frame $k+1$ the example assumes that the best ranked source-to-source correlation produces two source-to-source correlated tracks. The first source-correlated track of this hypothesis associates track 1 from source 1 (denoted in the figure as 1-1), track 1 from source 2 (1-2), and track 1 from source 3 (1-3). The second source-correlated track of this hypothesis associates track 2 from source 1 and track 2 from source 2. The 2nd highest ranking hypothesis correlates track 1 from source 3 with track 2 from source 2 and track 2 from source 1. The 3rd highest ranking hypothesis considers track 1 from source 3 a singleton track. For simplicity the example assumes that the correlation of time frame $k + 2$ produces the same hypotheses as the correlation of time frame $k + 1$ (this will in typical situations not be the case).

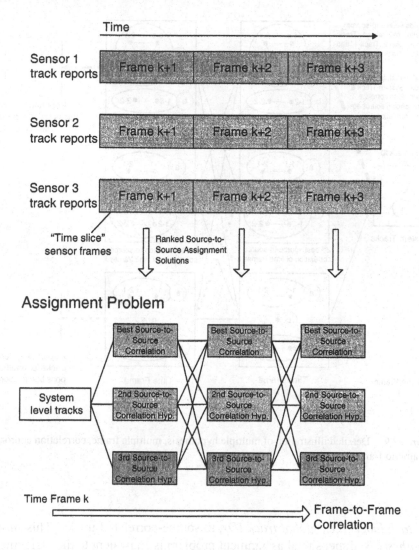

Figure 12.8. Illustration of multiple hypothesis, multiple frame, correlation approach to frame-to-frame matching

The frame-to-frame matching attempts to connect the source-correlated tracks to a given set of global or system tracks at time frame k. If only the highest ranking hypothesis would be considered per time frame, the frame-to-frame matching over N frames can be formulated as an N-dimensional assignment problem, as discussed in Section 12.3.2. The solution of the N-dimensional assignment problem allows one to connect the set of global tracks to the source-correlated tracks. One can say that the frame-to-frame matching allows one

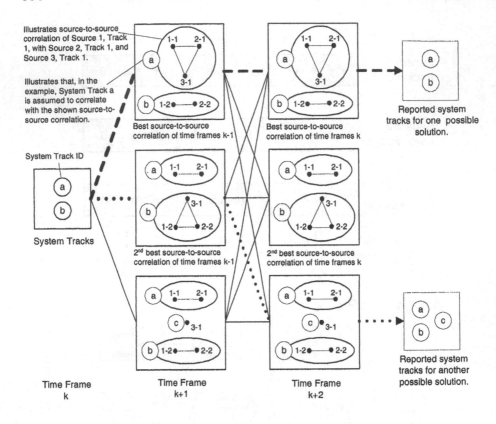

Figure 12.9. Detailed illustration of multiple hypothesis, multiple frame, correlation approach to frame-to-frame matching

to *attach labels, i.e., global track IDs*, to source-correlated tracks. This *single hypothesis N*-dimensional assignment problem is equivalent to the assignment problem illustrated by the thick dashed arrow in Figure 12.9. The more robust multiple hypothesis approach will consider alternative source-to-source correlation hypotheses. This requires the solution (and scoring) of $\prod_{l=1}^{N-1} n_l$, N-dimensional, assignment problems. The example in Figure 12.9 allows for 9 different frame-to-frame matchings that are given through the 9 possible paths through the trellis. The thick dotted line shows one of these alternative solutions. By making decisions over multiple (time) frames of data, this approach has the potential to make more robust frame-to-frame global track extension decisions. For computational efficiency it is important to note that the scoring of the assignments, which is the computationally expensive part, typically involves many equivalent computations within the scoring of alternative solutions. Thus, for efficiency, the multiple hypothesis correlation approach requires a *track tree* or *caching* component that avoids redundant computations.

Let

$$\mathcal{M}_{j_1 j_2 \ldots j_{N-1}}^k = \left\{ \mathcal{G}_k, \ \mathcal{S}_{j_1}^{k+1}, \ \mathcal{S}_{j_2}^{k+2}, \ldots, \mathcal{S}_{j_{N-1}}^{k+N-1} \right\}$$

denote the N-dimensional assignment problem for frame-to-frame matching of the global tracks at time frame k with the source-correlated tracks from the j_1th ranked hypothesis at time frame $k + 1$, the source-correlated tracks from the j_2th ranked hypothesis at time frame $k + 2$, etc. The solution of this N-dimensional assignment problem allows one to compute a frame-to-frame matching cost $m_{j_1 j_2 \ldots j_{N-1}}^k$ that is the sum of the costs of the individual global tracks (see Section 12.4) obtained as the solution of the respective assignment problem. The optimal frame-to-frame matching for an extension of global tracks from time frame k to time frame $k + N - 1$, given data from time frame k through time frame $k + N - 1$, is now given by the matching that has minimal cost:

$$\mathcal{M}_{k+N-1|k}^{\text{optimal}} := \mathcal{M}_{j_1^{\text{opt}} \ldots j_{N-1}^{\text{opt}}}^k,$$

with

$$j_1^{\text{opt}} \ldots j_{N-1}^{\text{opt}} = \arg \min_{j_1, \ldots, j_{N-1}} m_{j_1 \ldots j_{N-1}}^k, \ \ j_l = 1, \ldots, n_l, \ \ l = 1, \ldots, N - 1.$$

The multiple hypothesis, multiple frame, frame-to-frame matching requires a sliding window of N frames of data. Figure 12.10 illustrates the functioning of the sliding window for $N = 3$. Initially the system is assumed to have processed data up to time frame $k + 2$. This implies a set of established global tracks $\mathcal{G}_k$ at frame k and sets of source-correlated tracks at frames $k + 1$ and $k + 2$ respectively. In order to process the time frame $k + 3$ the window needs to be *shifted* to open a new slot in the window for the new data. This requires to make a *hard frame-to-frame matching decision* at the *back* of the window. This will produce a new set of established global tracks $\mathcal{G}_{k+1}$ at time frame $k+1$ that are firmly correlated to source tracks. Once the next time frame is received the new N-dimensional assignment problem needs to be solved to obtain the new frame-to-frame matching cost $m_{j_1 j_2 \ldots j_{N-1}}^{k+1}$. This allows one to compute the new, currently best, frame-to-frame matching estimate $\mathcal{M}_{k+N|k+1}^{\text{optimal}}$. Based on this estimate, global tracks can be reported at time frame $k + N$ by propagating the global tracks from frame $k+1$ through the set of of source-correlated tracks $\mathcal{S}_{j_1^{\text{opt}}}^{k+2}, \ldots, \mathcal{S}_{j_N^{\text{opt}}}^{k+N}$, e.g., in the example, $\mathcal{S}_2^{k+2}$ and $\mathcal{S}_1^{k+3}$. The process can then continue with sliding of the window and processing the next time frame of data.

The advantage of the multiple hypothesis, multiple frame, frame-to-frame matching approach is two-fold. First, we can make potentially *better current source-to-source and source-to-system track correlation decisions based on*

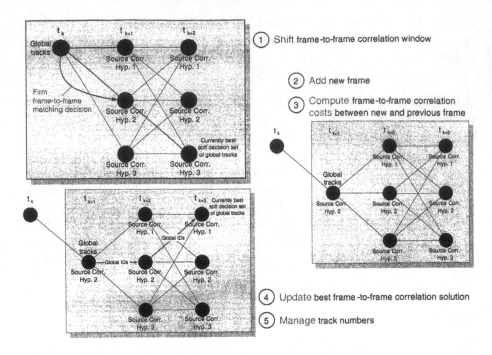

Figure 12.10.   Illustration of sliding window for frame-to-frame matching

the information from previous frames of data. Second we have the potential to *revise previously reported source-to-source and source-to-system track correlation decisions* based on current data. In addition, the multiple hypotheses allow one to assess *ambiguity* between source-to-source and source-to-system track correlations which can be used to avoid track fusion for ambiguous assignments.

4.   Computation of Cost Coefficients using a Batch Methodology.

This section introduces a novel concept for computing assignment scores for track fusion that is based on a *batch distance metric*. This methodology was specifically designed to address the problem of missing covariances in track fusion systems that occurs if some sensors from the set of sensors to be fused only report track states without full covariance information.

4.1   The Dynamical Model and Estimation Problem

It is known that, with Gaussian a priori statistics, the maximum a posteriori estimate is equivalent to an appropriate least-squares curve fit, using the inverses of the plant- and report-noise covariances as weight matrices. The preliminary development in this section follows closely that in the book by Jazwinski [Jazwinski, 1970] and the book by Sage and Melsa [Sage and Melsa, 1971].

The discrete message and observation models are given by

$$\mathbf{x}_{k+1} = \mathbf{f}_k(\mathbf{x}_k) + \mathbf{w}_k,$$
$$\mathbf{z}_k = \mathbf{h}_k(\mathbf{x}_k) + \mathbf{v}_k,$$

(12.4)

where

$\mathbf{x}_k = n$-dimensional state vector (system level track),

$\mathbf{f}_k(\mathbf{x}_k) = n$-dimensional vector-valued function,

$\mathbf{w}_k = r$-dimensional plant-noise vector,

$\mathbf{z}_k = m$-dimensional observation vector (source level track),

$\mathbf{h}_k(\mathbf{x}_k) = m$-dimensional vector-valued function,

$\mathbf{v}_k = m$-dimensional observation-noise vector.

In report based association and fusion, $\mathbf{w}_k$ and $\mathbf{v}_k$ are assumed to be independent zero-mean Gauss-Markov white sequences such that

$$E\{\mathbf{w}_k\mathbf{w}_k^T\} = \mathbf{Q}_k,$$
$$E\{\mathbf{v}_k\mathbf{v}_k^T\} = \mathbf{R}_k,$$
$$E\{(\mathbf{x}_0 - \bar{\mathbf{x}}_0)^T(\mathbf{x}_0 - \bar{\mathbf{x}}_0)\} = \mathbf{P}_0,$$
$$\mathbf{x}_0 - \bar{\mathbf{x}}_0 \stackrel{def}{=} \mu_0 \sim \mathcal{N}(0, \mathbf{P}_0),$$

where $\mathbf{Q}_k$ and $\mathbf{R}_k$ are nonnegative definite $m \times m$ and $r \times r$ covariance matrices, respectively.

The sequences $\mathbf{x}_0, \mathbf{x}_1, ..., \mathbf{x}_N$ and $\mathbf{z}_1, \mathbf{z}_2, ..., \mathbf{z}_N$ are denoted by X_N and Z_N, respectively. We further assume that $p(\mathbf{x}_0)$ is known and is normal with mean $\bar{\mathbf{x}}_0$ and covariance $\mathbf{P}_0$. We shall assume that all functions are sufficiently smooth so that up to two derivatives exist and are continuous.

The best estimate of $\mathbf{x}$ throughout an interval will, in general, depend on the criteria used to determine the best estimate. Here the term "best estimate" denotes that estimate derived from maximizing the conditional probability function $p(X \mid Z)$ as a function of $\mathbf{x}$ throughout that interval, and is known as the maximum a posteriori estimator. The following derivation is adapted from that found in the book by Sage and Melsa [Sage and Melsa, 1971] and basically assumes that the cross-correlations (within the batch of data) are negligible.

Applying Bayes' rule to $p(X_N \mid Z_N)$ results in

$$p(X_N \mid Z_N) = \frac{p(Z_N \mid X_N)\, p(X_N)}{p(Z_N)}. \qquad (12.5)$$

From Eqn. (12.4) it follows that if $\mathbf{x}_k$ is known, $p(\mathbf{z}_k \mid \mathbf{x}_k)$ is Gaussian, since $\mathbf{v}_k$ is assumed Gaussian. If X_N is given,

$$p(Z_N \mid X_N) = \prod_{k=1}^{N} \frac{\exp\left\{-\tfrac{1}{2}(\mathbf{z}_k - \mathbf{h}_k(\mathbf{x}_k))^T \mathbf{R}_k^{-1}(\mathbf{z}_k - \mathbf{h}_k(\mathbf{x}_k))\right\}}{\sqrt{(2\pi)^r \det(\mathbf{R}_k)}}. \qquad (12.6)$$

Using the chain rule for probabilities,

$$p(\alpha, \beta) = p(\alpha \mid \beta)\, p(\beta), \qquad (12.7)$$

results in

$$p(X_N) = p(\mathbf{x}_N \mid X_{N-1})\, p(\mathbf{x}_{N-1} \mid X_{N-2}) \cdots p(\mathbf{x}_1 \mid \mathbf{x}_0)\, p(\mathbf{x}_0). \qquad (12.8)$$

If $\mathbf{w}_k$ is a white Gauss-Markov sequence, $\mathbf{x}_k$ is Markov, and

$$p(\mathbf{x}_r \mid X_{r-1}) = p(\mathbf{x}_r \mid \mathbf{x}_{r-1}). \qquad (12.9)$$

Thus $p(X_N)$ is composed of Gaussian terms, and

$$p(X_N) = p(\mathbf{x}_0) \prod_{k=1}^{N} p(\mathbf{x}_k \mid \mathbf{x}_{k-1}), \qquad (12.10)$$

$p(Z_N)$ contains no terms in $\mathbf{x}_k$ since Z_N is the known conditioning variable. Thus $p(Z_N)$ can be considered a normalizing constant with respect to the intended maximization.

Summarizing, we have

$$p(X_N \mid Z_N) = \frac{p(Z_N \mid X_N)\, p(X_N)}{p(Z_N)}$$

$$= \frac{p(\mathbf{x}_0)}{p(Z_N)} \prod_{k=1}^{N} p(\mathbf{z}_k \mid \mathbf{x}_k) p(\mathbf{x}_k \mid \mathbf{x}_{k-1}).$$

Part of the estimation is to solve the optimization problem

$$\underset{\mathbf{x}_0, \mathbf{x}_1, \ldots, \mathbf{x}_N}{\text{Minimize}} \; J = \frac{1}{2}\|\{\mathbf{x}_0 - \bar{\mathbf{x}}_0, \mathbf{x}_1 - \mathbf{f}_0(\mathbf{x}_0), \mathbf{z}_1 - \mathbf{h}_1(\mathbf{x}_1), \ldots, \qquad (12.11)$$

$$\mathbf{x}_N - \mathbf{f}_{N-1}(\mathbf{x}_{N-1}), \mathbf{z}_N - \mathbf{h}_N(\mathbf{x}_N)\}^T\|_{\mathbf{P}^{-1}}^2, \qquad (12.12)$$

where

$$\Sigma = E\left\{ [\mu_0, \mathbf{v}_1, \mathbf{w}_1, \ldots, \mathbf{w}_N, \mathbf{v}_N]^T [\mu_0, \mathbf{v}_1, \mathbf{w}_1, \ldots, \mathbf{w}_N, \mathbf{v}_N] \right\}.$$

If there are no statistical cross correlations, then $\Sigma = \text{diag}\{\mathbf{P}_0, \mathbf{Q}_0, \mathbf{R}_1, \ldots, \mathbf{Q}_{N-1}, \mathbf{R}_N\}$ resulting in the *Maximum a Posteriori (MAP) batch estimation* problem,

$$\underset{\mathbf{x}_0, \mathbf{x}_1, \ldots, \mathbf{x}_N}{\text{Minimize}} \quad J = \frac{1}{2}\|\mathbf{x}_0 - \bar{\mathbf{x}}_0\|^2_{\mathbf{P}_0^{-1}} \tag{12.13}$$

$$+ \frac{1}{2}\sum_{k=1}^{N}\|\mathbf{z}_k - \mathbf{h}_k(\mathbf{x}_k)\|^2_{\mathbf{R}_k^{-1}} + \frac{1}{2}\sum_{k=1}^{N}\|\mathbf{x}_k - \mathbf{f}_{k-1}(\mathbf{x}_{k-1})\|^2_{\mathbf{Q}_{k-1}^{-1}},$$

$$\tag{12.14}$$

where $(\bar{\mathbf{x}}_0, \mathbf{P}_0)$ ($\bar{\mathbf{x}}_0 = \int_{-\infty}^{\infty} \mathbf{x}_0 p(\mathbf{x}_0) d\mathbf{x}_0$) is the prior state and covariance and where it is assumed that $\mathbf{Q}_{k-1}$ is independent of the state.

Now, however, statistical cross correlations are present and are represented generally by the covariance matrix Σ via

$$\Sigma = E\left\{ [\mu_0, \mathbf{v}_1, \mathbf{w}_1, \ldots, \mathbf{w}_N, \mathbf{v}_N]^T [\mu_0, \mathbf{v}_1, \mathbf{w}_1, \ldots, \mathbf{w}_N, \mathbf{v}_N] \right\}.$$

We denote the solution of the nonlinear least squares problem by $\tilde{\mathbf{x}}_0, \tilde{\mathbf{x}}_1, \ldots, \tilde{\mathbf{x}}_N$.

4.2   The States Used in Global Track Extension Score Computation

Section 12.3.2 introduced the N-dimensional assignment problem for frame-to-frame matching that required the computation of a cost coefficient $c_{i\,j_1 \ldots j_{N-1}}$ that represents the cost for extending a global, or system level, track G_i^k at time frame k, with source correlated tracks $S_{j_1}^{k+1}, \ldots, S_{j_{N-1}}^{k+N-1}$, respectively.

Recall that a global track G_i^k represents a set $\{ID_i, S_{l_i}^{k_i}, \mathcal{A}_i^k\}$, where $S_{l_i}^{k_i}$ represents the source correlated track that associated to the global track at the last update time k_i, and $\mathcal{A}_i^k$ represents a set that contains the source correlated tracks from previous updates. For computation of the cost coefficient $c_{i\,j_1 \ldots j_{N-1}}$ we will represent the global track G_i^k with a *prior* state $(\bar{\mathbf{x}}_i^k, \mathbf{P}_i^k)$ that is assumed to be of full covariance[2] and computed from the previous update $S_{l_i}^{k_i}$ and potentially a set of previous source correlated tracks.

Different options may be considered for computation of the prior state. The choice of prior is designed to *minimize the amount of correlation* between sys-

---

[2]This requirement is not necessary but ensures greater numerical stability of the batch if N is small.

tem track prior and source correlated tracks within the N-dimensional frame-to-frame matching problem. The prior may for example represent the fused state obtained from fusion of the source track reports correlated to the track $S_{l_i}^{k_i}$ (this represents fusion based on data from a single time frame). As an alternative we may *select* the most recent full state source update of best quality as representative of the global track state G_i^k.

Now recall from Section 12.2 that a source correlated track S_j^{k+l} represents a *set* of source track reports from possibly a number of different sensors within the time interval $[t_{k+l-1}, t_{k+l})$. Each source track report may contain multiple reported source track updates within the given time interval. In the following we will represent a source correlated track S_j^{k+l} with a set of measurements

$$S_j^{k+l} = \left\{ \mathbf{z}_{r_1^j}^{i_1^j, k+l}, \dots, \mathbf{z}_{r_{m_j}^j}^{i_{m_j}^j, k+l} \right\},$$

where $\mathbf{z}_{r_m^j}^{i_m^j, k+l}$ is assumed to be the most recent track update within the time interval $[t_{k+l-1}, t_{k+l})$ for source track r_m received from sensor i_m that is correlated to source track S_j^{k+l}.

The computation of the cost coefficient $c_{i\, j_1 \dots j_{N-1}}$ will now be based on the following set of states:

$$\left\{ \tilde{\mathbf{x}}_i^k, S_{j_1}^{k+1}, \dots, S_{j_{N-1}}^{k+N-1} \right\}$$
$$= \left\{ \tilde{\mathbf{x}}_i^k, \; \mathbf{z}_{r_1^{j_1}}^{i_1^{j_1}, k+1}, \dots, \mathbf{z}_{r_{m_{j_1}}^{j_1}}^{i_{m_{j_1}}^{j_1}, k+1}, \dots, \mathbf{z}_{r_1^{j_{N-1}}}^{i_1^{j_{N-1}}, k+N-1}, \dots, \mathbf{z}_{r_{m_{j_{N-1}}}^{j_{N-1}}}^{i_{m_{j_{N-1}}}^{j_{N-1}}, k+N-1} \right\},$$

$$(12.15)$$

and the respective covariances. For notational convenience we will drop indices and consider the following set of states

$$\{(t_0, \overline{\mathbf{x}}_0, \mathbf{P}_0), \; (t_1, \mathbf{z}_1, \mathbf{R}_1), \; (t_2, \mathbf{z}_2, \mathbf{R}_2), \; \dots, (t_m, \mathbf{z}_m, \mathbf{R}_m)\}, \quad (12.16)$$

in place of the set given in Eqn. (12.15). Notationally and computationally, the batch score computation then treats the source track updates $\mathbf{z}_j$ as measurements at time t_j with covariances $\mathbf{R}_j$ that will "fuse to" the established system track prior $(\overline{\mathbf{x}}_0, \mathbf{P}_0)$ at time t_0. Figure 12.11 illustrates this frame-to-frame matching for two sources and $N = 3$, i.e., the prior and two consecutive time frames.

We note that a source track update $\mathbf{z}_j$ can be replaced with an *equivalent* measurement [Frenkel, 1995, Drummond, 1997], i.e., a decorrelated track report or tracklet, if the set of state Eqn. (12.16) contains multiple track reports on the same source track from the same sensor.

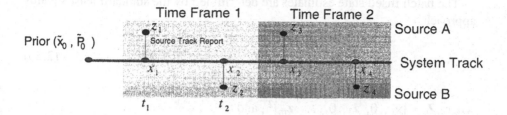

Figure 12.11. Illustration of the batch scoring for frame-to-frame matching

4.3 Batch Fusion

Before the distance metric can be computed that is the basis for the computation of the cost coefficient $c_{i\,j_1...j_{N-1}}$, fused estimates $\tilde{\mathbf{x}}_0, \ldots, \tilde{\mathbf{x}}_m$ at times $t_0, \ldots, t_m$, respectively, need to be computed. We assume the linearized version of the dynamics model, Eqn. (12.4), i.e., $\mathbf{f}_k(\mathbf{x}_k) := \mathbf{F}_k\mathbf{x}_k$ and $\mathbf{h}_k(\mathbf{x}_k) := \mathbf{H}_k\mathbf{x}_k$. The batch methodology is the same as that given in [Poore et al., 2003], and we summarize the approach here. The fused state estimates $\{\tilde{\mathbf{x}}_0, \ldots, \tilde{\mathbf{x}}_m\}$ are computed by finding the solution to the following least squares problem,

$$\{\tilde{\mathbf{x}}_0, \ldots, \tilde{\mathbf{x}}_m\} \leftarrow \arg\min_{X_m} \left\| \begin{array}{c} \mathbf{x}_0 - \bar{\mathbf{x}}_0 \\ \mathbf{x}_1 - \mathbf{F}(t_1, t_0)\mathbf{x}_0 \\ \mathbf{z}_1 - \mathbf{H}_1\mathbf{x}_1 \\ \vdots \\ \mathbf{x}_m - \mathbf{F}(t_m, t_{m-1})\mathbf{x}_{m-1} \\ \mathbf{z}_m - \mathbf{H}_m\mathbf{x}_m \end{array} \right\|^2_{\Sigma^{-1}}, \qquad (12.17)$$

where Σ is a normalization covariance matrix given by

$$\Sigma = E\left\{ \begin{bmatrix} \mu_0 \\ \mathbf{w}_0 \\ \mathbf{v}_1 \\ \vdots \\ \mathbf{w}_{m-1} \\ \mathbf{v}_m \end{bmatrix} \begin{bmatrix} \mu_0 \\ \mathbf{w}_0 \\ \mathbf{v}_1 \\ \vdots \\ \mathbf{w}_{m-1} \\ \mathbf{v}_m \end{bmatrix}^T \right\}. \qquad (12.18)$$

This covariance matrix will have cross-correlation terms, but when those terms are ignored the matrix is given by

$$\Sigma = \text{diag}(\mathbf{P}_0, \mathbf{Q}_0, \mathbf{R}_1, \ldots, \mathbf{Q}_{m-1}, \mathbf{R}_m). \qquad (12.19)$$

The batch fused state estimates are determined by the standard least squares approach,

$$\begin{bmatrix} \tilde{\mathbf{x}}_0 \\ \vdots \\ \tilde{\mathbf{x}}_m \end{bmatrix} = (\mathbf{A}^T \boldsymbol{\Sigma}^{-1} \mathbf{A})^{-1} \mathbf{A}^T \boldsymbol{\Sigma}^{-1} \mathbf{Z}, \tag{12.20}$$

where $\mathbf{Z} = [\bar{\mathbf{x}}_0, \ 0, \ \mathbf{z}_1, \ 0, \ldots, \ \mathbf{z}_m]^T$, and

$$\mathbf{A} = \begin{bmatrix} \mathbf{I} & 0 & 0 & 0 & 0 & \cdots & 0 & 0 \\ -\mathbf{F}(t_1, t_0) & \mathbf{I} & 0 & 0 & 0 & \cdots & 0 & 0 \\ 0 & \mathbf{H}_1 & 0 & 0 & 0 & \cdots & 0 & 0 \\ 0 & 0 & -\mathbf{F}(t_2, t_1) & \mathbf{I} & 0 & \cdots & 0 & 0 \\ \vdots & \vdots & \vdots & \vdots & \vdots & \vdots & \vdots & \vdots \\ 0 & 0 & 0 & 0 & 0 & \cdots & 0 & \mathbf{H}_m \end{bmatrix}. \tag{12.21}$$

If $\Delta t = t_i - t_j$ is very small, then $\boldsymbol{\Sigma}$ can become close to singular. To mitigate this, one must detect situations when Δt is small, and in that case modify $\mathbf{A}$ and $\mathbf{Z}$ to not include the dynamics portion, i.e., the transition $\mathbf{x}_i - \mathbf{F}(t_i, t_j)\mathbf{x}_j$. If all the tracks in the batch do not report the velocity covariance, then the number of unknowns to be estimated relative to the number of equations becomes too large and the batch estimate will produce poor results (or the estimation model will become underdetermined and (12.20) will fail). The only recourse is to use an approximate (estimated) covariance for the velocity.

4.4   Batch Distance Metric

Given the fused track states $\{\tilde{\mathbf{x}}_0, \ldots, \tilde{\mathbf{x}}_m\}$ we can compute the *batch distance metric*. We form the vector $\mathbf{y}$:

$$\mathbf{y} = \begin{bmatrix} \tilde{\mathbf{x}}_0 - \bar{\mathbf{x}}_0 \\ \tilde{\mathbf{x}}_1 - \mathbf{F}(t_1, t_0)\tilde{\mathbf{x}}_0 \\ \mathbf{z}_1 - \mathbf{H}_1 \tilde{\mathbf{x}}_1 \\ \vdots \\ \tilde{\mathbf{x}}_m - \mathbf{F}(t_m, t_{m-1})\tilde{\mathbf{x}}_{m-1} \\ \mathbf{z}_m - \mathbf{H}_m \tilde{\mathbf{x}}_m \end{bmatrix}. \tag{12.22}$$

The batch distance metric is given by the Mahalanobis distance

$$d_y^2 = \mathbf{y}^T \boldsymbol{\Sigma}^{-1} \mathbf{y}, \tag{12.23}$$

where $\boldsymbol{\Sigma}$ is the covariance matrix defined in (12.18) and (12.19).

The cross term values in the covariance, $\boldsymbol{\Sigma}_{ij}$, could be computed using the approximation technique of Bar-Shalom [Bar-Shalom and Li, 1995, pg. 455], refined more recently in [Chen et al., 2003].

$$[\Sigma_{ij}]_{(m,n)} = \rho \cdot \text{sign}(\rho^i_{mn}\rho^j_{mn})\sqrt{|\rho^i_{mn}\rho^j_{mn}|} \cdot [\Sigma_i]_{(m,n)} \cdot [\Sigma_j]_{(m,n)} \quad (12.24)$$

where $[\Sigma]_{(m,n)}$ is the (m, n) row-column element of the matrix Σ and

$$\rho^i_{mn} = \frac{[\Sigma_i]_{(m,n)}}{\sqrt{[\Sigma_i]_{(m,n)} \cdot [\Sigma_j]_{(m,n)}}} \quad (12.25)$$

The value of ρ is constant selected in $[0.1, 0.4]$ [Chen et al., 2003]. For a non-maneuvering target it should be selected closer to 0.1.

4.5   Assignment Score for Global Track Extension

We now describe how the desired cost coefficient $c_{i\,j_1...j_{N-1}}$ is obtained. Once we have computed the distance metric d_y^2 using Eqn. (12.23) and the corresponding covariance Σ from Eqn. (12.19), then the log-likelihood ratio can be computed. The ratio for track matching is given for the case of two tracks in [Blackman and Popoli, 1999]. The extension to the case of m simultaneous tracks is shown here. We wish to clarify that $m \geq N$ since the cost coefficient $c_{i\,j_1...j_{N-1}}$ describes the cost for the correlation of $N - 1$ source correlated tracks to a global track. However, since we use the "original" source tracks associated to a source correlated track we will in general have to consider $m \geq N$ tracks in the distance matrix and likelihood computation.

Generalizing the formulation of Blackman [Blackman and Popoli, 1999, p.629] to m track states, the likelihood function for the hypothesis that tracks $x_0, \ldots x_m$ go together is

$$\Lambda_{i_1...i_m} = \frac{\beta_T P_{T_{i_1}} P_{T_{i_2}} \cdot \ldots \cdot P_{T_{i_m}} \exp(-d_y^2/2)}{\sqrt{(2\pi)^{\dim(y)}|\Sigma|}}, \quad (12.26)$$

where β_T is the target density, P_{T_i} is the probability that the sensor producing the ith track has a track on a given target in the common field of view.

The likelihood function for the hypothesis that the tracks do not go together is the product of the m likelihoods,

$$\begin{aligned}
\Lambda_{i_10...0} &= \beta_T P_{T_{i_1}}(1 - P_{T_{i_2}}) \cdot \ldots \cdot (1 - P_{T_{i_m}}) + \beta_{FT_{i_1}} \\
\Lambda_{0i_20...0} &= \beta_T(1 - P_{T_{i_1}})P_{T_{i_2}} \cdot (1 - P_{T_{i_3}}) \cdot \ldots \cdot (1 - P_{T_{i_m}}) + \beta_{FT_{i_2}} \\
&\vdots \\
\Lambda_{0...0i_m} &= \beta_T(1 - P_{T_{i_1}}) \cdot \ldots \cdot (1 - P_{T_{i_{m-1}}})P_{T_{i_m}} + \beta_{FT_{i_m}},
\end{aligned}$$

where β_{FT_i} is the false track density for the sensor providing the ith track. The likelihood ratio for track matching is then given by

$$L_{i_1\ldots i_m} = \frac{\Lambda_{i_1\ldots i_m}}{\Lambda_{i_1 0\ldots 0} \cdot \Lambda_{0 i_2 0\ldots 0} \cdot \ldots \cdot \Lambda_{0\ldots 0 i_m}}. \qquad (12.27)$$

The assignment score is the negative log-likelihood ratio given by

$$
\begin{aligned}
c_{i\,j_1\ldots j_{N-1}} := c_{i_1\ldots i_m} &= -\ln\left(L_{i_1\ldots i_m}\right) \\
&= \frac{1}{2}\left(d_y^2 + \ln\left(\det(\boldsymbol{\Sigma})\right) + \dim(\mathbf{y})\ln(2\pi)\right) - \ln(\kappa),
\end{aligned}
$$

where

$$\kappa = \frac{\beta_T P_{T_{i_1}} P_{T_{i_2}} \cdot \ldots \cdot P_{T_{i_m}}}{\Lambda_{i_1 0\ldots 0} \cdot \Lambda_{0 i_2 0\ldots 0} \cdot \ldots \cdot \Lambda_{0\ldots 0 i_m}}. \qquad (12.28)$$

Note that the calculation of the score requires the availability of the parameters $(\beta_T, P_{T_i}, \beta_{FT_i})$ within κ, but that frequently these parameters are unknown for a given sensor. Further, extensive "tuning" of these parameters is often required to get the best performance from the assignment solver. Thus, in practice it is often sufficient to specify a single κ value for the application.

An extension of this likelihood function to include initiating and terminating track probabilities can be accomplished following the approach in [Poore, 1995].

5.   Summary

Multiple sensor tracking applications typically require either measurement-to-track or track-to-track fusion. Many advanced multiple hypothesis correlation approaches have been developed for measurement-to-track fusion, most noticeably multiple hypothesis tracking (MHT) and multiple frame assignment (MFA) tracking. Equally, or possibly more important, is the correlation of track states to track states. With respect to this problem to date all approaches have been limited to single frame correlation techniques. This chapter introduced a novel multiple hypothesis approach to track-to-track correlation.

Architectures for track-to-track correlation and fusion can be categorized into two fundamentally different processing architectures: (1) sensor-to-sensor track fusion and (2) sensor-to-system track fusion. The sensor-to-sensor track fusion architecture requires a robust mechanism for global, or system level, track data management. This chapter focused on a multiple hypothesis, multiple frame, correlation approach for the frame-to-frame matching problem in sensor-to-sensor track fusion.

We formulated the frame-to-frame matching problem as an N-dimensional assignment problem. Given this formulation, the approach was generalized toward a multiple hypothesis correlation approach that considers multiple source-to-source correlation hypotheses for a given time frame and selects the optimal correlation over multiple frames of data. We introduced a novel batch scoring mechanism for computation of the cost coefficients in the N-dimensional assignment problem.

The use of the proposed multiple hypothesis, multiple frame correlation system is expected to improve the fusion system performance in scenarios where significant track assignment ambiguity exists. In the same way that multiple frame processing has shown improvements in the tracking performance in measurement-to-track fusion applications, we expect to achieve improvements in the track-to-track fusion problem.

In this chapter, we have also noted that the multiple hypothesis correlation approach, as formulated, represents a special case of the general group assignment problem, which was previously introduced in [Poore and Gadaleta, 2003]. In future work we plan to formulate the multiple hypothesis track correlation within this more general framework. The advantage of this approach will be a more efficient method for solving the multiple hypothesis correlation assignment problem.

References

[Bar-Shalom, 1990] Bar-Shalom, Y., editor (1990). *Multitarget-Multisensor Tracking: Advanced Applications*. Artech House, Dedham, MA.

[Bar-Shalom, 1992] Bar-Shalom, Y., editor (1992). *Multitarget-Multisensor Tracking: Applications and Advances*. Artech House, Dedham, MA.

[Bar-Shalom and Blair, 2000] Bar-Shalom, Y. and Blair, W. D., editors (2000). *Multitarget-Multisensor Tracking: Applications and Advances, Volume III*. Artech House, Dedham, MA.

[Bar-Shalom and Li, 1995] Bar-Shalom, Y. and Li, X.-R. (1995). *Multitarget-Multisensor Tracking: Principles and Techniques*. YBS.

[Blackman and Popoli, 1999] Blackman, S. and Popoli, R. (1999). *Design and Analysis of Modern Tracking Systems*. Artech House, Norwood, MA.

[Chen et al., 2003] Chen, H., Kirubarajan, T., and Bar-Shalom, Y. (2003). Performance limits of track-to-track fusion versus centralized estimation: theory and application. *IEEE Transactions on Aerospace and Electronic Systems*, 39(2):386–398.

[Chong et al., 2000] Chong, C.-Y., Mori, S., Barker, W., and Chang, K.-C. (2000). Architectures and algorithms for track association and fusion. *IEEE AES Systems Magazine*, **15**:5–13.

[Chong et al., 1990] Chong, C. Y., Mori, S., and Chang, K. C. (1990). Distributed multitarget multisensor tracking. In Bar-Shalom, Y., editor, *Multitarget-Multisensor Tracking: Advanced Applications*, volume 1, chapter 8, pages 247–295. Artech House, Norwood, MA.

[Drummond, 1997] Drummond, O. E. (1997). A hybrid fusion algorithm architecture and tracklets. *Signal and Data Processing of Small Targets 1997, SPIE*, 3136:485–502.

[Frenkel, 1995] Frenkel, G. (1995). Multisensor tracking of ballistic targets. In *SPIE*, volume **2561**, pages 337–346.

[Gadaleta et al., 2002] Gadaleta, S., Klusman, M., Poore, A. B., and Slocumb, B. J. (2002). Multiple frame cluster tracking. In *SPIE Vol. 4728, Signal and Data Processing of Small Targets*, pages 275–289.

[Gadaleta et al., 2004] Gadaleta, S., Poore, A. B., Roberts, S., and Slocumb, B. J. (2004). Multiple hypothesis clustering, multiple frame assignment tracking. In *SPIE Vol. 5428, Signal and Data Processing of Small Targets*.

[Jazwinski, 1970] Jazwinski, A. (1970). *Stochastic processes and filtering theory*. Academic Press, New York.

[Liggens et al., 1997] Liggens, M. E., Chong, C. Y., Kadar, I., Alford, M. G., Vannicola, V., and Thomopoulos, S. (1997). Distributed fusion architectures and algorithms for target tracking. *Proceedings of the IEEE*, 85(1):95–107.

[Murty, 1968] Murty, K. (1968). An algorithm for ranking all the assignments in order of increasing cost. *Operations Research*, 16:682–687.

[Poore and Gadaleta, 2003] Poore, A. and Gadaleta, S. (2003). The group assignment problem. In *SPIE Vol. 5204, Signal and Data Processing of Small Targets*, pages 595–607.

[Poore, 1994] Poore, A. B. (1994). Multidimensional assignment formulation of data association problems arising from multitarget tracking and multisensor data fusion. *Computational Optimization and Applications*, 3:27–57.

[Poore, 1995] Poore, A. B. (1995). Multidimensional assignments and multitarget tracking. In Cox, I. J., Hansen, P., and Julesz, B., editors, *Partitioning Data Sets*, volume 19 of *DIMACS Series in Discrete Mathematics and Theoretical Computer Science*, pages 169–198, Providence, RI. American Mathematical Society.

[Poore et al., 2001a] Poore, A. B., Lu, S., and Suchomel, B. (2001a). Network MFA tracking architectures. In *Proceedings of SPIE Conference on Small Targets 2001, Oliver E. Drummond Editor*.

[Poore et al., 2001b] Poore, A. B., Lu, S., and Suchomel, B. J. (2001b). Data association using multiple frame assignments. In *Handbook of Multisensor Data Fusion*. CRC Press LLC.

[Poore et al., 2003] Poore, A. B., Slocumb, B. J., Suchomel, B. J., Obermeyer, F. H., Herman, S. M., and Gadaleta, S. M. (2003). Batch Maximum Likelihood (ML) and Maximum A Posteriori (MAP) estimation with process noise for tracking applications. In *SPIE Vol. 5204, Signal and Data Processing of Small Targets*, pages 188–199.

[Sage and Melsa, 1971] Sage, A. and Melsa, J. (1971). *Estimation theory with applications to communications and control*. McGraw-Hill, New York.

Chapter 13

COMPUTATIONAL MOLECULAR BIOLOGY

Giuseppe Lancia

Dipartimento di Matematica e Informatica
Università di Udine
Via delle Scienze 206, 33100 Udine
Italy

lancia@dei.unipd.it

1. Introduction

Computational (Molecular) Biology is a relatively young science, which has experienced a tremendous growth in the last decade. The seeds for the birth of Computational Biology were sowed in the end of the Seventies, when computers became cheaper, easily available and simpler to use, so that some biology labs decided to adopt them, mainly for storing and managing genomic data. The use of computers allowed the quick completion of projects that before would have taken years. With a snowball effect, these projects generated larger and larger amounts of data whose management required more and more powerful computers. Recently, the advent of the Internet has allowed all the laboratories to share their data, and make them available worldwide through some new genomic data banks (such as GenBank [12], EMBL [79], PDB [13]). Without any doubt, today a computer is a necessary instrument for doing research in molecular biology, and is invariably present, together with microscopes and other more classical instruments, in any biotechnology lab.

Currently, there is not a full agreement on what "Computational Biology" means. Some researchers use a very loose definition such as "Computer Science applied to problems arising in Molecular Biology". This definition encompasses at least two major types of problems: (i) Problems of storage, organization and distribution of large amounts of genomic data; (ii) Problems of interpretation and analysis of genomic data. We would rather call problems of the first type *Bioinformatics* problems, and reserve the term *Computational Biology* for the study of problems of the second type. Furthermore, although Computer Science plays a key role in the solution of such problems, other disciplines such as Discrete Mathematics, Statistics and Optimization, are as important. Hence, the following definition of Computational Biology will be adopted in this survey: "*Study of mathematical and computational problems of modeling biological processes in the cell, removing experimental errors from genomic data, interpreting the data and providing theories about their biological relations*".

The study of a typical problem in Computational Biology starts by a representation of the biological data in terms of mathematical objects, such as strings, graphs, permutations, etc. The biological relations of the data are mapped into mathematical relations, and the original question is expressed either as a feasibility or an optimization problem $\mathcal{P}$. From this point on, depending on the nature of $\mathcal{P}$, standard techniques can be adopted for its solution. A first step is usually the study of the computational complexity of the problem. This may possibly lead to a polynomial algorithm, a result that has been obtained, in particular, with Dynamic Programming for several problems in Computational Biology. This line of attack is conducted primarily by computer scientists. However, most times the problems turn out to be NP-hard optimization problems. Their exact solution becomes then work for scientists with a background in Combinatorial Optimization, and, particularly, Operations Research (OR). Standard OR techniques, such as Integer Linear or Quadratic Programming, solved by Branch-and-Cut, Branch-and-Price, or Lagrangian Relaxation, have been increasingly adopted for Computational Biology problems in the last years. Correspondingly, the recent years have witnessed an increasing number of scientists with OR background approaching the field of Computational Biology.

In this chapter, we intend to review some of the main results in Computational Biology that were obtained with the use of Mathematical Programming techniques. The problems will come from different areas, touching virtually every aspect of modern Computational Biology. However, many relevant results that do not employ Mathematical Programming will not be addressed in this survey. For a complete coverage of the many beautiful and important results that have been achieved in Computational Biology, the reader is referred to the classic textbooks by the pioneers of this field, i.e., Michael Waterman [82], Dan Gusfield [38] and Pavel Pevzner [72].

Perhaps the main contribution of Computational Biology so far lies in the key role it played in the completion of the Human Genome Project, which culminated with the 2001 announcement of the completion of the sequencing of the human genome [45, 80]. This result was obtained thanks to experimental techniques, such as *Shotgun Sequencing*, which posed challenging computational problems. Sophisticated algorithms for these problems (which are mostly variants of the famous *Shortest Superstring Problem*, see Garey and Johnson [31] problem [SR9]) were developed mainly by Myers et al. [67, 68, 84].

The material in this chapter is organized in sections. In each section, we describe one or more problems of the same nature, and the Mathematical Programming approaches that were proposed in the literature for their solution. Some of these problems are solved via a reduction to very well known optimization problems, such as the TSP and Set Covering. For these problems, the core of the analysis will focus on the modeling and the reduction, while the solving algorithm can be assumed to be a standard, state-of-the-art, code for the target optimization problem (e.g., CONCORDE [4] for the TSP). For the remaining problems, *ad hoc* algorithms were developed for their solution. Many of these algorithms are based on Integer Programming formulations with an exponential number of inequalities (or variables), that are solved by Branch-and-Cut (respectively, Branch-and-Price).

The chapter is organized as follows:

- **Alignment Problems.** In Section 13.3 we review several problems related with the alignment of genomic objects, such as DNA or RNA sequences and secondary or tertiary structures. We distinguish three types of alignment:

  i) Sequence vs Sequence Alignment. Section 13.3.1 is devoted to the problem of aligning genomic sequences. First, we discuss the classical multiple alignment problem, studied by Reinert et al. [75] and Kececioglu et al. [50], who developed a Branch-and-Price approach for its solution. Then, we describe a heuristic approach for the same problem, by Fischetti et al. [28]. The approach generalizes an approximation algorithm by Gusfield [37] and uses a reduction to the Minimum Routing Cost Tree problem, solved by Branch-and-Price. Closely related to the alignment problem, are the so called "center" problems, in which one tries to determine a sequence as close or as far as possible from a set of input sequences. For this problem, we describe an approximation algorithm based on LP relaxation and Randomized Rounding, by Ben Dor et al. [11]. We also discuss some works that, developing similar ideas, obtain Polynomial Time Approximation Schemes (PTAS) for the consensus and the farthest-sequence problems (Li et al. [61], Ma [64], Lanctot et al. [54]).

  ii) Sequence vs Structure Alignment. In Section 13.3.2 we describe the problem of aligning two RNA sequences, when for one of them the secondary structure is already known. For this problem, we describe a Branch-and-Price approach by Lenhof et al. [60] and Kececioglu et al. [50].

  iii) Structure vs Structure Alignment. In Section 13.3.3 we consider the problem of comparing two protein tertiary structures. For this problem, we review an Integer Linear Programming approach, proposed by

Lancia et al. [53], and another approach, based on a Quadratic Programming formulation and Lagrangian Relaxation, given by Caprara and Lancia [20].

- **Single Nucleotide Polymorphisms.** In Section 13.4 we describe some combinatorial problems related with human diversities (polymorphisms) at the genomic level. The *Haplotyping* problem consists in determining the values of a set of polymorphic sites in a genome. First, we discuss the single individual haplotyping problem, for which some Integer Programming techniques were employed by Lancia et al. [52] and Lippert et al. [63]. Then, we review the population version of the problem, as studied by Gusfield [39, 40]. The approach employs a reduction to a graph problem, which is then solved by Integer Programming.

- **Genome Rearrangements.** In Section 13.5, we describe a successful Branch-and-Price approach, by Caprara et al. [21, 22] for computing the evolutionary distance between two genomes evolved from a common ancestor.

- **Mapping problems and the TSP.** In Section 13.6 we review the *Physical Mapping* problem and its connections with the TSP problem and the Consecutive One property for 0-1 matrices. We first describe an Integer Programming approach for mapping a set of probes on a set of clones, proposed by Alizadeh et al. [3]. We then turn to the problem of mapping radiation hybrids, studied by Ben Dor et al. [9, 10] and by Agarwala et al [2]. These papers reduced the problem to a standard TSP, for which they used the best available software, based on Branch-and-Cut. Finally, we mention a Branch-and-Cut approach for the physical mapping of probes coming from the chromosomes ends, by Christof et al. [26].

- **Applications of Set Covering.** In Section 13.7 we mention a few problems of Computational Biology whose solution algorithm consisted mainly in a reduction to the Set Cover problem. In particular, we review the PCR primer design problem (Pearson et al. [71] and Nicodeme and Steyaert [70]) and the analysis of microarray expression data (Halldorsson et al. [41, 14]).

The elementary concepts of molecular biology needed to follow the material are introduced in Section 13.2 and in the context of the following sections. The web provides a wonderful source of information for further reading. In particular, we suggest visiting the sites of National Institute of Health (www.nih.gov) and the European Molecular Biology Laboratory (www.embl.de).

2. Elementary Molecular Biology Concepts

One of the major problems encountered by researchers coming from mathematical fields and approaching computational biology, is the lack of vocabulary and of understanding of the biological phenomena underlying the mathematical models of the problems. This section is intended to provide the reader with some, very basic, preliminary notions, and many additional biology concepts will be given in the context of the specific problems described in the following sections. A nice exposition of molecular biology at an introductory level can be found in the Human Genome Project Primer [24]. Alternatively, Fitch [29] has written an introductory paper aimed specifically at mathematicians and computer scientists. For a comprehensive and deeper analysis of the topic, the reader is referred to some of the standard textbooks in molecular biology, such as the one by Watson, Gilman, Witkowski and Zoller [83].

2.1   The DNA

A complete description of each living organism is contained in its *genome*. This can be thought of as a "program", in a very special language, describing the set of instructions to be followed by the organism in order to grow and fully develop to its final form. The language used by nature to encode life is represented by the *DNA*.

The *deoxyribonucleic acid* (DNA) is present in each of the organism's cells, and consists of two linear sequences (*strands*) of tightly coiled threads of *nucleotides*. Each nucleotide is a molecule composed by one sugar, one phosphate and one nitrogen-containing chemical, called a *base*. Four different bases are present in the DNA, namely *Adenine* (A), *Thymine* (T), *Cytosine* (C) and *Guanine* (G). The particular order of the bases is called the *DNA sequence* and varies among different organisms. The sequence specifies the precise genetic instructions to create a particular form of life with its own unique traits.

The two DNA strands are inter-twisted in a typical helix form, and held together by bonds between the bases on each strand, forming the so called *base pairs*. A complementarity law relates one strand to its opposite, the only admitted pairings being Adenine with Thymine (A $\leftrightarrow$ T) and Cytosine with Guanine (C $\leftrightarrow$ G). Thanks to this complementarity, the DNA has the ability to replicate itself. Each time a cell divides into two new cells, its full genome is duplicated: First, the DNA molecule unwinds and the bonds between the base pairs break (figure 13.2.1). Then, each strand directs the synthesis of a complementary new strand, by matching up free nucleotides floating in the cell with their complementary bases on each of the separated strands. The complementarity rules should ensure that the new genome is an exact copy of the old one. However, although very reliable, this process is not completely error-free and there is the possibility that some bases are lost, duplicated or simply changed. Variations

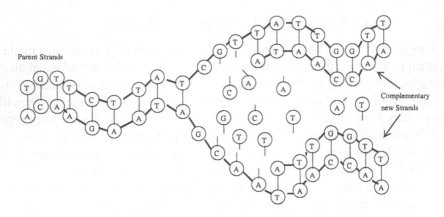

Figure 13.1.   Schematic DNA replication

to the original content of a DNA sequence are called *mutations* and can affect the resulting organism or its offspring. In many cases mutations are deadly. In some other cases, they can be completely harmless, or lead, in the long run, to the evolution of a species.

In humans the DNA in each cell contains about $3 \cdot 10^9$ base pairs. Human cells are *diploid*, i.e. their DNA is organized in *pairs* of *chromosomes*. The two chromosomes that form a pair are called *homologous*. One copy of each pair is inherited from the father and the other copy from the mother. The cells of some lower organisms, such as fungi and single-celled algae, may have only a single set of chromosomes. Cells of this type are called *haploid*.

2.2   Genes and proteins

The information describing an organism is encoded in its genome by means of a universal code, called the *genetic code*. This code is used to describe how to build the *proteins*, which provide the structural components of cells and tissues, as well as the *enzymes* needed for essential biochemical reactions.

Not all of the DNA sequence contains coding information (as a matter of fact, only a small fraction does). The information-coding DNA regions are organized into *genes*, where each gene is responsible for the coding of a different protein. Regions of the DNA sequences containing genes are called *exons*, since they are "expressed" by the mechanism which builds the proteins. Conversely, the *introns* are non-coding regions whose functions are still obscure.

The gene is the unit of heredity, which, together with many other such units, is transmitted from parents to offspring. Each gene, acting either alone or with other genes, is responsible for one or more characteristics of the resulting organism. Each characteristic can take many distinct values, called *alleles*. For a simplistic example, the alleles for the eye-color could be {black, blue, green, hazel, brown}.

Genes occur on the chromosomes, at specific positions called their *loci*. In diploid organisms, for each gene on one chromosome, there is a corresponding similar gene on the homologous chromosome. Thanks to this redundancy, sometimes an organism may survive even when one copy of a gene is defective, provided the other is not. A pair of corresponding genes can consist of the same allele on both chromosomes, or of different alleles. In the first case we say the organism is *homozygous* for the gene, while in the second it is *heterozygous*. The human genome is estimated to comprise between $30,000$ and $50,000$ genes, whose size ranges from a thousand to hundreds of thousands of base pairs.

A *protein* is a large molecule, consisting of one or more linear chains of *amino acids*, folded into a characteristic 3-dimensional shape called the protein's *native state*. The linear order of the amino acids is determined by the sequence of nucleotides in the gene coding for the protein. There exist 20 amino acids, each of which is identified by some triplets of DNA letters. The DNA triplets are also called *codons* and the correspondence of codons with amino acids is the *genetic code*. Since there are $4^3 = 64$ possible triplets of nucleotides, most amino acids are coded by more than one triplet (called *synonyms* for that amino acid), whereas a few are identified by a unique codon. For example, the amino acid Proline (Pro) is represented by the four triplets CCT, CCC, CCA and CCG. The code redundancy reduces the probability that random mutations of a single nucleotide can cause harmful effects. For instance, any mutation in the third base of Proline would result in a codon still correctly identifying the amino acid.

In order to build a protein, the DNA from a gene is first copied onto the *messenger RNA* (mRNA), which will serve as a template for the protein synthesis. This process is termed *transcription*. Only exons are copied, while the introns are spliced out from the molecule. Then the mRNA moves out of the cell's nucleus, and some cellular components named *ribosomes* start to read its triplets sequentially, identify the corresponding amino acids and link them so as to form the protein. The process of translation of genes into proteins, requires the presence of signals to identify the beginning and the end of each information-coding region. To this purpose, there are codons that specifically determine where a gene begins and where it terminates.

2.3 Some experiments in molecular biology

We now describe some common experiments performed in molecular biology labs, which are relevant to the problems described in this survey.

- **Cloning in-vivo and Polymerase Chain Reaction.**

Since most experiments cannot be conducted on a single copy of a DNA region, the preliminary step for further analysis consists in making a large quantity of copies (*clones*) of the DNA to be studied. This process is referred to as *cloning* or *DNA amplification*. There are two basic cloning procedures, namely *cloning in-vivo* and *Polymerase Chain Reaction* (PCR). Cloning in-vivo, consists in inserting the given DNA fragments into a living organism, such as bacteria and yeast cells. After the insertion, when the host cells are are naturally duplicated, the foreign DNA gets duplicated as well. Finally, the clones can be extracted from the new cells.

PCR is an automatic technique by which a given DNA sequence can be amplified hundreds of millions of times within a few hours. In order to amplify a double-stranded DNA sequence by PCR, one needs two short DNA fragments (*primers*) from the ends of the region of interest. The primers are put in a mixture containing the region of interest, free DNA bases and a specialized polymerase enzyme. The mixture is heated and the two strands separate. Later, the mixture is cooled and the primers bind to their complementary sequences on the separated strands. Finally, the polymerase enzyme synthesizes new complementary strands by extending the primers. By repeating this cycle of heating and cooling, an exponential number of copies of the target DNA sequence can be readily obtained.

- **Gel Electrophoresis.** Gel electrophoresis is an experiment by which DNA fragments can be separated according to their size, which can then be estimated with high precision. A large amount of DNA fragments are put in an agarose gel, to which an electric field is applied. Under the field, the fragments migrate in the gel, moving with a speed inversely proportional to their size. After some time, one can compare the position of the fragments in the gel to the position of a sample molecule of known size, and derive their size by some simple computation.

- **(Fragment) Sequencing.** *Sequencing* is the process of reading out the ordered list of bases from a DNA fragment. Due to technological limitations, it is impossible to sequence fragments longer than a few hundred base pairs. One technique for sequencing is as follows. Given a fragment which has been amplified, say, by PCR, the copies are cut randomly at all positions. This way one obtains a set of nested fragments, differing in length by exactly one nucleotide. The specific base at the end of each successive fragment is then detected, after the fragments have been separated by gel electrophoresis. The machine to sequence DNA is called a *sequencer*. To each output base, the sequencer attaches a value (*confidence level*) which represents the probability that the base has been determined correctly.

- **(Genome) Shotgun Sequencing.**

In order to sequence a long DNA molecule (e.g., a whole genome), this must first be amplified into many copies, and then be broken, at random, into several fragments, of about 1,000 nucleotides each, which are individually fed to a sequencer. The cloning phase is necessary so that the fragments can have nonempty overlap. From the overlap of two fragments one may infer a longer fragment, and so on, until the original DNA sequence has been reconstructed. This is, in essence, the principle of *Shotgun Sequencing*, in which the fragments are *assembled* back into the original sequence by using sophisticated algorithms and powerful computers.

The *assembly* (i.e. overlap and merge) phase is complicated by the fact that in a genome there exist many regions with identical content (*repeats*) scattered all around and due to replicating events during evolution. The repeats may be confused by the assembler to be all copies of a same, unique, region. To partly overcome the problems of repeats, some fragments used in shotgun sequencing may have some extra information attached. In particular, some fragments can be obtained by a process that generates pairs (called *mate pairs*) of fragments instead of individual ones. Each pair is guaranteed to come from the same copy of a chromosome and to have a given distance between its elements. A pair of mates can help since, even if one of them comes from a repeat region, there is a good chance that the other does not.

3. Alignment Problems

Oftentimes, in Computational Biology, one must compare objects which consist of a set of elements arranged in a linearly ordered structure. A typical example is the genomic sequence, which, as we saw, is equivalent to a string. Another example is the protein, when this is regarded as a linear chain of amino acids (and hence with a *first* and a *last* amino acid).

Aligning two (or more) such objects consists in determining subsets of corresponding elements in each. The correspondence must be order-preserving, i.e., if the i-th element of object 1 corresponds to the k-th element of object 2, no element following i in object 1 can correspond to an element preceding k in object 2.

The situation can be described graphically as follows. Given two objects, where the first has n elements, numbered $1, \ldots, n$ and the second has m ele-

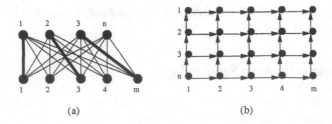

Figure 13.2.   (a) A noncrossing matching (alignment). (b) The directed grid.

ments, numbered $1, \ldots, m$, we consider the complete bipartite graph

$$W_{nm} := ([n], [m], L) \tag{13.1}$$

where $L = [n] \times [m]$ and $[k] := \{1, \ldots, k\} \; \forall k \in Z^+$.

We hereafter call a pair (i, j) with $i \in [n]$ and $j \in [m]$ a *line* (since it *aligns* i with j), and we denote it by $[i, j]$. L is the set of all lines. Two lines $[i, j]$ and $[i', j']$, with $[i, j] \neq [i', j']$, are said to *cross* if either $i' \geq i$ and $j' \leq j$ or $i' \leq i$ and $j' \geq j$ (by definition, a line does not cross itself). Graphically, crossing lines correspond to lines that intersect in a single point. A matching is a subset of lines no two of which share an endpoint. An alignment is identified by a *noncrossing matching*, i.e., a matching for which no two lines cross, in W_{nm}. Figure 13.2(a) shows (in bold) an alignment of two objects.

In the following Sections 13.3.1, 13.3.2 and 13.3.3, we will describe Mathematical Programming approaches for three alignment problems, i.e., Sequences vs. Sequences (Reinert et al. [75], Kececioglu et al. [50]), Sequences vs. Structures (Lenhof et al. [60], Kececioglu et al. [50]), and Structures vs. Structures (Lancia et al. [53], Caprara and Lancia [20]) respectively. All Integer Programming formulations for these problems contain binary variables x_l to select which lines $l \in L$ define the alignment, and constraints to insure that the selected lines are noncrossing. Therefore, it is appropriate to discuss here this part of the models, since the results will apply to all the following formulations. We will consider here the case of two objects only. This case can be generalized to more objects, as mentioned in Section 13.3.1.

Let X_{nm} be the set of incidence vectors of all noncrossing matchings in W_{nm}. A noncrossing matching in W_{nm} corresponds to a stable set in a new graph, G_L (the *Line Conflict Graph*), defined as follows. Each line $l \in L$ is a vertex of G_L, and two vertices l and h are connected by an edge if the lines l and h cross.

The graph G_L has been studied in Lancia et al. [53], where the following theorem is proved:

THEOREM 13.1 *([53]) The graph G_L is perfect.*

A complete characterization of conv(X_{nm}) in terms of linear inequalities, is given –besides non-negativity– by the following *clique inequalities*:

$$\sum_{l \in Q} x_l \leq 1 \qquad \forall Q \in \text{clique}(L). \qquad (13.2)$$

where clique(L) denotes the set of all cliques in G_L. Given weights x_l^* for each line l, the separation problem for the clique inequalities (13.2) consists in finding a clique Q in G_L with $x^*(Q) > 1$. From Theorem 13.1, we know that this problem can be solved in polynomial time by finding the maximum x^*-weight clique in G_L. Instead of resorting to slow and complex algorithms for perfect graphs, the separation problem can be solved directly by Dynamic Programming, as described in Lenhof et al. [60] and, independently and with a different construction, in Lancia et al. [53]. Both algorithms have complexity $O(nm)$. We describe here the construction of [60].

Consider L as the vertex set of a directed grid, in which vertex $[1, n]$ is put in the lower left corner and vertex $[m, 1]$ in the upper right corner (see Figure 13.2(b)).

At each internal node $l = [i, j]$, there are two incoming arcs, one directed from the node below, i.e. $[i, j + 1]$, and one directed from the node on the left, i.e. $[i - 1, j]$. Each such arc has associated a length equal to x_l^*.

THEOREM 13.2 *([60, 53]) The nodes on a directed path P from $[1, n]$ to $[m, 1]$ in the grid correspond to a clique Q, of maximal cardinality, in G_L and vice versa.*

Then the most violated clique inequality can be found by taking the longest $[1, n]$-$[m, 1]$ path in the grid. There is a violated clique inequality if and only if the length of the path (plus x_{1n}^*) is greater than 1.

Closely related to the problem of finding a largest-weight clique in G_L is the problem of finding a largest-weight stable set in G_L (i.e., a maximum weighted noncrossing matching in W_{nm}). Also for this problem, there is a Dynamic Programming solution, of complexity $O(nm)$. The basic idea is to compute the matching recursively. Let $V(a, b)$ be the value of the maximum w-weight noncrossing matching of nodes $\{1, \ldots, a\}$ with nodes $\{1, \ldots, b\}$, where each line $[i, j]$ has a weight w_{ij}. Then, in the optimal solution, either a is matched with b, and $V(a, b) = w_{ab} + V(a - 1, b - 1)$, or one of a and b is unmatched, in which case $V(a, b) = \max\{V(a - 1, b), V(a, b - 1)\}$. With a look-up table, $V(n, m)$ can be computed in time $O(nm)$. Incidentally, this algorithm coincides with the basic algorithm for computing the *edit distance* of two strings (a problem described in the next section), rediscovered independently by many authors, among which Smith and Waterman [78]. In [20], Caprara and Lancia solve the Lagrangian Relaxation of a model for aligning two protein structures, by reducing it to finding the largest weighted noncrossing matching in a suitable graph. This is described in Section 13.3.3.

3.1   Sequence vs Sequence Alignments

Comparing genomic sequences drawn from individuals of the same or different species is one of the fundamental problems in molecular biology. In fact, such comparisons are needed to identify highly conserved (and therefore presumably functionally relevant) DNA regions, spot fatal mutations, suggest evolutionary relationships, and help in correcting sequencing errors.

A genomic sequence can be represented as a string over an alphabet Σ consisting of either the 4 nucleotide letters or the 20 letters identifying the 20 amino acids. Aligning a set of sequences (i.e., computing a *Multiple Sequence Alignment*) consists in arranging them in a matrix having each sequence in a row. This is obtained by possibly inserting gaps (represented by the '−' character) in each sequence so that they all result of the same length. The following is a simple example of an alignment of the sequences ATTCCGAC, TTCCCTG and ATCCTC. The example highlights that the pattern TCC is common to all sequences.

```
A   T   T   C   C   G   A   -   C
-   T   T   C   C   C   -   T   G
A   -   T   C   C   -   -   T   C
```

By looking at a column of the alignment, we can reconstruct the events that have happened at the corresponding position in the given sequences. For instance, the letter G in sequence 1, has been *mutated* into a C in sequence 2, and *deleted* in sequence 3.

The multiple sequence alignment problem has been formalized as an optimization problem. The most popular objective function for multiple alignment generalizes ideas from optimally aligning two sequences. This problem, called *pairwise alignment*, is formulated as follows: Given symmetric costs (or, alternatively, profits) $\gamma(a, b)$ for replacing a letter a with a letter b and costs $\gamma(a, -)$ for deleting or inserting a letter a, find a minimum-cost (respectively, maximum-profit) set of symbol operations that turn a sequence S' into a sequence S''. For genomic sequences, the costs $\gamma(\cdot, \cdot)$ are usually specified by some widely used *substitution matrices* (e.g., PAM and BLOSUM), which score the likelihood of specific letter mutations, deletions and insertions. The pairwise alignment problem can be solved by Dynamic Programming in time and space $O(l^2)$, where l is the maximum length of the two sequences, by Smith and Waterman's algorithm [78]. The value of an optimal solution is called the *edit distance* of S' and S'' and is denoted by $d(S', S'')$.

Formally, an *alignment* $\mathcal{A}$ of two or more sequences is a bidimensional array having the (gapped) sequences as rows. The value $d_{\mathcal{A}}(S', S'')$ of an alignment of two sequences S' and S'' is obtained by adding up the costs for the pairs of characters in corresponding positions, and it is easy to see that $d(S', S'') = \min_{\mathcal{A}} d_{\mathcal{A}}(S', S'')$. This objective is generalized, for k sequences $\{S_1, \ldots, S_k\}$, by the *Sum–of–Pairs* (SP) score, in which the cost of an alignment is obtained by adding up the costs of the symbols matched up at the same positions, over all the pairs of sequences,

$$SP(\mathcal{A}) := \sum_{1 \leq i < j \leq k} d_{\mathcal{A}}(S_i, S_j) = \sum_{1 \leq i < j \leq k} \sum_{l=1}^{|\mathcal{A}|} \gamma(\mathcal{A}[i][l], \mathcal{A}[j][l]) \quad (13.3)$$

where $|\mathcal{A}|$ denotes the length of the alignment, i.e., the number of its columns.

Finding the optimal SP alignment was shown to be NP-hard by Wang and Jiang [81]. A straightforward generalization of the Dynamic Programming from 2 to k sequences, of length l, leads to an exponential-time algorithm of complexity $O(2^k l^k)$. In typical real-life instances, while k can be possibly small (e.g., around 10), l is in the order of several hundreds, and the Dynamic Programming approach turns out to be infeasible for all but tiny problems.

As an alternative approach to the ineffective Dynamic Programming algorithm, Kececioglu et al. [50] (see also Reinert et al. [75] for a preliminary version) proposed an approach based on Mathematical Programming to optimize an objective function called the Maximum Weight Trace (MWT) Problem. The MWT generalizes the SP objective as well as many other alignment problems in the literature. The problem was introduced in 1993 by Kececioglu [49], who described a combinatorial Branch-and-Bound algorithm for its solution.

3.1.1 Trace and Mathematical Programming approach.

The *trace* is a graph theoretic generalization of the noncrossing matching to the case of the alignment of more than two objects. Suppose we want to align k objects, of $n_1, n_2, \ldots, n_k$ elements. We define a complete k-partite graph $W_{n_1, \ldots, n_k} = (V, L)$, where $V = \bigcup_{i=1}^{k} V_i$, the set $V_i = \{v_{i1}, \ldots, v_{in_i}\}$ denotes the nodes of level i (corresponding to the elements of the i-th object), and $L = \bigcup_{1 \leq i < j \leq k} V_i \times V_j$. Each edge $[i, j] \in L$ is still called a line. Given an alignment $\mathcal{A}$ of the objects, we say that the alignment *realizes* a line $[i, j]$ if i and j are put in the same column of $\mathcal{A}$. A trace T is a set of lines such that there exists an alignment $\mathcal{A}$ that realizes all the lines in T. The Maximum Weight Trace (MWT) Problem is the following: given weights $w_{[i,j]}$ for each line $[i, j]$, find a trace T of maximum total weight.

Let A be the set of directed arcs pointing from each element to the elements that follow it in an object (i.e., arcs of type (v_{is}, v_{it}) for $i = 1, \ldots, k, 1 \leq s < t \leq n_i$). Then a trace T is characterized as follows:

PROPOSITION 13.3 *([75]) T is a trace if and only if there is no directed mixed cycle in the mixed graph $(V, T \cup A)$.*

A directed mixed cycle is a cycle containing *both* arcs and edges, in which the arcs (elements of A) are traversed according to their orientation, while the undirected edges (lines of T) can be traversed in either direction.

Using binary variables x_l for the lines in L, the MWT problem can be modeled as follows [50]:

$$\max \sum_{l \in L} w_l x_l \tag{13.4}$$

subject to

$$\sum_{l \in C \cap L} x_l \leq |C \cap L| - 1 \qquad \forall \text{ directed mixed cycles } C \text{ in } (V, L \cup A) \tag{13.5}$$

$$x_l \in \{0, 1\} \qquad \forall l \in L. \tag{13.6}$$

The polytope $\mathcal{P}$ defined by (13.5) and $0 \leq x_l \leq 1$, $l \in L$, is studied in [50], and some classes of facet-defining inequalities are described. Among them, are the clique inequalities (13.2), relative to each subgraph $W_{n_i, n_j} := (V_i, V_j, V_i \times V_j)$ of $W_{n_1, \ldots, n_k}$, and the following subset of (13.5), called *mixed-cycle* inequalities. Given a directed mixed cycle C, a line $l = [i, j]$ is a *chord* of C if $C_1 \cup \{l\}$ and $\{l\} \cup C_2$ are directed mixed cycles, where C_1 and C_2 are obtained by splitting C at i and j.

LEMMA 13.4 *([50]) Let C be a directed mixed cycle of $(V, L \cup A)$. Then the inequality*

$$x(C \cap L) \leq |C \cap L| - 1$$

is facet-defining for $\mathcal{P}$ if and only if C has no chord.

The mixed-cycle inequalities can be separated in polynomial time, via the computation of at most $O(\sum_{i=1}^{k} n_i)$ shortest paths in $(V, L \cup A)$, [50]. A Branch-and-Cut algorithm based on the mixed-cycle inequalities and the clique inequalities (as well as other cuts, separated heuristically) allowed Kececioglu et al. [50] to compute, for the first time, an optimal alignment for a set of 15 proteins of about 300 amino acids each from the SWISSPROT database, whereas the limit for the combinatorial version of Kececioglu's algorithm for this problem was 6 sequences of size 200 [49]. We remark the the optimal alignment of 6 sequences of size 200 is already out of reach for Dynamic Programming-based algorithms. We must also point out, however, that the length of the sequences is still a major limitation to the applicability of the Integer Programming solution, and the method is not suitable for sequences longer than a few hundred letters.

3.1.2   Heuristics for SP alignment.   Due to the complexity of the alignment problem, many heuristic algorithms have been developed over time, some of which provide an approximation-guarantee on the quality of the solution found. In this section we describe one such heuristic procedure, which uses ideas from Network Design and Integer Programming techniques. This heuristic is well suited for long sequences.

A popular approach for many heuristics is the so called "progressive" alignment, in which the solution is incrementally built by considering the sequences one at a time. The most effective progressive alignment methods proceed by first finding a tree whose node set spans the input sequences, and then by using the tree as a guide for aligning the sequences iteratively. One such algorithm is due to Gusfield [37], who suggested to use a *star* (i.e., a tree in which at most one node -the *center*- is not a leaf), as follows. Let $S = \{S_1, \ldots, S_k\}$ be the set of input sequences. Fix a node (sequence) S_c as the center, and compute $k - 1$ pairwise alignments $\mathcal{A}_i$, one for each each S_i aligned with S_c. These alignments can then be merged into a single alignment $\mathcal{A}(c)$, by putting in the same column two letters if they are aligned to the same letter of S_c. The following is an example of the alignments of $S_1 = $ ATGTC and $S_2 = $ CACGG with $S_c = $ ATCGC and their merging.

				S_1	-AT-GTC
S_1	AT-GTC	S_2	CA-CGG	S_2	CA-CG-G
S_c	ATCG-C	S_c	-ATCGC	S_c	-ATCG-C

Note that each $\mathcal{A}_i$ realizes the edit distance of S_i to S_c, and it is true also in $\mathcal{A}(c)$ that each sequence is aligned optimally with S_c, i.e.

$$d_{\mathcal{A}(c)}(S_i, S_c) = d_{\mathcal{A}_i}(S_i, S_c) = d(S_i, S_c) \quad \forall i. \tag{13.7}$$

Furthermore, for all pairs i, j, the triangle inequality $d_{\mathcal{A}(c)}(S_i, S_j) \leq d(S_i, S_c) + d(S_j, S_c)$ holds, since the distance in an alignment is a metric over the sequences (assuming the cost function γ is a metric over Σ). Let c^* be such that $SP(\mathcal{A}(c^*)) = \min_r SP(\mathcal{A}(r))$. Call OPT be the optimal SP value, and HEUR $:= SP(\mathcal{A}(c^*))$. By comparing $\sum_{1 \leq i < j \leq k} d(S_i, S_j)$ (a lower bound to OPT) to $\sum_{1 \leq i < j \leq k} (d(S_i, S_{c^*}) + d(S_j, S_{c^*}))$ (an upper bound to HEUR), Gusfield proved the following 2-approximation result:

THEOREM 13.5 *([37]) HEUR $\leq \left(2 - \frac{2}{k}\right)$ OPT.*

In his computational experiments, Gusfield showed that the above approximation ratio is overly pessimistic on real data, and that his star-alignment heuristic was finding solutions within 15% from optimum on average.

Gusfield's algorithm can be generalized so as to use *any* tree, instead of just a star, and still maintain the 2-approximation guarantee. The idea was first described in Wu et al. [85], in connection to a Network Design problem, i.e., the Minimum Routing Cost Tree (MRCT). Given a complete weighted graph of k vertices, and a spanning tree T, the *routing cost* of T is defined as the sum of the weights of the $\binom{k}{2}$ paths, between all pairs of nodes, contained in T. This value is denoted by $r(T)$. Computing the minimum routing cost tree is NP-hard, but a Polynomial Time Approximation Scheme (PTAS) is possible, and is described in [85].

To relate routing cost and alignments, we start by considering the set of sequences S as the vertex set of a complete weighted graph G, where the weight of an edge $S_i S_j$ is $d(S_i, S_j)$. The following procedure, attributed by folklore to Feng and Doolittle, shows that, for any spanning tree T, we can build an alignment $\mathcal{A}(T)$ which is optimal for $k - 1$ of the the $\binom{k}{2}$ pairs of sequences. A sketch of the procedure is as follows: (i) pick an edge $S_i S_j$ of the tree and align recursively the two sets of sequences induced by the cut defined by the edge, obtaining two alignments, say $\mathcal{A}_i$ and $\mathcal{A}_j$; (ii) align optimally S_i with S_j; (iii) merge $\mathcal{A}_i$ and $\mathcal{A}_j$ into a complete solution, by aligning the columns of $\mathcal{A}_i$ and $\mathcal{A}_j$ in the same way as the letters of S_i are optimally aligned to the letters of S_j.

From this procedure, the validity of the next theorem follows.

THEOREM 13.6 *For any spanning tree $T = (S, E)$ of G, there exists an alignment $\mathcal{A}(T)$ such that $d_{\mathcal{A}(T)}(S_i, S_j) = d(S_i, S_j)$ for all pairs of sequences $S_i S_j \in E$.*

Furthermore, the alignment $\mathcal{A}(T)$ can be easily computed, as outlined above. By triangle inequality, it follows

COROLLARY 13.7 *For any spanning tree $T = (S, E)$ of G, there exists an alignment $\mathcal{A}(T)$ such that $SP(\mathcal{A}(T)) \leq r(T)$.*

From Corollary 13.7, it follows that a good tree to use for aligning the sequences is a tree T^* such that $r(T^*) = \min_T r(T)$. This tree does not guarantee an optimal alignment, but guarantees the *best possible upper bound* to the alignment value, for an alignment obtained from a tree (such as Gusfield's alignment). In [28], Fischetti et al. describe a Branch-and-Price algorithm for finding the minimum routing cost tree in a graph, which is effective for up to $k = 30$ nodes (a large enough value for alignment problems). The algorithm is based on a formulation with binary variables x_P for each path P in the graph, and constraints that force the variables set to 1 to be the $\binom{k}{2}$ paths induced by a tree. The pricing problem is solved by computing $O(k^2)$ nonnegative shortest paths. Fischetti et al. [28] report on computational experiments of their algorithm applied to alignment problems, showing that using the best routing cost tree instead of the best star produces alignments that are within 6% of optimal on average.

3.1.3 Consensus Sequences.

Given a set $\mathcal{S} = \{S_1, \ldots, S_k\}$ of sequences, an important problem consists in determining their *consensus* i.e., a new sequence which agrees with all the given sequences as much as possible. The consensus, in a way, represents all the sequences in the set. Let C be the consensus. Two possible objectives for C are either *min-sum* (the total distance of C from all sequences in S is minimum) or *min-max* (the maximum distance of C from any sequence in S is minimum). Both objectives are NP-hard ([30, 77]).

Assume all input sequences have length n. For most consensus problems instead of using the edit distance, the *Hamming distance* is preferable (some biological reasons for this can be found in [54]). This distance is defined for sequences of the same length, and weighs all the positions at which there is a mismatch. If C is the consensus, of length n, the distance of a sequence S from C is defined as $\sum_{i=1}^{n} \gamma(S[i], C[i])$, where $X[i]$ denotes the i–th symbol of a sequence X.

One issue with the min-sum objective is that, when the data are biased, the consensus tends to be biased as well. Consider the following situation. A biologist searches a genomic data base for all sequences similar to a newly sequenced protein. Then, he computes a consensus of these sequences to highlight their common properties. However, since genomic data bases contain mostly human sequences, the search returns human sequences more than any other species, and they tend to dominate the definition of the consensus. In this case, an unbiased consensus should optimize the min-max objective. For this problem, an approximation algorithm based on an Integer Programming formulation and Randomized Rounding has been proposed by Ben-Dor et al [11].

The Integer Programming formulation for the consensus problem has a binary variable $x_{i,\sigma}$, for every symbol $\sigma \in \Sigma$, and every position i, $1 \leq i \leq n$, to indicate whether $C[i] = \sigma$. The model reads:

$$\min \ r \tag{13.8}$$

subject to

$$\sum_{\sigma \in \Sigma} x_{i,\sigma} = 1 \qquad \forall i = 1, \ldots, n \tag{13.9}$$

$$\sum_{i=1}^{n} \sum_{\sigma \in \Sigma} \gamma(\sigma, S_j[i]) x_{i,\sigma} \leq r \qquad \forall j = 1, \ldots, k \tag{13.10}$$

$$x_{i,\sigma} \in \{0, 1\} \qquad \forall i = 1, \ldots, n, \ \forall \sigma \in \Sigma. \tag{13.11}$$

Let OPT denote the value of the optimum solution to the program above. An approximation to OPT can be obtained by Randomized Rounding [74]. Consider the Linear Programming relaxation of (13.8)-(13.11), obtained by replacing (13.11) with $x_{i,\sigma} \geq 0$. Let r^* be the LP value and $x^*_{i,\sigma}$ be the LP optimal solution, possibly fractional. An integer solution $\bar{x}$ can be obtained from x^* by *randomized rounding*: independently, at each position i, choose a letter $\sigma \in \Sigma$ (i.e., set $\bar{x}_{i,\sigma} = 1$ and $\bar{x}_{i,\tau} = 0$ for $\tau \neq \sigma$) with probability of $\bar{x}_{i,\sigma} = 1$ given by $x^*_{i,\sigma}$. Let HEUR be the value of the solution $\bar{x}$, and $\Gamma = \max_{a,b \in \Sigma} \gamma(a, b)$. The first approximation result for the consensus was given in [11], and states that, for any constant $\epsilon > 0$,

$$Pr\left(\text{HEUR} > \text{OPT} + \Gamma\sqrt{3\,\text{OPT}\log\frac{k}{\epsilon}}\right) < \epsilon. \qquad (13.12)$$

This probabilistic algorithm can be de-randomized using standard techniques of conditional probabilities ([73]). Experiments with the exact solution of (13.8)-(13.11) are described in Gramm et al. [35].

The above IP model is utilized in a string of papers on consensus as well as other, related, problems [54, 61, 64], such as finding the *farthest* sequence (i.e., a sequence "as dissimilar as possible" from each sequence of a given set S).

The main results obtained on these problems are PTAS based on the above IP formulation and Randomized Rounding, coupled with random sampling. In [61] and [64], Li et al. and Ma respectively, describe PTAS for finding the consensus, or a consensus substring. The main idea behind the approach is the following. Given a subset of r sequences from S, line them up in an $r \times n$ array, and consider the positions where they all agree. Intuitively, there is a high likelihood that the consensus should also agree with them at these positions. Hence, all one needs to do is to optimize on the positions where they do not agree, which is done by LP relaxation and randomized rounding.

A PTAS for the farthest sequence problem, by Lanctot et al., is described in [54]. Also this result is obtained by LP relaxation and Randomized Rounding of an IP similar to (13.8)-(13.11).

3.2   Sequence vs Structure Alignments

Contrary to DNA molecules, which are double-stranded, RNA molecules are single-stranded and the exposed bases tend to form hydrogen bonds within the same molecule, according to Watson-Crick pairing rules. This bonds lead to structure formation, also called *secondary structure* of the RNA sequence. As an example, consider the following RNA sequence, $S = \text{UCGUGCGGUAACUUCCACGA}$. Since the two ends of S are self-complementary, part of the sequence may self-hybridize, leading to the formation of a so-called *loop*, illustrated below:

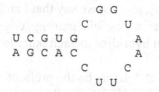

```
            G  G
         C        U
   U C G U G            A
   A G C A C            A
         C        C
            U  U
```

Determining the secondary structure from the nucleotide sequence is not an easy problem. For instance, complementary bases may not hybridize if they are not sufficiently apart in the molecule. It is also possible that not all the feasible Watson-Crick pairings can be realized at the same time, and Nature chooses the most favorable ones according to an energy minimization which is not yet well understood.

All potential pairings can be represented by the edge set of a graph $G_S = (V_S, E_S)$. The vertex set $V_S = \{v_1, \ldots, v_n\}$ represents the nucleotides of S, with v_i corresponding to the i-th nucleotide. Each edge $e \in E_S$ connects two nucleotides that may possibly hybridize to each other in the secondary structure. For instance, the graph of Figure 13.3 describes the possible pairings for sequence S, assuming a pairing can be achieved only if the bases are at least 8 positions apart from each other. Only a subset $E'_S \subseteq E_S$ of pairings is realized by the actual secondary structure. This subset is drawn in bold in Figure 13.3. The graph $G'_S = (V_S, E'_S)$ describes the secondary structure. Note that each node in G'_S has degree ≤ 1.

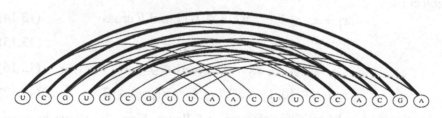

Figure 13.3. Graph of RNA secondary structure

Two RNA sequences that look quite different as strings of nucleotides, may have similar secondary structure. A generic sequence alignment algorithm, such as those described in the previous section, would not spot the structure similarity, and hence a different model for comparison is needed for this case.

One possible model was proposed in Lenhof et al. [60] (see also Kececioglu et al. [50]). In this model, a sequence U of unknown secondary structure is compared to a sequence K of known secondary structure. This is done by comparing the graph $G_U = (V_U, E_U)$ to $G'_K = (V_K, E'_K)$.

Assume $V_U = [n]$ and $V_K = [m]$. An alignment of G_U and G'_K is defined by a noncrossing matching of W_{nm} (which, recalling (13.1), is the complete bipartite graph (V_U, V_K, L)) Given an edge $e = ij \in E_U$ and two noncrossing lines $l = [i, u] \in L$, $h = [j, v] \in L$, we say that l and h *generate* e if $uv \in E'_K$. Hence, two lines generate a possible pairing (edge of E_U) if they align its endpoints to elements that hybridize to each other in the secondary structure of K.

For each line $l = [i, j] \in L$ let w_l be the profit of aligning the two endpoints of the line (i.e., $w_l = \gamma(U[i], K[j])$). Furthermore, for each edge $e \in E_U$, let w_e be a nonnegative weight associated to the edge, measuring the "strength" of the corresponding bond. The objective of the RNA Sequence/Structure Alignment (RSA) problem is to determine an alignment which has maximum *combined* value: the value is given by the sum of the weights of the alignment lines (as for a typical alignment) and the weights of the edges in E_U that these lines generate.

Let $\prec$ be an arbitrary total order defined over the set of all lines L. Let $\mathcal{G}$ be the set of pairs of lines (p, q) with $p \prec q$, such that each pair $(p, q) \in \mathcal{G}$ generates an edge in E_U. For each $(p, q) \in \mathcal{G}$, define $w_{pq} := w_e$, where $p = [i, u]$, $q = [j, v]$ and $e = ij \in E_U$.

The Integer Programming model in [60] for this alignment problem has variables x_l for lines $l \in L$ and y_{pq} for pairs of lines $(p, q) \in \mathcal{G}$:

$$\max \sum_{l \in L} w_l\, x_l + \sum_{(p,q) \in \mathcal{G}} w_{pq}\, y_{pq} \tag{13.13}$$

subject to

$$x_l + x_h \leq 1 \qquad \forall\, l, h \in L \mid l \text{ and } h \text{ cross} \tag{13.14}$$

$$y_{pq} \leq x_p \qquad \forall\, (p, q) \in \mathcal{G} \tag{13.15}$$

$$y_{pq} \leq x_q \qquad \forall\, (p, q) \in \mathcal{G} \tag{13.16}$$

$$x_l, y_{pq} \in \{0, 1\} \qquad \forall l \in L, \quad \forall (p, q) \in \mathcal{G}. \tag{13.17}$$

The model can be readily tightened as follows. First, the clique inequalities (13.2) can be replaced for the weaker constraints (13.14). Secondly, for a line $l \in L$, let $\mathcal{G}_l = \{(p, q) \in \mathcal{G} \mid p = l \vee q = l\}$. Each element (i, j) of $\mathcal{G}_l$ identifies a pairing $e \in E_U$ such that l is one of the generating lines of e. Note that if an edge $e \in E_U$ can be generated by l and $a \in L$ and, alternatively, by l and $b \neq a \in L$, then a and b must cross (in particular, they share an endpoint). Hence, of all the generating pairs in $\mathcal{G}_l$, at most one can be selected in a feasible solution. The natural generalization for constraints (13.15) and (13.16) is therefore

$$\sum_{(a,b) \in \mathcal{G}_l} y_{ab} \leq x_l \qquad \forall l \in L. \tag{13.18}$$

These inequalities are shown in [60] to be facet-defining for the polytope

$$\mathrm{conv}\{(x, y) \mid (x, y) \text{ satisfies } (13.14)\text{-}(13.17)\}.$$

A Branch-and-Cut procedure based on the above inequalities is described in [60, 50]. The method was validated by aligning RNA sequences of known structure to RNA sequences of known structure, but without using the structure information for one of the two sequences. In all cases tested, the algorithm retrieved the correct alignment, as computed by hand by the biologists. For comparison, a "standard" sequence alignment algorithm (which ignores structure in both sequences) was used, and failed to retrieve the correct alignments.

In a second experiment, the method was used to align RNA sequences of unknown secondary structure, of up to 1,400 nucleotides each, to RNA sequences of known secondary structure. The optimal solutions found were compared with alternative solutions found by "standard" alignment, and their merits were discussed. The results showed that structure information is essential in retrieving biologically relevant alignments.

It must however be remarked that, in aligning large sequences, not all the variables were present in the model. In particular, there were variables only for a relatively small subset of all possible lines (which would otherwise be more than a million). This subset was determined heuristically in a pre-processing phase meant at keeping only lines that have a good chance of being in an optimal solution.

3.3 Structure vs Structure Alignments

As described in Section 13.2.2, a protein is a chain of molecules known as amino acids, or *residues*, which folds into a peculiar 3-D shape (called its *tertiary* structure) under several molecular forces (entropic, hydrophobic, thermodynamic). A protein's fold is perhaps the most important of all protein's features, since it determines how the protein functions and interacts with other molecules. In fact, most biological mechanisms at the protein level are based on shape-complementarity, and proteins present particular concavities and convexities that allow them to bind to each other and form complex structures, such as skin, hair and tendon.

The comparison of protein structures is a problem of paramount importance in structural genomics, and an increasing number of approaches for its solution have been proposed over the past years (see Lemmen and Lengauer [59] for a survey). Several protein structure classification servers (the most important of which is the Protein Data Bank, PDB [13]) have been designed based on them, and are extensively used in practice.

Loosely speaking, the structure comparison problem is the following: *Given two 3-D protein structures, determine how similar they are.* Some of the reasons motivating the problem are:

1. *Clustering.* Proteins can be clustered in families based on structure similarity. Proteins within a family are functionally and evolutionarily closely related.

2. *Function determination.* The function of a protein can oftentimes be determined by comparing its structure to some known ones, whose function is already understood.

3. *Fold Prediction Assessment.* This is the problem faced by the CASP (Critical Assessment of Structure Prediction [65, 66]) jurors, in a bi-annual competition in which many research groups try to predict a protein structure from its amino acid sequence. Given a set of "tentative" folds for a protein, and a correct one (determined experimentally), the jurors need to decide which guess comes closest to the true answer.

Comparing two structures implies to "align" them in some way. Since, by their nature, three-dimensional computational problems are inherently more complex than the similar one-dimensional ones, there is a dramatic difference between the complexity of two-sequences alignment and two-structures alignment. Not surprisingly, various simplified versions of the structure comparison problems were shown NP-hard [34].

Pairwise structure comparison requires a structure similarity scoring scheme that captures the biological relevance of the chemical and physical constraints involved in molecular recognition. Determining a satisfactory scoring scheme is still an open question. The most used schemes, follow mainly three themes: *RMSD (Root Mean Square Deviation) of rigid-body superposition* [47], *distance map similarity* [44] and *contact map overlap* (CMO) [32]. All these similarity measures use distances between amino acids and raise computational issues that at present do not have effective solutions.

Due to the inherent difficulty of the problem, most algorithms for structure comparison in the literature are heuristics of some sort, with a notable exception, as far as the the CMO measure is concerned. In fact, the CMO measure is the only one for which performance-guarantee approximation algorithms, as well as exact algorithms based on Mathematical Programming techniques, were proposed over the last few years. In the remainder of this subsection we will overview the main ideas underlying the exact algorithms for the CMO problem.

3.3.1   Contact Maps.   A *contact map* is a 2-D representation of a 3-D protein structure. When a proteins folds, two residues that were not adjacent in the protein's linear sequence, may end up close to each other in the 3-D space (Figure 13.4 (a) and (b)). The contact map of a protein with n residues is defined as a 0-1, symmetric, $n \times n$ matrix, whose 1–elements correspond to pairs of residues that are within a "small" distance (typically around 5Å) in

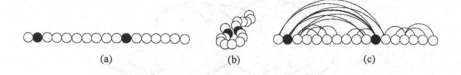

Figure 13.4. (a) An unfolded protein. (b) After folding. (c) The contact map graph.

the protein's fold, but are not adjacent in the linear sequence. We say that such a pair of residues are in *contact*. It is helpful to regard the contact map of a protein as the adjacency matrix of a graph G. Each residue is a node of G, and there is an edge between two nodes if the corresponding residues are in contact (see Figure 13.4 (c)).

The CMO problem tries to evaluate the similarity in the 3-D folds of two proteins by determining the similarity of their contact maps (the rationale being that a high contact map similarity is a good indicator of high 3-D similarity). This measure was introduced in [33], and its optimization was proved NP-hard in [34], thus justifying the use of sophisticated heuristics or enumerative methods.

Given two folded proteins, the CMO problem calls for determining an alignment between the residues of the first protein and of the second protein. The alignment specifies the residues that are considered equivalent in the two proteins. The goal is to find the alignment which highlights the largest set of common contacts. The value of an alignment is given by the number of pairs of residues in contact in the first protein which are aligned with pairs of residues that are also in contact in the second protein. Given the graph representation for contact maps, we can phrase the CMO problem in graph–theoretic language. The input consists of two undirected graphs $G_1 = (V_1, E_1)$ and $G_2 = (V_2, E_2)$, with $V_1 = [n]$ and $V_2 = [m]$. For each edge ij, with $i < j$, we distinguish a left endpoint (i) and a right endpoint (j), and therefore we denote the edge by the ordered pair (i, j). A solution of the problem is an alignment of V_1 and V_2, i.e., a noncrossing matching $L' \subseteq L$ in W_{nm}. Two edges $e = (i, j) \in E_1$ and $f = (i', j') \in E_2$ contribute a *sharing* to the value of an alignment L' if $l = [i, i'] \in L'$ and $h = [j, j'] \in L'$. In this case, we say that l and h *generate* the sharing (e, f). The CMO problem consists in finding an alignment which maximizes the number of sharings. Figure 13.5 shows two contact maps and an alignment with 5 sharings.

From the compatibility of lines one can derive a notion of compatible sharings. Let $e = (i, j), e' = (i', j') \in E_1$ and $f = (u, v), f' = (u', v') \in E_2$. The sharings (e, f) and (e', f') are said *compatible* if the following lines are non-crossing: $[i, u], [j, v], [i', u']$ and $[j', v']$. After defining a new graph G_S (the Sharings Conflict Graph), the CMO problem can be reduced to the STABLE

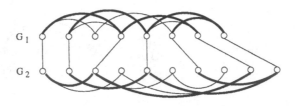

Figure 13.5.   An alignment of value 5

SET (SS) problem. In G_S, there is a node N_{ef} for each $e \in E_1$ and $f \in E_2$ and two nodes N_{ef} and $N_{e'f'}$ are connected by an edge if the sharings (e, f) and (e', f') are not compatible. A feasible set of sharings is a stable set in G_S, and the optimal solution to CMO is identified by the largest stable set in G_S.

3.3.2   Mathematical Programming approaches to the CMO problem.

Attacking the CMO problem as a SS problem is not viable, since the graph G_S is very large for real proteins, even of moderate size. In fact, a small protein has about 50 residues and 100 contacts, giving rise to a graph G_S with about 10,000 nodes, beyond the capabilities of the best codes for the SS problem [46].

The most successful approaches for the CMO solution rely on formulating the problem as an Integer Program and solving it by Branch-and-Bound. In two different approaches, the bound has been obtained either from the Linear Programming relaxation [53] of an Integer Linear Programming formulation, or from a Lagrangian relaxation of a Integer Quadratic Programming formulation [20]. Both formulations are based on binary variables x_l for each line l, and y_{pq}, for pairs of lines p, q which generate a sharing.

The first formulation of the problem was given by Lancia et al. [53]. This formulation is very similar to the formulation for the RSA problem, described in Section 13.3.2. For completeness, we restate it here. We denote with $\mathcal{G}$ the set of all pairs of lines that generate a sharing. For each pair $(l, t) \in \mathcal{G}$, where $l = [i, i']$, and $t = [j, j']$, the edge $(i, j) \in E_1$ and the edge $(i', j') \in E_2$. For a line $l \in L$, let $\mathcal{G}_l^- = \{(p, l) \in \mathcal{G}\}$ and $\mathcal{G}_l^+ = \{(l, p) \in \mathcal{G}\}$. The model in [53] reads

$$\max \sum_{(l,t) \in \mathcal{G}} y_{lt} \qquad (13.19)$$

subject to the clique inequalities (13.2) and the constraints

$$\sum_{(p,l) \in \mathcal{G}_l^-} y_{pl} \leq x_l \qquad \forall l \in L \qquad (13.20)$$

$$\sum_{(l,p) \in \mathcal{G}_l^+} y_{lp} \leq x_l \qquad \forall l \in L \qquad (13.21)$$

$$x_l, y_{pq} \in \{0, 1\} \qquad \forall l \in L, \quad \forall (p, q) \in \mathcal{G}. \qquad (13.22)$$

In fact, the CMO problem can be seen as a special case of the RSA problem (13.13)-(13.17), in which each line l has weight $w_l = 0$ and each pair (p, q) of generating lines have weight $w_{pq} = 1$. Another difference between the two problems is that in the RSA problem the degree of each node in G_1 is at most 1, while for the CMO problem this is not necessarily true.

A Branch-and-Cut procedure based on the above constraints, as well as other, weaker, cuts, was used in [53] to optimally align, for the first time, real proteins from the PDB. The algorithm was run on 597 protein pairs, with sizes ranging from 64 to 72 residues. Within the time limit of 1 hour per instance, 55 problems were solved optimally and for 421 problems (70 percent) the gap between the best solution found and the upper bound was ≤ 5, thereby providing a certificate of near-optimality. The feasible solutions, whose value was used as a lower bound to prune the search tree, were obtained via heuristics such as *Genetic Algorithms* and fast *Steepest Ascent Local Search*, using several neighborhoods. The solutions found by the genetic algorithms resulted close to optimum on many runs, and in general their quality prevailed over the local search solutions.

In a second experiment, the CMO measure was validated from a biological point of view. A set of 33 proteins, known to belong to four different families, were pairwise compared. The proteins were then re-clustered according to their computed CMO value. In the new clustering, no two proteins were put in the same cluster which were not in the same family before (a 0% *false positives* rate) and only few pairs of proteins that belonged to the same family were put in different clusters (a 1.3% *false negatives* rate).

3.3.3 The Lagrangian approach.

Although the approach in [53] allowed to compute, for the first time, the optimal structure alignment of real proteins from the PDB, the method had several limitations. In particular, due to the time required to solve many large and expensive LPs, the method could only be applied to problems of moderate size (proteins of up to 80 residues and 150 contacts each). To overcome these limitations, in [20] Caprara and Lancia described an alternative formulation for the problem, based on a Quadratic Programming formulation and Lagrangian Relaxation.

The theory of Lagrangian Relaxation is a well established branch of Combinatorial Optimization, and perhaps the most successful approach to tackle very large problems (such as large instances of the well known Set Covering Problem, the Optimization problem most frequently solved in real-world applications [19, 69]).

The CMO algorithm proposed in [20] is based on an approach which was successfully used for Binary Quadratic Programming problems, such as the Quadratic Assignment Problem [23]. This problem bears many similarities with the structure alignment problem. In particular, there are profits p_{ij} in the objective function which are attained when two binary variables x_i and x_j are *both* set to 1 in a solution. Analogously, in the alignment problem, there may be a profit in aligning two specific residues of the proteins *and* some other two.

For $(p, q) \in L \times L$, let b_{pq} denote a profit achieved if the lines p and q are both in the solution. Two lines p and q such that $x_p = 1$ and $x_q = 1$ contribute $b_{pq} + b_{qp}$ to the objective function. Hence, the value of an alignment can be computed as $\sum_{p \in L} \sum_{q \in L} b_{pq} x_p x_q$, provided that $b_{pq} + b_{qp} = 1$ when either $(p, q) \in \mathcal{G}$ or $(q, p) \in \mathcal{G}$, and $b_{pq} + b_{qp} = 0$ otherwise. Given such a set of profits b, the problem can be stated as

$$\max \sum_{p \in L} \sum_{q \in L} b_{pq}\, x_p\, x_q \qquad (13.23)$$

subject to

$$\sum_{l \in Q} x_l \leq 1 \qquad \forall Q \in \text{clique}(L) \qquad (13.24)$$

$$x_l \in \{0, 1\} \qquad \forall l \in L. \qquad (13.25)$$

Problem (13.23), (13.24), (13.25) is a *Binary Quadratic Program*. The formulation shows how the problem is closely related to the Quadratic Assignment Problem. The difference is that the CMO problem is in maximization form, the matching to be found does not have to be perfect and it must be noncrossing.

The objective function (13.23) can be linearized by introducing variables y_{pq}, for $p, q \in L$, and replacing the product $x_p x_q$ by y_{pq}. By adopting a standard procedure used in the convexification of ILPs [1, 7], a linear model for the problem is obtained as follows. Each constraint (13.24) associated with a set Q is multiplied by x_p for some $p \in L$ and $x_l x_p$ is replaced by y_{lp}, getting

$$\sum_{l \in Q} y_{lp} \leq x_p.$$

By doing the above for all constraints in (13.24) and variables x_m, $m \in L$, we get the model in [20]:

$$\max \sum_{p \in L} \sum_{q \in L} b_{pq}\, y_{pq} \qquad (13.26)$$

subject to

$$\sum_{l \in Q} x_l \leq 1 \qquad \forall Q \in \text{clique}(L) \qquad (13.27)$$

$$\sum_{l \in Q} y_{lp} \le x_p \qquad \forall Q \in \text{clique}(L), \quad \forall p \in L \qquad (13.28)$$

$$y_{pq} = y_{qp} \qquad \forall p, q \in L \,|\, p \prec q \qquad (13.29)$$

$$x_l, y_{pq} \in \{0, 1\} \qquad \forall l, p, q \in L. \qquad (13.30)$$

The constraints (13.29) can be relaxed in a *Lagrangian* way, by associating a *Lagrangian multiplier* λ_{pq} to each constraint (13.29), and adding to the objective function (13.26) a linear combination of constraints (13.29), each weighed by its Lagrangian multiplier.

By defining for convenience $\lambda_{qp} := -\lambda_{pq}$ for $p \prec q$, and by letting $c_{pq} := b_{pq} + \lambda_{pq}$, the objective function of the Lagrangian relaxation becomes

$$U(\lambda) := \max \sum_{p \in L} \sum_{q \in L} c_{pq}\, y_{pq} \qquad (13.31)$$

and we see that, intuitively, the effect of the Lagrangian relaxation is to redistribute the profit $b_{pq} + b_{qp}$ between the two terms in the objective function associated with y_{pq} and y_{qp}.

In [20] it is shown how the Lagrangian relaxed problem (13.31) subject to (13.27), (13.28) and (13.30) can be effectively solved, via a decomposition approach. In this problem, besides the integrality constraints (13.30), each variable y_{lp} appears only in the constraints (13.28) associated with x_p. For each $p \in L$, this implies that, if $x_p = 0$, all variables y_{lp} take the same value, whereas, if variable $x_p = 1$, the optimal choice of y_{lp} for $l \in L$ amounts to solving the following:

$$\max \sum_{l \in L} c_{lp} y_{lp} \qquad (13.32)$$

subject to

$$\sum_{l \in Q} y_{lp} \le 1 \qquad \forall Q \in \text{clique}(L) \,|\, p \notin Q \qquad (13.33)$$

$$\sum_{l \in Q} y_{lp} \le 0 \qquad \forall Q \in \text{clique}(L) \,|\, p \in Q \qquad (13.34)$$

$$y_{lp} \in \{0, 1\} \qquad \forall l \in L. \qquad (13.35)$$

In other words, the profit achieved if $x_p = 1$ is given by the optimal solution of (13.32)–(13.35). Let d_p denote this profit. Interpreting c_{lp} as the weight of line l and y_{lp} as the selection variable for line l, we see that the problem (13.32)–(13.35) is simply a *maximum weight noncrossing matching* problem. This problem can be solved in quadratic time by Dynamic Programming, as explained in the introduction of Section 13.3.

Once the profits d_p have been computed for all $p \in L$, the optimal solution to the overall Lagrangian problem can be obtained by solving one more max-weight noncrossing matching problem, in which d are the weights of the lines and x are the selection variables.

As far as the overall complexity, we have the following result:

PROPOSITION 13.8 *([20]) The Lagrangian relaxation defined by (13.31) subject to (13.27), (13.28) and (13.30) can be solved in $O(|E_1||E_2|)$ time.*

Let $U(\lambda^*) := \min_\lambda U(\lambda)$ be the best upper bound that can be obtained by Lagrangian relaxation. By denoting with U_0 the upper bound obtained in [53], corresponding to the LP relaxation of (13.19)-(13.22) and (13.2), it can be shown that

$$U_0 \geq U(\lambda^*) \tag{13.36}$$

and the inequality can be tight.

In order to determine the optimal multipliers λ^*, or at least near-optimal multipliers, the approach of [20], employs a standard *subgradient optimization* procedure (see Held and Karp [43]).

The exact algorithm for CMO proposed in [20] is a Branch-and-Bound procedure based on the above Lagrangian relaxation. Furthermore, a greedy heuristic procedure is also described in [20], which performed very well in practice. The procedure constructs a solution by choosing a line which maximizes a suitable score, fixing it in the solution and iterating. As to the score, the procedure chooses the line p such that the value of the Lagrangian relaxed problem, with the additional constraint $x_p = 1$, is maximum.

In their computational experiments, Caprara and Lancia showed how the Lagrangian Relaxation approach can be orders of magnitude faster than the Branch-and-Cut approach of [53]. For instance, a set of 10,000 pairs of moderate-size proteins were aligned optimally or near-optimally in about a day. The optimal solution can be readily found if the two proteins have a similar enough structure. In this case, optimal alignments can be computed even for proteins of very large size, within few seconds. For instance, in less than one minute, the algorithm in [20] could find, for the first time, an optimal alignment of two large proteins from the same family, with about 900 residues and 2000 contacts each.

It must be remarked, however that the optimal solution of instances associated with substantially different proteins appears to be completely out of reach not only for this algorithm, but also for the other methods method currently known in Combinatorial Optimization. Such a situation is analogous to the case of the Quadratic Assignment Problem, for which instances with 40 nodes are quite far from being solved to optimality.

```
Chrom.c, paternal:ataggtccCtatttccaggcgcCgtatacttcgacgggActata
Chrom.c, maternal:ataggtccGtatttccaggcgcCgtatacttcgacgggTctata
```

```
Haplotype 1 →            C               C               A
Haplotype 2 →            G               C               T
```

Figure 13.6. A chromosome and the two haplotypes

4.   Single Nucleotide Polymorphisms

The recent whole-genome sequencing efforts [80, 45] have confirmed that the genetic makeup of humans is remarkably well conserved, with different people sharing some 99% identity at the DNA level. A DNA region whose content varies in a statistically significant way within a population is called a *genetic polymorphism.* The smallest possible region consists of a single nucleotide, and hence is called a *Single Nucleotide Polymorphism,* or SNP (pronounced "snip"). A SNP is a nucleotide site, in the middle of a DNA region which is otherwise identical for everybody, at which we observe a statistically significant variability in a population. For some reasons still unclear, the variability is restricted to only two alleles out of the four possible (A, G, C, G). The alleles can be different for different SNPs. It is believed that SNPs are the predominant form of human genetic variation [25], so that their importance cannot be overestimated for medical, drug-design, diagnostic and forensic applications.

Since DNA of diploid organisms is organized in pairs of chromosomes, for each SNP one can either be *homozygous* (same allele on both chromosomes) or *heterozygous* (different alleles). The values of a set of SNPs on a particular chromosome copy define a *haplotype. Haplotyping* consists in determining a pair of haplotypes, one for each copy of a given chromosome, that provides full information of the SNP fingerprint for an individual at that chromosome. In Figure 13.6 we give a simplistic example of a chromosome with three SNP sites. This individual is heterozygous at SNPs 1 and 3 and homozygous at SNP 2. The haplotypes are CCA and GCT.

In recent years, several optimization problems have been defined for SNP data. Some of these were tackled by Mathematical Programming techniques, which we recall in this section. Different approaches were also followed, aimed, e.g., at determining approximation algorithms, dynamic programming procedures, and computational complexity results. However, these approaches do not fall within the scope of this survey.

We consider the haplotyping problems relative to a single individual and to a set of individuals (a population). In the first case, the input is inconsistent haplotype data, coming from sequencing, with unavoidable sequencing errors. In the latter case, the input is ambiguous *genotype* data, which specifies only the multiplicity of each allele for each individual (i.e., it is known if individual i is homozygous or heterozygous at SNP j, for each i and j).

4.1   Haplotyping a single individual

As discussed in Section 13.2.3, sequencing produces either single fragments or pairs of fragments (mate pairs) from the same chromosome copy. As of today, even with the best possible technology, sequencing errors are unavoidable. These consist in bases which have been miscalled or skipped altogether. Further, there can be the presence of *contaminants*, i.e. DNA coming from another organism which was wrongly mixed with the one that had to be sequenced. In this framework, the *Haplotyping Problem For an Individual* can be informally stated as: "given inconsistent haplotype data coming from fragment sequencing, find and correct the errors so as to retrieve a consistent pair of SNPs haplotypes." The mathematical framework for this problem is as follows. Independently of what the actual alleles at a SNP are, in the sequel we denote the two values that each SNP can take by the letters A and B. A haplotype, i.e. a chromosome content projected on a set of SNPs, is then a string over the alphabet $\{A, B\}$. Let $S = \{1, \ldots, n\}$ be a set of SNPs and $\mathcal{F} = \{1, \ldots, m\}$ be a set of fragments. Each SNP is covered by some of the fragments, and can take the values A or B. For a pair $(f, s) \in \mathcal{F} \times S$, denote with the symbol "$-$" the fact that fragment f does not cover SNP s. Then, the the data can be represented by an $m \times n$ matrix over the alphabet $\{A, B, -\}$, called the *SNP matrix*, where each row represents a fragment and each column a SNP.

A *gapless* fragment is one covering a set of consecutive SNPs (i.e., in the row corresponding to the fragment, between any two entries in $\{A,B\}$ there is no "$-$" entry), otherwise, the fragment has *gaps*. There can be gaps for two reasons: (i) thresholding of low-quality reads (i.e., when the sequencer cannot call a SNP A or B with enough confidence, it is marked with a -); (ii) mate-pairing in shotgun sequencing (one pair of mates is the same as a single fragment with one gap in the middle).

Two fragments i and j are said to be *in conflict* if there exists a SNP k such that $M[i, k] = $ A and $M[j, k] = $ B or $M[i, k] = $ B and $M[j, k] = $ A. This implies that either i and j are not from the same chromosome copy, or there are errors in the data. Given a SNP matrix M, the *fragment conflict graph* is the graph $G_{\mathcal{F}}(M) = (\mathcal{F}, E_{\mathcal{F}})$ with an edge for each pair of fragments in conflict (see Figure 13.7).

The data are consistent with the existence of two haplotypes (one for each chromosome copy) if and only if $G_{\mathcal{F}}(M)$ is a bipartite graph. In the presence of contaminants, or of some "bad" fragments (fragments with many se-

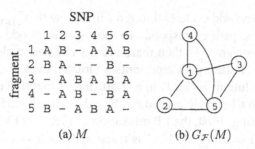

	SNP
fragment	1 2 3 4 5 6
1	A B – A A B
2	B A – – B –
3	– A B A B A
4	– A B – B A
5	B – A B A –

(a) M (b) $G_{\mathcal{F}}(M)$

Figure 13.7. A SNP matrix M and its fragment conflict graph

quencing errors), to correct the data one must face the problem of removing the fewest number of rows from M so that $G_{\mathcal{F}}(M)$ becomes bipartite. This problem is called the *Minimum Fragment Removal* (MFR) problem. The problem was shown to be polynomial when all fragments are gapless, by Lancia et al. [52]. When the fragments contains gaps, however, it was shown that the problem is NP-hard. In order to solve the problem in the presence of gaps [63], the following IP formulation can be adopted so as to minimize the number of fragments (nodes) removed from $G_{\mathcal{F}}(M)$ until it is bipartite. Introduce a 0-1 variable x_f for each fragment. The variables for which $x_f = 1$ in an optimal solution define the nodes to remove. Let $\mathcal{C}$ be the set of all odd cycles in $G_{\mathcal{F}}(M)$. Since a graph is bipartite if and only if there are no odd cycles, one must remove at least one node from each cycle in $\mathcal{C}$. For a cycle $C \in \mathcal{C}$, let $V(C)$ denote the nodes in C. We obtain the following Integer Programming formulation:

$$\min \sum_{f \in \mathcal{F}} x_f \tag{13.37}$$

s.t.

$$\sum_{f \in V(C)} x_f \geq 1 \qquad \forall C \in \mathcal{C} \tag{13.38}$$

$$x_f \in \{0, 1\} \qquad \forall f \in \mathcal{F}. \tag{13.39}$$

To solve the LP relaxation of (13.37)-(13.39), one must be able to solve in polynomial time the following separation problem: Given fractional weights x^* for the fragments, find an odd cycle C, called *violated*, for which $\sum_{f \in V(C)} x_f^* < 1$.

This problem can be solved as follows. First, build a new graph $Q = (W, F)$ made, roughly speaking, of two copies of $G_{\mathcal{F}}$. For each node $i \in \mathcal{F}$ there are two nodes $i', i'' \in W$, and for each edge $ij \in E_{\mathcal{F}}$, there are two edges $i'j'', i''j' \in F$. Edges in F are weighted, with weights $w(i'j'') = w(i''j') := x_i^*/2 + x_j^*/2$. By a parity argument, a path P in Q of weight $w(P)$, from i' to

i'', corresponds to an odd cycle C through i in $G_{\mathcal{F}}$, with $\sum_{j \in V(C)} x_j^* = w(P)$. Hence, the *shortest* path corresponds to the "smallest" odd cycle, and if the shortest path has weight ≥ 1 then there are no violated cycles (through node i). Otherwise, the shortest path represents a violated cycle.

To find good solutions of MFR in a short time, Lippert et al. [63] have also experimented with a heuristic procedure, based on the above LP relaxation and randomized rounding. First, the LP relaxation of (13.37)-(13.39) is solved, obtaining an optimal solution x^*. If x^* is fractional, each variable x_f is rounded to 1 with probability x_f^*. If the solution is feasible, stop. Otherwise, fix to 1 all the variables that were rounded so far (i.e., add the constraints $x_f = 1$) and iterate a new LP. This method converged very quickly on all tests on real data, with a small gap in the final solution (an upper bound to the optimum) and the first LP value (a lower bound to the optimum).

4.2   Haplotyping a population

Haplotype data is particularly sought after in the study of complex diseases (those affected by more than one gene), since it can give complete information about which set of gene alleles are inherited together. However, because polymorphism screens are conducted on large populations, in such studies it is not feasible to examine the two copies of each chromosome separately, and *genotype*, rather than haplotype, data is usually obtained. A genotype describes the multiplicity of each SNP allele for the chromosome of interest. At each SNP, three possibilities arise: either one is homozygous for the allele A, or homozygous for the allele B, or heterozygous (a situation denoted by the symbol X). Hence a genotype is a string over the alphabet $\{A, B, X\}$, where each position of the letter X is called an *ambiguous* position. A genotype $g \in \{A, B, X\}^n$ is *resolved* by the pair of haplotypes $h, q \in \{A, B\}^n$, written $g = h \oplus q$, if for each SNP j, $g[j] = A$ implies $h[j] = q[j] = A$, $g[j] = B$ implies $h[j] = q[j] = B$, and $g[j] = X$ implies $h[j] \neq q[j]$. A genotype is called *ambiguous* if it has at least two ambiguous positions (a genotype with at most one ambiguous positions can be resolved uniquely). A genotype g is said to be *compatible* with a haplotype h if h agrees with g at all unambiguous positions. The following inference rule, given a genotype g and a compatible haplotype h, defines q such that $g = h \oplus q$:

Inference Rule: Given a genotype g and a compatible haplotype h, obtain a new haplotype q by setting $q[j] \neq h[j]$ at all ambiguous positions and $q[j] = h[j]$ at the remaining positions.

The *Haplotyping Problem For a Population* is the following: given a set $\mathcal{G}$ of m genotypes over n SNPs, find a set $\mathcal{H}$ of haplotypes such that each genotype is resolved by one pair of haplotypes in $\mathcal{H}$.

To turn this problem into an optimization problem, one has to specify the objective function, i.e., the cost of each solution. One such objective has been studied by Gusfield [39, 40], and is based on a greedy algorithm for haplotype inference, also known as *Clark's rule*.

This rule was was proposed by the biologist Andy Clark in 1990 [27], with arguments from theoretical population genetics in support of its validity. The goal is to derive the haplotypes in $\mathcal{H}$ by successive applications of the inference rule, starting from the set of haplotypes obtained by resolving the unambiguous genotypes (of which it is assumed here there is always at least one).

In essence, Clark proposed the following, nondeterministic, algorithm: Let $\mathcal{G}'$ be the set of non-ambiguous genotypes. Start with setting $\mathcal{G} := \mathcal{G} - \mathcal{G}'$ and $\mathcal{H}$ to the set of haplotypes obtained unambiguously from $\mathcal{G}'$. Then, repeat the following: take a $g \in \mathcal{G}$ and a compatible $h \in \mathcal{H}$ and apply the inference rule, obtaining q. Set $\mathcal{G} := \mathcal{G} - \{g\}$, $\mathcal{H} := \mathcal{H} \cup \{q\}$ and iterate. When no such g and h exist, the algorithm has succeeded if $\mathcal{G} = \emptyset$ and failed otherwise.

For example, suppose $\mathcal{G} = \{\texttt{XAAA}, \texttt{XXAA}, \texttt{BBXX}\}$. The algorithm starts by setting $\mathcal{H} = \{\texttt{AAAA}, \texttt{BAAA}\}$ and $\mathcal{G} = \{\texttt{XXAA}, \texttt{BBXX}\}$. The inference rule can be used to resolve $\texttt{XXAA}$ from $\texttt{AAAA}$, obtaining $\texttt{BBAA}$, which can, in turn, be used to resolve $\texttt{BBXX}$, obtaining $\texttt{BBBB}$. However, one could have started by using $\texttt{BAAA}$ to resolve $\texttt{XXAA}$ obtaining $\texttt{ABAA}$. At that point, there would be no way to resolve $\texttt{BBXX}$. The non-determinism in the choice of the pair g, h to which we apply the inference rule can be settled by fixing a deterministic rule based on the initial sorting of the data. Clark in [27] used a large (but tiny with respect to the total number of possibilities) set of random initial sortings to run the greedy algorithm on real and simulated data sets, and report the best solution overall. To find the best possible order of application of the inference rule, Gusfield considered the following optimization problem: find the ordering of application of the inference rule that leaves the fewest number of unresolved genotypes in the end. Gusfield showed the problem to be APX-hard in [40].

As for practical algorithms, Gusfield [39, 40] proposed a graph-theoretic formulation of the problem and an Integer Programming approach for its solution. The problem is first transformed (by an exponential-time reduction) into a problem on a digraph $G = (N, A)$, defined as follows. Let $N = \bigcup_{g \in \mathcal{G}} N(g)$, where $N(g) := \{(h, g) \mid h \text{ is compatible with } g\}$. $N(g)$ is (isomorphic to) the set of possible haplotypes obtainable by setting each ambiguous position of a genotype g to one of the 2 possible values. Let $N' = \bigcup_{g \in \mathcal{G}'} N(g)$ be (isomorphic to) the subset of haplotypes unambiguously determined from the set $\mathcal{G}'$ of unambiguous genotypes. For each pair $v = (h, g')$, $w = (q, g)$ in N, there is an arc $(v, w) \in A$ iff g is ambiguous, $g' \neq g$ and $g = h \oplus q$ (i.e., q can be inferred from g via h). Then, any directed tree rooted at a node $v \in N'$ specifies a feasible history of successive applications of the inference rule starting at node $v \in N'$. The problem can then be stated as: Find the largest number of nodes in $N - N'$ that can be reached by a set of node–disjoint directed trees, where each tree is rooted at a node in N' and where for every ambiguous genotype g, *at most* one node in $N(g)$ is reached.

The above graph problem was also shown to be NP-hard [40] (note that the reduction of the haplotyping problem to this one is exponential–time, and hence it does not imply NP–hardness trivially). For its solution Gusfield proposed the following formulation. Let x_v be a 0-1 variable associated to node $v \in N$.

$$\max \sum_{v \in N - N'} x_v$$

subject to

$$\sum_{v \in N(g)} x_v \leq 1 \qquad \forall g \in \mathcal{G} - \mathcal{G}'$$

$$x_w \leq \sum_{v \in \delta^-(w)} x_v \qquad \forall w \in N - N'$$

with $x_v \in \{0, 1\}$ for $v \in N$.

Gusfield [39] used this model to solve the Haplotyping problem. To reduce the problem dimension and make the algorithm practical, he actually defined variables only for nodes in a subgraph of G, i.e. the nodes reachable from nodes in N'. He observed that the LP solution, on real and simulated data, was almost always integer, thus requiring no branching in the branch and bound search. He also pointed out that, although there is a possibility of an integer solution containing directed cycles, this situation never occurred in the experiments, and hence he did not need to add *subtour elimination*-type inequalities to the model.

5.   Genome Rearrangements

Thanks to the large amount of genomic data that has become available in the past years, it is now possible to compare the genomes of different species, in order to find their differences and similarities. This is a very important problem because, when developing new drugs, we typically test them on animals before humans. This motivates questions such as determining, e.g., how close to a human a mouse is, or, in a way, how much evolution separates us from mice.

In principle, a genome can be thought of as a (very long) sequence and hence one may want to compare two genomes via a sequence alignment algorithm. Besides the large time needed to obtain an optimal alignment, there is a biological reason why sequence alignment is not the right model for large-scale genome comparisons. In fact, at the genome level, differences should be measured not in terms of insertions/deletions/mutations of single nucleotides, but

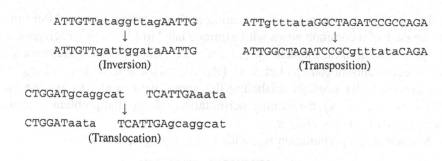

Figure 13.8. Evolutionary events

rather rearrangements (misplacements) of long DNA regions which occurred during evolution.

Among the main evolutionary events known are *inversions, transpositions,* and *translocations.* Each of these events affects a long fragment of DNA on a chromosome. When an inversion or a transposition occurs, a DNA fragment is detached from its original position and then is reinserted, on the same chromosome. In an *inversion,* it is reinserted at the same place, but with opposite orientation than it originally had. In a *transposition,* it keeps the original orientation but ends up in a different position. A *translocation* causes a pair of fragments to be exchanged between the ends of two chromosomes. Figure 13.8 illustrates these events, where each string represents a chromosome.

Since evolutionary events affect long DNA regions (several thousand bases) the basic unit for comparison is not the nucleotide, but rather the *gene.*

The general *Genome Comparison Problem* can be informally stated as: Given two genomes (represented by two sets of ordered lists of genes, one list per chromosome) find a sequence of evolutionary events that, applied to the first genome, turn it into the second. Under a common parsimony principle, the solution sought is the one requiring the minimum possible number of events. Pioneering work in the definition of genome rearrangement problems is mainly due to Sankoff and colleagues [76].

In the past decade, people have concentrated on evolution by means of some specific event alone, and have shown that even these special cases can be already very hard to solve. Since inversions are considered the predominant of all types of rearrangements, they have received the most attention. For historical reasons, they have become known as *reversals* in the computer science community.

In the remainder of this section, we will outline a successful Mathematical Programming approach for the solution of computing the evolutionary distance between two genomes evolved by reversals only, a problem known as *Sorting by Reversals.* The approach is due to Caprara et al [21, 22].

Two genomes are compared by looking at their common genes. After numbering each of n common genes with a unique label in $\{1, \ldots, n\}$, each genome is a permutation of the elements $\{1, \ldots, n\}$ (we assume here to focus on a single, specific chromosome). Let $\pi = (\pi_1 \ldots \pi_n)$ and $\sigma = (\sigma_1 \ldots \sigma_n)$ be two genomes. By possibly relabeling the genes, we can always assume that $\sigma = \iota := (1\ 2\ \ldots\ n)$, the identity permutation. Hence, the problem becomes turning π into ι, i.e., *sorting* π.

A *reversal* is a permutation ρ_{ij}, with $1 \leq i < j \leq n$, defined as

$$\rho_{ij} = (1 \ldots i-1 \quad \boxed{j\ j-1 \ldots i+1\ i} \quad j+1 \ldots n).$$
$$\text{reversed}$$

By applying ρ_{ij} to π, one obtains $(\pi_1 \ldots \pi_{i-1},\ \pi_j\ \pi_{j-1} \ldots \pi_i,\ \pi_{j+1} \ldots \pi_n)$, i.e., the order of the elements $\pi_i, \ldots, \pi_j$ has been reversed. Let $\mathcal{R} = \{\rho_{ij}, 1 \leq i < j \leq n\}$. Each permutation π can be expressed (non-uniquely) as a product $\iota\rho^1\rho^2 \ldots \rho^D$ with $\rho^i \in \mathcal{R}$ for $i = 1 \ldots D$. The minimum value D such that $\iota\rho^1\rho^2 \ldots \rho^D = \pi$ is called the *reversal distance* of π, and denoted by $d(\pi)$. *Sorting by Reversals* (SBR) is the problem of finding $d(\pi)$ and a sequence $\rho^1\rho^2 \ldots \rho^{d(\pi)}$ satisfying the previous equation.

The first optimization algorithm for computing the reversal distance was a branch-and-bound procedure, due to Kececioglu and Sankoff [51], based on a combinatorial bound. This algorithm is only suitable for small problems ($n \leq 30$), and becomes quickly impractical for larger values of n (note that, as of this writing, the data for real-life genome comparison instances involve a number n of common genes in the range $50 \leq n \leq 200$). The algorithm was developed at a time when the complexity of the problem was still unknown, although it was conjectured that SBR was NP-hard. Settling the complexity of SBR became a longstanding open question, eventually answered by Caprara [18] who showed that in fact, SBR is NP-hard.

A major step towards the practical solution of the problem was made by Bafna and Pevzner [6], who found a nice combinatorial characterization of π in terms of its breakpoints. A *breakpoint* is given by a pair of adjacent elements in π that are not adjacent in ι — that is, there is a breakpoint at position i, if $|\pi_i - \pi_{i-1}| > 1$. Let $b(\pi)$ denote the number of breakpoints. A trivial bound is $d(\pi) \geq \lceil b(\pi)/2 \rceil$, since a reversal can remove at most two breakpoints, and ι has no breakpoints. However, Bafna and Pevzner showed how to obtain a much better bound from the *breakpoint graph* $G(\pi)$. $G(\pi)$ has a node for each element of π and edges of two colors, say red and blue. There is a red edge between π_i and π_{i-1} for each position i at which there is a breakpoint, and a blue edge between each h and k such that $|h - k| = 1$ and h, k are not adjacent in π. Each node has the same number of red and blue edges incident on it (0, 1 or 2), and $G(\pi)$ can be decomposed into a set of edge-disjoint color-alternating cycles. Note that this cycles are not necessarily simple, i.e., they can repeat nodes. Let $c(\pi)$ be the maximum number of edge-disjoint alternating cycles in $G(\pi)$. Bafna and Pevzner [6] proved the following

THEOREM 13.9 *([6]) For every permutation* π, $d(\pi) \geq b(\pi) - c(\pi)$.

The lower bound $b(\pi) - c(\pi)$ turns out to be very tight, as observed first experimentally by various authors and then proved to be almost always the case by Caprara [17], who showed in [18] that determining $c(\pi)$ is essentially the same problem as determining $d(\pi)$, and hence, NP-hard as well.

Since computing $c(\pi)$ is hard, the lower bound $b(\pi) - c(\pi)$ should not be used directly in a branch-and-bound procedure for SBR. However, for any upper bound $c'(\pi)$ to $c(\pi)$, also the value $b(\pi) - c'(\pi)$ is a lower bound to $d(\pi)$, which may be quite a bit easier to compute. Given an effective Integer Linear Programming (ILP) formulation to find $c(\pi)$, a good upper bound to $c(\pi)$ can be obtained by LP relaxation. Let $\mathcal{C}$ denote the set of all the alternating cycles of $G(\pi) = (V, E)$, and for each $C \in \mathcal{C}$ define a binary variable x_C. The following is the ILP formulation of the maximum cycle decomposition:

$$c(\pi) := \max \sum_{C \in \mathcal{C}} x_C \qquad (13.40)$$

subject to

$$\sum_{C \ni e} x_C \leq 1 \qquad \forall e \in E \qquad (13.41)$$

$$x_C \in \{0, 1\} \qquad \forall C \in \mathcal{C}. \qquad (13.42)$$

The LP relaxation to the above problem is obtained by replacing the constraints $x_C \in \{0, 1\}$ with $x_C \geq 0, \forall C \in \mathcal{C}$ (it is easy to see that the constraints $x_C \leq 1$ can be omitted). Caprara et al showed in [21] how to solve the exponentially large LP relaxation of (13.40)-(13.42) in polynomial time by column-generation techniques. To price-in the variables, the solution of some non-bipartite min-cost perfect matching problems is required.

Consider the dual of the LP relaxation of (13.40)-(13.42), which reads

$$c'(\pi) := \min \sum_{e \in E} y_e \qquad (13.43)$$

subject to

$$\sum_{e \in C} y_e \geq 1 \qquad \forall C \in \mathcal{C} \qquad (13.44)$$

$$y_e \geq 0, \quad \forall e \in E. \qquad (13.45)$$

The separation problem for (13.43)-(13.45) is the following: given a weight y_e^* for each $e \in E$, find an alternating cycle $C \in \mathcal{C}$ such that

$$\sum_{e \in C} y_e^* < 1, \qquad (13.46)$$

or prove that none exists. Call an alternating cycle satisfying (13.46) a *violated cycle*. To solve (13.43)-(13.45) in polynomial time, one must be able to identify a violated cycle in polynomial time, i.e., an alternating cycle of $G(\pi)$ having weight < 1, where each edge $e \in E$ is given the weight y_e^*.

We illustrate the idea for doing this, under the simplifying assumption that the decomposition contains only cycles that are *simple* (i.e., they do not go through the same node twice). Consider the following construction, analogous to one used by Grötschel and Pulleyblank for finding minimum-weight odd and even paths in an undirected graph [36]. Define the graph H, depending on $G(\pi)$, with $H = (V_R \cup V_B, L_R \cup L_B \cup L_V)$, as follows. For each node i of $G(\pi)$, H contains two *twin* nodes $i_R \in V_R, i_B \in V_B$. For each red edge ij in $G(\pi)$, H has an edge $i_R j_R \in L_R$; for each blue edge ij in $G(\pi)$, H has an edge $i_B j_B \in L_B$. Each edge in $L_R \cup L_B$ is given the same weight as its counterpart in $G(\pi)$. Finally twin nodes are connected in H by means of edges $i_R i_B \in L_V$, for each $i \in V$, each having weight 0.

The following proposition justifies the use of H.

PROPOSITION 13.10 *([21]) There is a bijection between perfect matchings of H and (possibly empty) sets of node-disjoint simple alternating cycles of $G(\pi)$.*

From this proposition it follows

PROPOSITION 13.11 *There is a simple alternating cycle in G of weight < 1 if and only if there exists a perfect matching $M \neq L_V$ in H of weight < 1.*

Accounting for alternating cycles that are not simple requires to first break up the nodes of degree 4 in G into pairs of nodes, and then apply the above reduction to the resulting graph (technical details can be found in [21]). Note that the graph H obtained is, in general, non-bipartite.

Although polynomial, the solution of the weighted matching problem on general graphs is computationally expensive to find. In [22] Caprara et al. describe a weaker bound, obtained by enlarging the set $\mathcal{C}$ in (13.40)-(13.42) to include also the *pseudo-alternating* cycles, i.e., cycles that alternate red and blue edges but may possibly use an edge twice. With this new set of variables, the pricing becomes much faster, since only *bipartite* matching problems must be solved.

Let $\mathcal{P}$ be the set of pseudo-alternating cycles. Similarly to before, one can now formulate the problem of finding the decomposition of $G(\pi)$ into a set of edge-disjoint pseudo-alternating cycles. The associated column generation problem requires finding, given a weighting u^* for the edges, a pseudo-alternating cycle $C \in \mathcal{P}$ of weight $u^*(C) < 1$. The weight of C is the sum of the weights of its edges, where each edge e which is traversed twice by the cycle, contributes $2u^*(e)$ to the sum. We next show how to solve this problem.

Construct an arc-weighted directed graph $D = (V, A)$ from $G(\pi)$ and u^* as follows. D has the same node set as $G(\pi)$ and, for each node pair $i, j \in V$, D has an arc $(i, j) \in A$ if there exists $k \in V$ such that ik is a blue edge of $G(\pi)$ and kj is a red edge. The arc weight for (i, j) is given by $u_{ik}^* + u_{kj}^*$ (if there exist two such k, consider the one yielding the minimum weight of arc (i, j)). Call *dicycle* a simple (i.e. without node repetitions) directed cycle of D. Then, each dicycle of D corresponds to a pseudo-alternating cycle of $G(\pi)$ of the same weight. Vice versa, each pseudo-alternating cycle of $G(\pi)$ corresponds to a dicycle of D whose weight is not larger.

It follows that, for a given u^*, $G(\pi)$ contains a pseudo-alternating cycle $C \in \mathcal{P}$ of weight < 1 if and only if D contains a dicycle of weight < 1.

To find such a cycle, introduce *loops* in D, i.e. arcs of the form (i, i) for all $i \in V$, where weight of the loops is initially set to 0. The *Assignment Problem* (i.e., the perfect matching problem on bipartite graphs) can then be used to find a set of dicycles in D, spanning all nodes, and with minimum total weight.

From this proposition it follows

PROPOSITION 13.12 *There is a pseudo-alternating cycle in $G(\pi)$ of weight < 1 if and only there is a solution to the assignment problem on D not containing only loops and with weight < 1.*

Although weaker than the bound based on alternating cycles, the bound based on pseudo-alternating cycles turns out to be still very tight. Based on this bound, Caprara et al. developed a branch-and-price algorithm [22] which can routinely solve, in a matter of seconds, instances with $n = 200$ elements, a large enough size for all real-life instances available so far. Note that no effective ILP formulation has ever been found for modeling SBR directly. Finding such a formulation constitutes an interesting theoretical problem.

With this branch and price algorithm, SBR is regarded as practically solved, being one of the few NP-hard problems for which a (worst-case) exponential algorithm is fairly good on most instances.

We end this section by mentioning a closely related problem, for which an elegant and effective solution has been found by Hannenhalli and Pevzner. However, this solution does not employ Mathematical Programming techniques and hence falls only marginally within the scope of this survey.

Since a DNA molecule has associated a concept of "reading direction" (from the so-called *5'-end* to the *3'-end*), after a reversal, not only the order of some genes is inverted, but also of the nucleotide sequence of each such gene. To account for this situation, a genome can be represented by a *signed* permutation, i.e. a permutation in which each element is signed either '+' or '−'. For a signed permutation, the effect of a reversal is not only to flip the order of some consecutive elements, but also to complement their sign. For instance, the reversal ρ_{24} applied to the signed permutation $(+1 \ -4 \ +3 \ -5 \ +2)$ yields the signed permutation $(+1 \ +5 \ -3 \ +4 \ +2)$. The *signed* version of Sorting by Reversals (SSBR) consists in determining a minimum number of reversals that turn a signed permutation π into $(+1 \ +2 \ \ldots \ +n)$.

With a deep combinatorial analysis of the cycle decomposition problem for the permutation graph of a signed π, SSBR was shown to be polynomial, by Hannenhalli and Pevzner [42]. The original $O(n^4)$ algorithm of Hannenhalli and Pevzner for SSBR was improved over the years to an $O(n^2)$ algorithm for finding the optimal solution [48] and an $O(n)$ algorithm [5] for finding the optimal value only (i.e., the signed reversal distance).

6.   Genomic Mapping and the TSP

In the last years (especially prior to the completion of the sequencing of the human genome), a great deal of effort has been devoted to determining *genomic maps*, of various detail.

Similarly to a geographic map, the purpose of a genomic map is to indicate the location of some landmarks in a genome. Each landmark (also called a *marker*) corresponds to a specific DNA region, whose content may or may not be known. Typical markers are genes, or short DNA sequences of known content (called STSs, for Sequence Tagged Sites), or restriction-enzyme sites (which are short, palindrome, DNA sequences).

At the highest resolution, to locate a marker one needs to specify the chromosome on which it appears and the distance in base pairs, from one specific end of the chromosome. A map of lower resolution, instead of the absolute position of the markers, may give their relative positions, e.g., in the form of the order with which the markers appear on a chromosome.

The construction of genomic maps gives us the opportunity to describe a Computational Biology problem for which the solution proposed consists in a reduction to a well-known optimization problem, in this case the TSP. In particular, we will discuss the *mapping* problem for either *clones vs. probes* or *radiation hybrids vs. markers*. For both problems, the input data consists of a 0-1, $m \times n$, matrix M.

In the first problem considered, there are m *clones* and n *probes*. Both clones and probes are DNA fragments, but of quite different length. In particular, a clone is a long DNA fragment (several thousand bases), from a given chromosome, while a probe is a short DNA fragment (a few bases), also from the same chromosome. An *hybridization* experiment can tell, for each clone i

and probe j, if j is contained in i. The experiment is based on the Watson-Crick pairing rules, according to which j finds its complement in i, if present, and forms strong bonds with it. By attaching to it a fluorescent label, it is possible to determine if j has hybridized with its complement in i.

The results of the experiments are stored in the 0-1 matrix M, which can be interpreted as follows: $M_{ij} = 1$ if and only if probe j appears in clone i. Due to experimental errors, some clones may be *chimeric*. A chimeric clone consists of two or more distinct fragments from the chromosome instead of only one. Other experimental errors are *false positives* ($M_{ij} = 1$ but i does not contain j) and *false negatives* ($M_{ij} = 0$ but i contains j). The *Probe Mapping Problem* consists in: given a clone vs probe matrix M, determine in which order the probes occur on the chromosome. The solution corresponds to a permutation of the columns of M. Note that, when there are no chimeric clones and no false positive/negatives, the correct ordering of the probes is such that, after rearranging the columns, in each clone (row of M) the elements "1" appear consecutively. Such a matrix is said to have the *consecutive 1 property*. Testing if a matrix has the consecutive 1 property, and finding the order that puts the 1s consecutively in each row, is a well known polynomial problem [16].

Under the assumption of possible errors in the data, the problem of determining the correct probe order is NP-hard, and Alizadeh et al. [3] have described an optimization model in which the best solution corresponds to the order minimizing

$$v(M') = \sum_{j=1}^{n-1} c(M'_j, M'_{j+1}). \tag{13.47}$$

Here, M' is a matrix obtained by permuting the columns of M, M'_k denotes the k-th column of M' and $c(x, y)$ is the cost of following a column x with a column y, both viewed as binary vectors. Assuming there are no false positive/negatives, and that there is a low probability for a clone to be chimeric, $c(x, y)$ can be taken as the Hamming distance of x and y. Then, the objective becomes that of minimizing the total number of times that a block of consecutive 1s is followed by some 0s and then by another block of consecutive 1s, on the same row (in which case, the row would correspond to a chimeric clone).

It is easy to see that the optimization of $v(M')$ is a TSP problem (in fact, a *Shortest Hamiltonian Path* problem). In their code, Alizadeh et al. used some standard TSP heuristics, such as 2-OPT and 3-OPT neighborhood search, and a Simulated Annealing approach. The procedure was tested on simulated data, for which the correct solution was known, as well as real data for chromosome 21. The experiments showed that, for several cost functions c, depending on the simulation model adopted, the TSP solutions approximate very well the correct solutions.

A very similar problem concerns the construction of *Radiation Hybrid* (RH) maps. Radiation hybrids are obtained by (i) breaking, by radiation, a human chromosome into several random fragments (the markers) (ii) fusing the radiated cells with rodent normal cells. The resulting hybrid cells, may retain one, none or many of the fragments from step (i).

For each marker j and each radiation hybrid i, an experiment can tell if j was retained by i or not. Hence, the data for m radiation hybrids and n markers can be represented by a 0-1 matrix M. As before, the problem consists in ordering the markers, given M. The problem is studied in Ben Dor et al. [9, 10]. Two different measures are widely used to evaluate the quality of a possible solution. The first is a combinatorial measure, called Obligate Chromosome Breaks (OCB), while the second is a statistically based, parametric method of maximum likelihood estimation (MLE). For a solution π, let M' be the matrix obtained from M after rearranging the columns according to π. The first measure is simply given by the number of times that, in some row of M', a 0 is followed by a 1, or vice versa. As we already saw, the objective of minimizing this measure is achieved by minimizing $v(M')$, defined in (13.47), when $c(x, y)$ is the Hamming distance.

In the MLE approach, the goal is to find a permutation of the markers and estimate the distance (breakage probability) between adjacent markers, such that the likelihood of the resulting map is maximized. The likelihood of a map (consisting of a permutation of the markers, and distances between them) is the probability of observing the RH data, given the map.

The reduction of MLE to TSP is carried out in three steps, as described in Ben Dor [8]. First, the retention probability is estimated. Then, the breakage probability between each pair of markers is estimated. Finally, the costs $c(x, y)$ are defined such that the total value of (13.47) is equal to the negative logarithm of the likelihood of the corresponding permutation. Therefore, minimizing $v(M')$ is equivalent to maximizing the logarithm of the likelihood, and hence, to maximizing the likelihood.

Although the OCB and MLE approaches are different, it can be shown that, under the assumption of evenly spaced markers, these two criteria are roughly equivalent. As a result, in this case it is enough to optimize the OCB objective, and the order obtained will optimize MLE as well. However, if the assumption is not met, the two objectives differ. At any rate, the optimization of both objectives corresponds to the solution of a TSP problem. In their work, Ben Dor at al. [9, 10] used *Simulated Annealing*, with the 2-OPT neighborhood structure of Lin and Kernighan [62], for the solution of the TSP.

The software package RHO (Radiation Hybrid Optimization), developed by Ben Dor et al., was used to compute RH maps for the 23 human chromosomes. For 18 out of 23 chromosomes, the maps computed with RHO were identical or nearly identical to the the corresponding maps computed at the Whitehead

Institute, by experimental, biological, techniques. For the remaining 5 chromosomes, the maps computed by using RHO were superior to the Whitehead maps with respect to both optimization criteria [10].

In [2], Agarwala et al. describe an improvement of the RHO program, obtained by replacing the simulated annealing module for the TSP with the state-of-the-art TSP software package, CONCORDE [4]. In a validating experiment, public data (i.e., the panels Genebridge 4 and Stanford G3) were gathered from a radiation hybrid databank, and new maps were computed by the program. The quality of a map was accessed by comparing how many times two markers, that appear in some order in a published sequence deposited on GenBank, are put in the opposite order by the solution (i.e., the map is inconsistent with the sequence). It was shown that the public maps for Genebridge 4 and G3 were inconsistent for at least 50% of the markers, while the maps computed by the program were far more consistent.

We end this section by mentioning one more mapping problem, the *Physical Mapping with end-probes*, for which a Branch-and-Cut approach was proposed by Christof et al. [26]. The problem has a similar definition that the problems discussed above. The main difference is that some probes are known to be *end-probes*, i.e., coming from the ends of some clone.

7.   Applications of Set Covering

Together with the TSP, the Set Covering (SC) problem is perhaps the only other notorious Combinatorial Optimization problem to which a large number of Computational Biology problems have been reduced. In this section we will review a few applications of this problem to areas such as *primer design* for PCR reactions and *tissue classification in microarrays.*

As described in Section 13.2.3, PCR is an experiment by which a DNA region can be amplified several thousand of times. In order to do so, two short DNA fragments are needed, one which must precede, and the other which must follow, the region to be amplified. These fragments are called *primers*, and their synthesis is carried on by specialized laboratories. The preparation of primers is delicate and expensive and hence, in the past years, an optimization problem concerning primer design has emerged. The goal is to determine the least expensive set of primers needed to perform a given set of reactions. The problem, in its most simplified version, can be described as follows.

Given a set $S = \{S_1, \ldots, S_n\}$ of strings (DNA sequences), find a set $\mathcal{P} = \{p_1, \ldots, p_n\}$ of pairs of strings, where each $p_i = (s_i, e_i)$. For each $i \in \{1, \ldots, n\}$, s_i and e_i must be substrings of S_i, of length at least k (a given input parameter). The objective requires to minimize $|\mathcal{P}|$.

The problem can be readily recognized as a SC, where the strings in S are the elements to be covered. Let $\mathcal{A}$ be the set of all substrings of length k.

Viewing each pair $(a_i, a_j) \in \mathcal{A} \times \mathcal{A}$ as a set A_{ij}, consisting of all strings which contain both a_i and a_j, the primer design problem corresponds the SC problem with sets A_{ij} and universe $\mathcal{S}$.

A heuristic approach to this problem was proposed by Pearson et al. [71] who, instead of solving once the SC described above, solved two times a different SC, to find both ends of each primer-pair. In their SC reduction, each element $a \in \mathcal{A}$ is viewed as a set, consisting of all strings which contain it. One of the main uses of PCR is that it provides an effective membership test for a sequence (e.g., a protein) and a certain family. The idea is that, given a set of primers designed for a family $\mathcal{F}$ and a new sequence s, if one of the primer-pairs amplifies part of s, then s belongs to $\mathcal{F}$, and otherwise it does not. In many cases (e.g., by performing a multiple alignment of the known family sequences) it is possible to find a short sequence common to all the elements of $\mathcal{F}$ (but perhaps also to elements not in $\mathcal{F}$), which can be used as one of the primers in each primer-pair. In this case, the SC reduction described by Pearson et al. correctly minimizes the total number of primers to synthesize. Finally, a heuristic, greedy, approach to minimize the number of primers to use in multiplex PCR reactions was employed by Nicodeme and Steyaert [70].

A second application of the SC problem arises in the context of microarrays, also called *DNA chips*. A microarray is a device which allows for a massively parallel experiment, in which several samples (targets) are probed at once. The experiment can determine, for each sample, the amount of production of mRNA for each of a given set of genes. The production of mRNA is directly proportional to the level of expression of a gene. In an application of this technology, one may compare samples from healthy cells and cancer cells, and highlight an excess in the expression of some genes, which may hint at the cause of the disease.

Physically, a microarray is an array on a chip, in which the rows (up to some hundreds) are indexed by a set of samples (e.g. tissues), while the columns (some thousands) are indexed by a set of genes. To each cell of the chip are attached a large number of identical short DNA sequences (probes). The probes along each column (corresponding to a gene g) are designed specifically to bind (under Watson-Crick complementarity) to the mRNA sequence encoded by the gene g.

After letting a set of fluorescently-labeled samples hybridize to each of the chip cells, an image processing software detects the amount of DNA that has been "captured" in each cell. The result is a matrix M of numbers, where $M[s, g]$ represents the amount of expression of gene g in sample s. The matrix is usually made binary by using a threshold, so that $M[s, g] = 1$ if and only if a gene g is expressed (at a sufficiently high level) by a sample s.

Given a binary matrix M containing the results of microarray hybridization, several optimization problems have been studied which aim at retrieving useful information from M. One such problem concers the use of gene expression to distinguish between several types of tissues (e.g., brain cells, liver cells, etc.). The objective is to find a subset G of the genes such that, for any two tissues $s \neq s'$, there exists at least a gene $g \in G$ with $M[s, g] \neq M[s', g]$, and the cardinality of G is minimum. This problem is equivalent to a well known NP-hard problem, the Minimum Test Collection (problem [SP96], Garey and Johnson [31]), and has been studied by Halldorsson et al. and De Bontridder et al.[41, 15, 14]. In order to solve the problem, Haldorsson et al. propose the following reduction to the SC. Each pair of tissues (s, s') is an element of the universe set, and each gene g corresponds to the subset of those pairs (s, s') for which $M[s, g] \neq M[s', g]$.

A variant of this problems, that is also of interest to Computational Biology, is the following: the input tissues are partitioned in k classes, and the problem consists again in finding a minimum subset of genes G such that, whenever two samples s and s' are not from the same class, there is a gene $g \in G$ with $M[s, g] \neq M[s', g]$. The special case of $k = 2$ is very important in the following situation: Given a set of healthy cells and a set of tumor cells, we want to find a set of genes G such that there is always always a gene $g \in G$ whose expression level can distinguish any healthy sample from any tumor one. The knowledge of the expression levels for such a set of genes would constitute a valid diagnostic test to determine the presence/absence of a tumor in a new input sample.

8. Conclusions

In this survey we have reviewed some of the most successful Mathematical Programming approaches to Computational Biology problems of the last few years. The results described show, once more, how profitably the sophisticated optimization techniques of Operations Research can be applied to a non-mathematical domain, provided the difficult phase of problem modeling has been successfully overcome. Clearly, the help of a biology expert is vitally needed in the modeling phase. However, we cannot stress hardly enough how important it is for the optimization expert to learn the language of molecular biology, in order to become a more active player during the critical problem-analysis phase.

Although this survey has touched many problems from different application areas, the number of new results which employ Mathematical Programming techniques is steadily increasing, and it is easy to foresee applications of these techniques to more and more new domains of Computational Biology. For instance, very recently a Branch-and-Cut approach was proposed for the solution of a protein fold prediction problem. The approach, by Xu et al. [86], tries to determine the best "wrapping"(called *threading*) of a protein sequence to a known 3-D protein structure, and bears many similarities to the models described in Section 13.3.2 and Section 13.3.3 for structure and sequence alignments.

Finally, besides the problems described in this survey, we would like to mention that Mathematical Programming techniques have been also successfully applied to related problems, arising in clinical biology and medicine. Some of these problems are concerned with the optimization of an instrument's use, or the speed-up of an experiment. Among the most effective approaches for this class of problems, we recall the works of Eva Lee et al., who developed Integer Programming models for the optimal schedule of clinical trials, and for the radiation treatment of prostate cancer [58, 56, 55, 57].

References

[1] W.P. Adams and H.D. Sherali. A tight linearization and an algorithm for zero-one quadratic programming problems. *Management Science*, 32:1274–1290, 1986.

[2] R. Agarwala, D. Applegate, D. Maglott, G. Schuler, and A. Schaffer. A fast and scalable radiation hybrid map construction and integration strategy. *Genome Research*, 10:230–364, 2000.

[3] F. Alizadeh, R. Karp, D. Weisser, and G. Zweig. Physical mapping of chromosomes using unique probes. In *Proceedings of the Annual ACM-SIAM Symposium on Discrete Algorithms (SODA)*, pages 489–500, New York, NY, 1994. ACM press.

[4] D. Applegate, R. Bixby, V. Chvatal, and W.Cook, editors. *CONCORDE: A code for solving Traveling Salesman Problems*. World Wide Web, http://www.math.princeton.edu/tsp/concorde.html.

[5] D. A. Bader, B. M. Moret, and M. Yan. A linear-time algorithm for computing inversion distances between signed permutations with an experimental study. *Journal of Computational Biology*, 8(5):483–491, 2001.

[6] V. Bafna and P.A. Pevzner. Genome rearrangements and sorting by reversals. *SIAM Journal on Computing*, 25:272–289, 1996.

[7] E. Balas, S. Ceria, and G. Cornuejols. A lift-and-project cutting plane algorithm for mixed 0-1 programs. *Mathematical Programming*, 58:295–324, 1993.

[8] A. Ben-Dor. *Constructing Radiation Hybrid Maps of the Human Genome*. PhD thesis, Technion, Israel Institute of Technology, Haifa, 1997.

[9] A. Ben-Dor and B. Chor. On constructing radiation hybrid maps. *Journal of Computational Biology*, 4:517–533, 1997.

[10] A. Ben-Dor, B. Chor, and D. Pelleg. Rho—Radiation Hybrid Ordering. *Genome Research*, 10:365–378, 2000.

[11] A. Ben-Dor, G. Lancia, J. Perone, and R. Ravi. Banishing bias from consensus sequences. In *Proceedings of the Annual Symposium on Combinatorial Pattern Matching (CPM)*, volume 1264 of *Lecture Notes in Computer Science*, pages 247–261. Springer, 1997.

[12] D. Benson, I. Karsch-Mizrachi, D. J. Lipman, J. Ostell, B. Rapp, and D. L. Wheeler DL. Genbank. *Nucleic Acids Research*, 30(1):17–20, 2002.

[13] H.M. Berman, J. Westbrook, Z. Feng, G. Gilliland, T.N. Bhat, H. Weissig, I.N. Shindyalov, and P.E. Bourne. The protein data bank. *Nucleic Acids Research*, 28:235–242, 2000. The PDB is at http://www.rcsb.org/pdb/.

[14] K. M. De Bontridder, B. Halldorsson, M. Halldorsson, C. A. J. Hurkens, J. K. Lenstra, R. Ravi, and L. Stougie. Approximation algorithms for the test cover problem. *Mathematical Programming-B*, To appear.

[15] K. M. De Bontridder, B. J. Lageweg, J. K. Lenstra, J. B. Orlin, and L. Stougie. Branch-and-bound algorithms for the test cover problem. In *Proceedings of the Annual European Symposium on Algorithms (ESA)*, volume 2461 of *Lecture Notes in Computer Science*, pages 223–233. Springer, 2002.

[16] K. S. Booth and G. S. Lueker. Testing for the consecutive ones property, intervals graphs and graph planarity testing using PQ-tree algorithms. *J. Comput. System Sci.*, 13:335–379, 1976.

[17] A. Caprara. On the tightness of the alternating-cycle lower bound for sorting by reversals. *Journal of Combinatorial Optimization*, 3:149–182, 1999.

[18] A. Caprara. Sorting permutations by reversals and eulerian cycle decompositions. *SIAM Journal on Discrete Mathematics*, 12:91–110, 1999.

[19] A. Caprara, M. Fischetti, and P. Toth. A heuristic method for the set covering problem. *Operations Research*, 47:730–743, 1999.

[20] A. Caprara and G. Lancia. Structural alignment of large-size proteins via lagrangian relaxation. In *Proceedings of the Annual International Conference on Computational Molecular Biology (RECOMB)*, pages 100–108, New York, NY, 2002. ACM Press.

[21] A. Caprara, G. Lancia, and S.K. Ng. A column-generation based branch-and-bound algorithm for sorting by reversals. In M. Farach-Colton, F.S. Roberts, M. Vingron, and M. Waterman, editors, *Mathematical Support for Molecular Biology*, volume 47 of *DIMACS Series in Discrete Math-*

ematics and Theoretical Computer Science, pages 213–226. AMS Press, 1999.

[22] A. Caprara, G. Lancia, and S.K. Ng. Sorting permutations by reversals through branch and price. *INFORMS journal on computing*, 13(3):224–244, 2001.

[23] P. Carraresi and F. Malucelli. A reformulation scheme and new lower bounds for the QAP. In P.M. Pardalos and H. Wolkowicz, editors, *Quadratic Assignment and Related Problems*, DIMACS Series in Discrete Mathematics and Theoretical Computer Science, pages 147–160. AMS Press, 1994.

[24] D. Casey. *The Human Genome Project Primer on Molecular Genetics*. U.S. Department of Energy, World Wide Web, http://www.ornl.gov/hgmis/publicat/primer/intro.html, 1992.

[25] A. Chakravarti. It's raining SNP, hallelujah? *Nature Genetics*, 19:216–217, 1998.

[26] T. Christof, M. Junger, J. Kececioglu, P. Mutzel, and G. Reinelt. A branch-and-cut approach to physical mapping with end-probes. In *Proceedings of the Annual International Conference on Computational Molecular Biology (RECOMB)*, New York, NY, 1997. ACM Press.

[27] A. Clark. Inference of haplotypes from PCR–amplified samples of diploid populations. *Molecular Biology Evolution*, 7:111–122, 1990.

[28] M. Fischetti, G. Lancia, and P. Serafini. Exact algorithms for minimum routing cost trees. *Networks*, 39(3):161–173, 2002.

[29] W. M. Fitch. An introduction to molecular biology for mathematicians and computer programmers. In M. Farach-Colton, F.S. Roberts, M. Vingron, and M. Waterman, editors, *Mathematical Support for Molecular Biology*, volume 47 of *DIMACS Series in Discrete Mathematics and Theoretical Computer Science*, pages 1–31. AMS Press, 1999.

[30] M. Frances and A. Litman. On covering problems of codes. *Theory of Computing Systems*, 30(2):113–119, 1997.

[31] M.R. Garey and D.S. Johnson. *Computers and Intractability, a Guide to the Theory of NP-Completeness*. W.H. Freeman and Co., San Francisco, CA, 1979.

[32] A. Godzik and J. Skolnick. Flexible algorithm for direct multiple alignment of protein structures and sequences. *CABIOS*, 10(6):587–596, 1994.

[33] A. Godzik, J. Skolnick, and A. Kolinski. A topology fingerprint approach to inverse protein folding problem. *Journal of Molecular Biology*, 227:227–238, 1992.

[34] D. Goldman, S. Istrail, and C. Papadimitriou. Algorithmic aspects of protein structure similarity. In *Proceedings of the Annual IEEE Symposium*

on Foundations of Computer Science (FOCS), pages 512–522, New York, NY, 1999. IEEE.

[35] J. Gramm, F. Huffner, and R. Niedermeier. Closest strings, primer design, and motif search. In L. Florea, B. Walenz, and S. Hannenhalli, editors, *Posters of Annual International Conference on Computational Molecular Biology (RECOMB)*, Currents in Computational Molecular Biology, pages 74–75, 2002.

[36] M. Grötschel and W. R. Pulleyblank. Weakly bipartite graphs and the max–cut problem. *Operations Research Letters*, 1:23–27, 1981.

[37] D. Gusfield. Efficient methods for multiple sequence alignment with guaranteed error bounds. *Bulletin of Mathematical Biology*, 55:141–154, 1993.

[38] D. Gusfield. *Algorithms on Strings, Trees, and Sequences: Computer Science and Computational Biology*. Cambridge University Press, Cambridge, UK, 1997.

[39] D. Gusfield. A practical algorithm for optimal inference of haplotypes from diploid populations. In R. Altman, T.L. Bailey, P. Bourne, M. Gribskov, T. Lengauer, I.N. Shindyalov, L.F. Ten Eyck, and H. Weissig, editors, *Proceedings of the Annual International Conference on Intelligent Systems for Molecular Biology (ISMB)*, pages 183–189, Menlo Park, CA, 2000. AAAI Press.

[40] D. Gusfield. Inference of haplotypes from samples of diploid populations: Complexity and algorithms. *Journal of Computational Biology*, 8(3):305–324, 2001.

[41] B. V. Halldrsson, M. M. Halldrsson, and R. Ravi. On the approximability of the minimum test collection problem. In *Proceedings of the Annual European Symposium on Algorithms (ESA)*, volume 2161 of *Lecture Notes in Computer Science*, pages 158–169. Springer, 2001.

[42] S. Hannenhalli and P. A. Pevzner. Transforming cabbage into turnip (polynomial algorithm for sorting signed permutations by reversals). In *Proceedings of the Annual ACM Symposium on Theory of Computing (STOC)*, pages 178–189. ACM press, 1995 (full version appeared in Journal of the ACM, 46, 1–27, 1999).

[43] M. Held and R. M. Karp. The traveling salesman problem and minimum spanning trees: Part II. *Mathematical Programming*, 1:6–25, 1971.

[44] L. Holm and C. Sander. 3-D lookup: fast protein structure searches at 90% reliability. In *Proceedings of the Annual International Conference on Intelligent Systems for Molecular Biology (ISMB)*, pages 179–187, Menlo Park, CA, 1995. AAAI Press.

[45] International Human Genome Sequencing Consortium. Initial sequencing and analysis of the human genome. *Nature*, 409:860–921, 2001.

[46] D. S. Johnson and M. A. Trick, editors. *Cliques, Coloring, and Satisfiability*, volume 26 of *DIMACS Series in Discrete Mathematics and Theoretical Computer Science*. AMS Press, 1996.

[47] W. Kabash. A solution for the best rotation to relate two sets of vectors. *Acta Cryst.*, A32:922–923, 1978.

[48] H. Kaplan, R. Shamir, and R. E. Tarjan. Faster and simpler algorithm for sorting signed permutations by reversals. *SIAM Journal on Computing*, 29(3):880–892, 1999.

[49] J. Kececioglu. The maximum weight trace problem in multiple sequence alignment. In *Proceedings of the Annual Symposium on Combinatorial Pattern Matching (CPM)*, volume 684 of *Lecture Notes in Computer Science*, pages 106–119. Springer, 1993.

[50] J. Kececioglu, H.-P. Lenhof, K. Mehlhorn, P. Mutzel, K. Reinert, and M. Vingron. A polyhedral approach to sequence alignment problems. *Discrete Applied Mathematics*, 104:143–186, 2000.

[51] J. Kececioglu and D. Sankoff. Exact and approximation algorithms for sorting by reversals, with application to genome rearrangement. *Algorithmica*, 13:180–210, 1995.

[52] G. Lancia, V. Bafna, S. Istrail, R. Lippert, and R. Schwartz. SNPs problems, complexity and algorithms. In *Proceedings of the Annual European Symposium on Algorithms (ESA)*, volume 2161 of *Lecture Notes in Computer Science*, pages 182–193. Springer, 2001.

[53] G. Lancia, R. Carr, B. Walenz, and S. Istrail. 101 optimal PDB structure alignments: A branch-and-cut algorithm for the maximum contact map overlap problem. In *Proceedings of the Annual International Conference on Computational Biology (RECOMB)*, pages 193–202, New York, NY, 2001. ACM Press.

[54] J. Lanctot, M. Li, B. Ma, S. Wang, and L. Zhang. Distinguishing string selection problems. In *Proceedings of the Annual ACM-SIAM Symposium on Discrete Algorithms (SODA)*, pages 633–642, 1999.

[55] E. Lee, T. Fox, and I. Crocker. Optimization of radiosurgery treatment planning via mixed integer programming. *Medical Physics*, 27(5):995–1004, 2000.

[56] E. Lee, T. Fox, and I. Crocker. Integer programming applied to intensity-modulated radiation treatment planning optimization. *Annals of Operations Research, Optimization in Medicine*, 119(1-4):165–181, 2003.

[57] E. Lee, R. Gallagher, and M. Zaider. Planning implants of radionuclides for the treatment of prostate cancer: an application of mixed integer pro-

gramming. *Optima (Mathematical Programming Society Newsletter)*, 61:1–10, 1999.

[58] E. Lee and M Zaider. Mixed integer programming approaches to treatment planning for brachytherapy – application to permanent prostate implants. *Annals of Operations Research, Optimization in Medicine*, 119(1-4):147–163, 2003.

[59] C. Lemmen and T. Lengauer. Computational methods for the structural alignment of molecules. *Journal of Computer–Aided Molecular Design*, 14:215–232, 2000.

[60] H.-P. Lenhof, K. Reinert, and M. Vingron. A polyhedral approach to RNA sequence structure alignment. *Journal of Computational Biology*, 5(3):517–530, 1998.

[61] M. Li, B. Ma, and L. Wang. On the closest string and substring problems. *Journal of the ACM*, 49(2):157–171, 2002.

[62] S. Lin and B. Kernigan. An efficient heuristic algorithm for the traveling-salesman problem. *Operations Research*, 21(2), 1973.

[63] R. Lippert, R. Schwartz, G. Lancia, and S. Istrail. Algorithmic strategies for the SNPs haplotype assembly problem. *Briefings in Bioinformatics*, 3(1):23–31, 2002.

[64] B. Ma. A polynomial time approximation scheme for the closest substring problem. In *Proceedings of the Annual Symposium on Combinatorial Pattern Matching (CPM)*, volume 1848 of *Lecture Notes in Computer Science*, pages 99–107. Springer, 2000.

[65] J. Moult, T. Hubbard, S. Bryant, K. Fidelis, J. Pedersen, and Predictors. Critical assessment of methods of proteins structure prediction (CASP): Round II. *Proteins*, Suppl.1:dedicated issue, 1997.

[66] J. Moult, T. Hubbard, K. Fidelis, and J. Pedersen. Critical assessment of methods of proteins structure prediction (CASP): Round III. *Proteins*, Suppl.3:2–6, 1999.

[67] E. Myers. Whole-genome DNA sequencing. *IEEE Computational Engineering and Science*, 3(1):33–43, 1999.

[68] E. Myers and J. Weber. Is whole genome sequencing feasible? In S. Suhai, editor, *Computational Methods in Genome Research*, pages 73–89, New York, 1997. Plenum Press.

[69] G. L. Nemhauser and L. Wolsey. *Integer and Combinatorial Optimization*. John Wiley and Sons, 1988.

[70] P. Nicodeme and J. M. Steyaert. Selecting optimal oligonucleotide primers for multiplex PCR. In *Proceedings of the Annual International Conference on Intelligent Systems for Molecular Biology (ISMB)*, pages 210–213, Menlo Park, CA, 1997. AAAI Press.

[71] W. R. Pearson, G. Robins, D. E. Wrege, and T. Zhang. On the primer selection problem in polymerase chain reaction experiments. *Discrete Applied Mathematics*, 71:231–246, 1996.

[72] P.A. Pevzner. *Computational Molecular Biology*. MIT Press, Cambridge, MA, 2000.

[73] P. Raghavan. A probabilistic construction of deterministic algorithms: Approximating packing integer programs. *Journal of Computer and System Sciences*, 37:130–143, 1988.

[74] P. Raghavan and C. D. Thompson. Randomized rounding: a technique for provably good algorithms and algorithmic proofs. *Combinatorica*, 7:365–374, 1987.

[75] K. Reinert, H.-P. Lenhof, P. Mutzel, K. Mehlhorn, and J. Kececioglu. A branch-and-cut algorithm for multiple sequence alignment. In *Proceedings of the Annual International Conference on Computational Molecular Biology (RECOMB)*, pages 241–249, New York, NY, 1997. ACM Press.

[76] D. Sankoff, R. Cedergren, and Y. Abel. Genomic divergence through gene rearrangement. In *Molecular Evolution: Computer Analysis of Protein and Nucleic Acid Sequences*, pages 428–438. Academic Press, 1990.

[77] J. S. Sim and K. Park. The consensus string problem for a metric is NP-complete. In *Proceedings of the Annual Australasian Workshop on Combinatorial Algorithms (AWOCA)*, pages 107–113, 1999.

[78] T. Smith and M. Waterman. Identification of common molecular subsequences. *Journal of Molecular Biology*, 147:195–197, 1981.

[79] G. Stoesser, W. Baker, A. van den Broek, , E. Camon, M. Garcia-Pastor, C. Kanz, T. Kulikova, R. Leinonen, Q. Lin, V. Lombard, R. Lopez, N. Redaschi, P. Stoehr, M. Tuli, K. Tzouvara, and R. Vaughan. The EMBL nucleotide sequence database. *Nucleic Acids Research*, 30(1):21–26, 2002.

[80] J.C. Venter *et al.* The sequence of the human genome. *Science*, 291:1304–1351, 2001.

[81] L. Wang and T. Jiang. On the complexity of multiple sequence alignment. *Journal of Computational Biology*, 1:337–348, 1994.

[82] M.S. Waterman. *Introduction to Computational Biology: Maps, Sequences, and Genomes*. Chapman Hall, 1995.

[83] J. D. Watson, M. Gilman, J. Witkowski, and M. Zoller. *Recombinant DNA*. Scientific American Books. W.H. Freeman and Co, 1992.

[84] J. Weber and E. Myers. Human whole genome shotgun sequencing. *Genome Research*, 7:401–409, 1997.

[85] B.Y. Wu, G. Lancia, V. Bafna, K.M. Chao, R. Ravi, and C.Y. Tang. A polynomial–time approximation scheme for minimum routing cost spanning trees. *SIAM Journal on Computing*, 29(3):761–778, 1999.

[86] J. Xu, M. Li, D. Kim, and Y. Xu. RAPTOR: Optimal protein threading by linear programming. *Journal of Bioinformatics and Computational Biology*, 1(1):95–117, 2003.

Index